Micropollutants and Challenges

Emerging in the Aquatic Environments and Treatment Processes

Micropollutants and Challenges

Emerging in the Aquatic Environments and Treatment Processes

Afsane Chavoshani

Department of Environmental Health Engineering, Isfahan University of Medical Sciences, Isfahan, Iran

Majid Hashemi

Department of Environmental Health Engineering, School of Public Health, Kerman University of Medical Sciences, Kerman, Iran; Environmental Health Engineering Research Center, Kerman University of Medical Sciences, Kerman, Iran

Mohammad Mehdi Amin

Department of Environmental Health Engineering, School of Health, Isfahan University of Medical Sciences, Isfahan, Iran; Environment Research Center, Research Institute for Primordial Prevention of Non-communicable Disease, Isfahan University of Medical Sciences, Isfahan, Iran

Suresh C. Ameta

Department of Chemistry, PAHER University, Udaipur, India

Elsevier
Radarweg 29, PO Box 211, 1000 AE Amsterdam, Netherlands
The Boulevard, Langford Lane, Kidlington, Oxford OX5 1GB, United Kingdom
50 Hampshire Street, 5th Floor, Cambridge, MA 02139, United States

British Library Cataloguing-in-Publication Data
A catalogue record for this book is available from the British Library

Library of Congress Cataloging-in-Publication Data
A catalog record for this book is available from the Library of Congress

ISBN: 978-0-12-818612-1

For Information on all Elsevier publications
visit our website at https://www.elsevier.com/books-and-journals

Publisher: Candice Janco
Acquisitions Editor: Louisa Munro
Editorial Project Manager: Lena Sparks
Production Project Manager: Kiruthika Govindaraju
Cover Designer: Alan Studholme

Typeset by MPS Limited, Chennai, India

Contents

List of contributors

Rakshit Ameta
Department of Chemistry, J. R. N. Rajasthan Vidyapeeth (Deemed to be University), Udaipur, India

Suresh C. Ameta
Department of Chemistry, PAHER University, Udaipur, India

Mohammad Mehdi Amin
Department of Environmental Health Engineering, School of Health, Isfahan University of Medical Sciences, Isfahan, Iran; Environment Research Center, Research Institute for Primordial Prevention of Non-Communicable Disease, Isfahan University of Medical Sciences, Isfahan, Iran

Jayesh Bhatt
Department of Chemistry, PAHER University, Udaipur, India

Afsane Chavoshani
Department of Environmental Health Engineering, School of Health, Isfahan University of Medical Sciences, Isfahan, Iran

Bahare Dehdashti
Department of Environmental Health Engineering, School of Health, Isfahan University of Medical Sciences, Isfahan, Iran; Environment Research Center, Research Institute for Primordial Prevention of Non-Communicable Disease, Isfahan University of Medical Sciences, Isfahan, Iran

Meghavi Gupta
Department of Chemistry, PAHER University, Udaipur, India

Majid Hashemi
Environmental Health Engineering, School of Public Health, Kerman University of Medical Sciences, Kerman, Iran; Environmental Health Engineering Research Center, Kerman University of Medical Sciences, Kerman, Iran

Avinash Kumar Rai
Department of Chemistry, PAHER University, Udaipur, India

Shubhang Vyas
Department of Chemistry, PAHER University, Udaipur, India

Preface

The worldwide trend toward urbanization leads to increasing contamination of aquatic environments by thousands of synthetic and natural compounds which are known as micropollutants. Although most of these chemicals occur at low concentrations, due to persistent, bioaccumulative, and toxic features, many of them show the considerable toxicological concerns and health side effects. Because of their partial removal during conventional wastewater treatment processes and lack of international safety and environmental standards, a large number of micropollutants and their metabolites still remained there and released in the nearby aquatic environments. In recent years, there has been a growing tendency to research about micropollutants' impacts on the receiving environment and human health. These compounds are more significantly sensitive in detecting processes than classic compounds. Therefore, using micropollutants as environmental indicators for anthropogenic activities is a common method and frequently applied today. This book on micropollutants and challenges emerging in the aquatic environments and treatment processes contains comprehensive information on the fate and removal methods of the various emerging micropollutants from water and wastewater plants and their human health threats. This book addresses the needs of both researchers and graduate students in fields of environmental health engineering, environmental engineering, civil engineering, chemistry, etc. Any suggestions from readers are welcome to further improve this book.

Afsane Chavoshani, Majid Hashemi, Suresh C. Ameta and Mohammad Mehdi Amin

CHAPTER

Introduction

1

Afsane Chavoshani[1], Majid Hashemi[2,3], Mohammad Mehdi Amin[1,4] and Suresh C. Ameta[5]

[1]Department of Environmental Health Engineering, School of Health, Isfahan University of Medical Sciences, Isfahan, Iran

[2]Environmental Health Engineering, School of Public Health, Kerman University of Medical Sciences, Kerman, Iran

[3]Environmental Health Engineering Research Center, Kerman University of Medical Sciences, Kerman, Iran

[4]Environment Research Center, Research Institute for Primordial Prevention of Non-Communicable Disease, Isfahan University of Medical Sciences, Isfahan, Iran

[5]Department of Chemistry, PAHER University, Udaipur, India

1.1 Emerging micropollutants

The term "emerging contaminants" called trace organic contaminants (TrOCs) (Grandclement et al., 2017; Rizzo et al., 2019) "defines a variety of chemical compounds that are currently used and are released into the environment" (Pablos et al., 2015). The emerging micropollutants (EMPs) group consists of substances which significantly vary in terms of toxicity, behavior, remediation/treatment technique, and so on. Release of EMPs to the environment may be occurred for a long time, but may not have been recognized until new detection methods were developed. In addition new chemical production or changes in use and disposal of compounds can create the new sources of emerging pollutants (Boxall, 2012). EMPs are a global concern in the aquatic environments and have the serious risk for species survival.

Easier access to EMP compounds such as pharmaceuticals, pesticides, detergents, and personal care products has significantly increased the loading of such compounds in both the natural and manmade environments (Wanda et al., 2017; Bunke et al., 2019). The presence of these compounds in sources of water varies from location to location and concentrations are considerably higher in groundwater and surface water than in drinking water (Benotti and Brownawell, 2009).

The most micropollutants are not included into routine monitoring for surface and drinking water yet (Wanda et al., 2017), therefore, compared to other anthropogenic contaminants, the EMPs have largely been outside the scope of monitoring and worldwide regulations. Also there is not enough data on their levels, occurrence, and

Micropollutants and Challenges. DOI: https://doi.org/10.1016/B978-0-12-818612-1.00001-5

fate in environment (Wanda et al., 2017). Today wastewater effluents reuse for agricultural applications and land amendment is one of the main challenges among scientists, policy makers, and stakeholders (Rizzo et al., 2018). Both emission and environmental monitoring data do not give a comprehensive image of the EMP situation in the aquatic environments. Also because of different temporal and spatial scales, there are many data-gaps on EMPs in the conducted studies (Gavrilescu et al., 2015; Brunsch et al., 2019). The aim of this chapter is to review of these challenges to provide the best knowledge on management of these compounds.

1.2 Resources of emerging micropollutants

Sources of EMPs are: (1) industrial wastewater, (2) runoff from agriculture, livestock, and aquaculture; (3) landfill leachates; and (4) domestic and hospital effluents (Barbosa et al., 2016) (Fig. 1.1). The EMPs comprise a wide range of natural and synthetic organic compounds, which include pharmaceuticals and personal care products (PPCPs), detergents, steroid hormones, industrial chemicals, pesticides, and many others (Wanda et al., 2018).

More than 80 compounds and several metabolites have been found in the aquatic environment, which indicates that not all contaminants are removed during water treatment (Heberer, 2002). While their presence is not a new phenomenon, remarkable attention has been drawn to them within the last decade (Besha et al., 2017), and recent studies have been interested in conducting research on EMPs effects on human and aquatic life (Wanda et al., 2017). The presence of over 50 individual PPCPs occurs mainly in wastewater treatment plant (WWTP) effluents, surface and ground water, and much less frequently in drinking water (Talib and Randhir, 2017).

Global EMPs production, between 1930 and 2000, has increased from 1 million to 400 million tons per each year (Gavrilescu et al., 2015). In the European Union (EU) more than 100,000 compounds have been registered of which 30,000 to 70,000 of them are used daily. Most of them end up to the aquatic environments. Natural waters receive about 300 million tons of synthetic compounds annually from industrial and consumer products effluents (Talib and Randhir, 2017).

During the last 10 years, the NORMAN Network group has detected around 970 emerging pollutants (Bunke et al., 2019). NORMAN Network is an international institution that improves the exchange of information and knowledge on emerging environmental compounds with purpose of support and management of measurement methods and monitoring tools (Bunke et al., 2019).

The list of some micropollutants presented in the watch list of EU Commission Decision 495/2015 are 17-alphaethinylestradiol (EE2), 17-beta-estradiol (E2), estrone (E1), diclofenac, 2,6-di-tert-butyl-4-methylphenol, 2-ethylhexyl-4-methoxycinnamate, macrolide antibiotics, methiocarb, neonicotinoids, oxadiazon, and triallate (Barbosa et al., 2016).

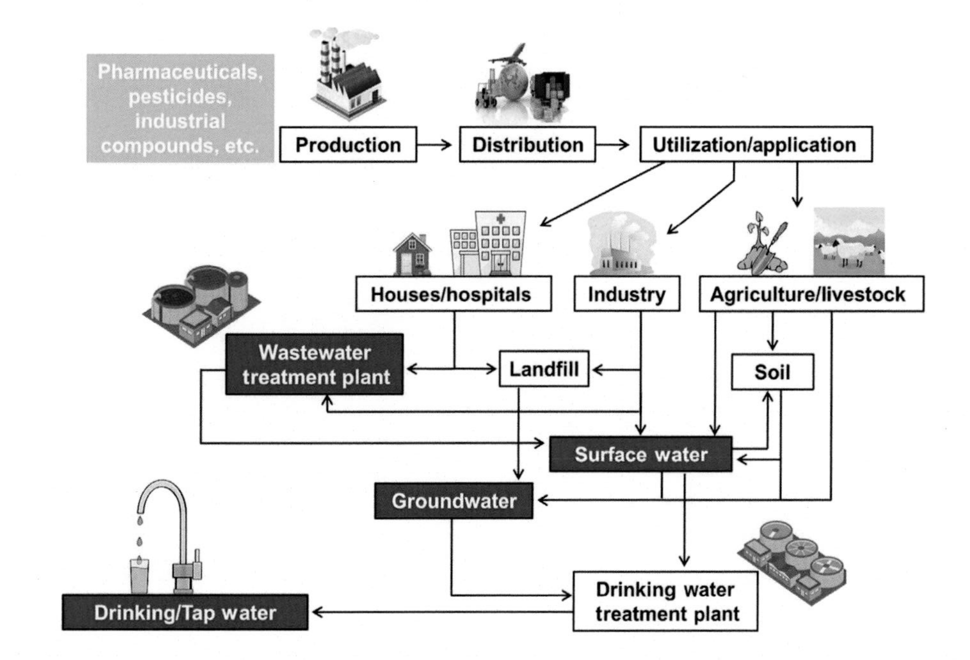

FIGURE 1.1

Representative sources and routes of micropollutants in the aquatic environments (Barbosa et al., 2016).

With permission.

At present different government and nongovernment organizations including the EU, the North American Environmental Protection Agency (EPA), the World Health Organization (WHO), or the International Program of Chemical Safety (IPCS) are considering these problems and setting up directives and legal frameworks to protect and improve the quality of freshwater resources (Hecker and Hollert, 2011).

According to the varieties in EMPs resources, contamination of water media with these chemicals is not a surprise (Besha et al., 2017; Pablos et al., 2015). It is cleared that EMPs discharge in water bodies is influenced by disposal of municipal, industrial, and agricultural wastes, excretion of pharmaceuticals, and accidental spills (Bunke et al., 2019; Talib and Randhir, 2017). In addition, future developments in society can result in the emission of new substances to the environment (Bunke et al., 2019).

Insufficient removal of EMPs by conventional wastewater treatment processes and lack of precautions and monitoring actions for EMPs have been caused that many of these compounds act as a threat for human and wildlife in the aquatic environments. The occurrence of EMPs at concentration between few nanogram per liter and several microgram per liter (Wanda et al., 2017) in the aquatic environment has been commonly associated with a number of negative effects such as chronic and acute toxicity, endocrine disrupting effects, and antibiotic resistance of microorganisms (Luo et al., 2014). Fig. 1.2 shows some human diseases associated with EMPs.

The presence of EMPs in the water media has recently been widely reported (Wanda et al., 2017; Kim and Zoh, 2016; Ebrahimi and Barbieri, 2019; Geissen et al., 2015), demonstrating an increasing concern about them. For instance a series of periodic review articles focusing on occurrence, fate, transport, and treatment of EMPs were published annually, and several original articles have suggested different methods for micropollutants treatment in water and wastewater (Jiang et al., 2013; Luo et al., 2014; Kwon et al., 2015; Virkutyte et al., 2010; Tröger et al., 2018).

Although the effects of micropollutants in aquatic environments are not very well known yet, there are clear indications that they have long-term impacts on ecosystem. Reasons for this are: (1) their potential to accumulate into aquatic organisms and human bodies (bioaccumulation), (2) their toxicity, and (3) their resistance to degradation in the environment (persistency). Regulations on their emission and discharge are thus vital for improving the aquatic environment and surface water quality (Chau et al., 2018; Antakyali et al.; Williamson et al., 1993).

The occurrence of the EMPs in aquatic environments (wastewater, surface water, groundwater, and drinking water) was reported in some regions such as Austria, China, EU-wide, France, Germany, Greece, Italy, Korea, Spain, Sweden, Switzerland, Western Balkan Region, United Kingdom, and United States (Luo et al., 2014).

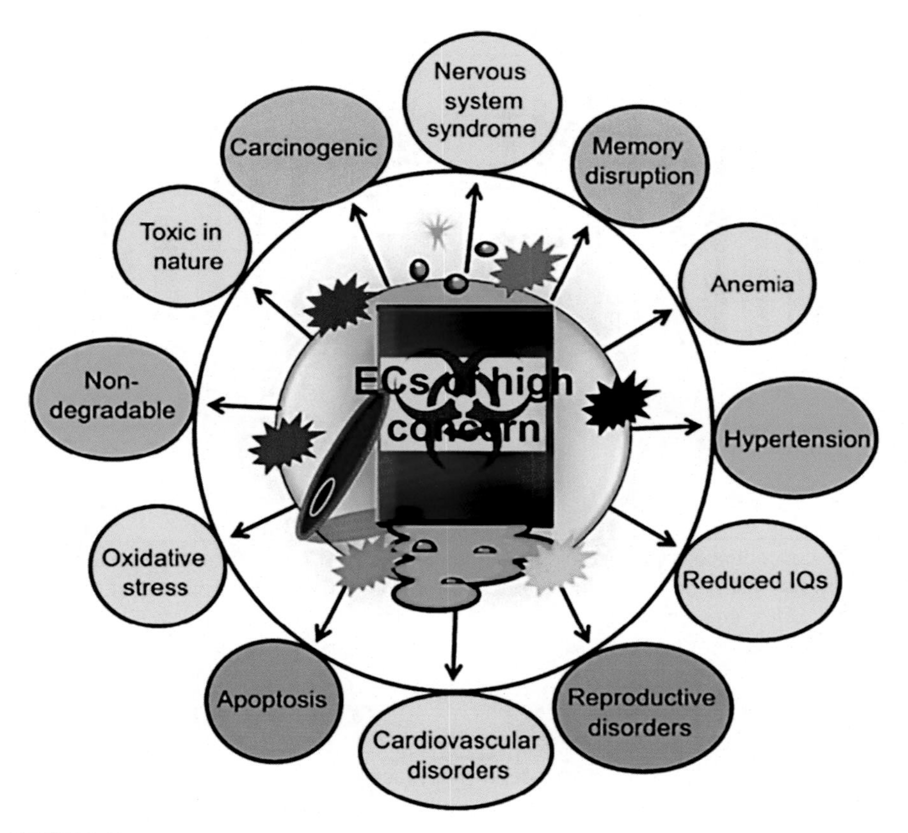

FIGURE 1.2

Major consequences and adverse effects of EMPs of high concern on human's health and the environment (Rasheed et al., 2018). *EMPs*, Emerging micropollutants.
With permission.

1.3 Occurrence of emerging micropollutants in wastewater treatment plants

Available water sources globally are limited because of several factors, such as domestic, agricultural and industrial uses, drought, global climate change, the increase of population density, and the continuous extraction of water from groundwater resources. Therefore for reuse of water, EMPs removal during water and wastewater treatment processes is necessary (Besha et al., 2017).

The EMPs concentration in WWTPs or WTPs is influenced by factors including EMPs distribution in a region, EMPs extraction rate (by urine and feces),

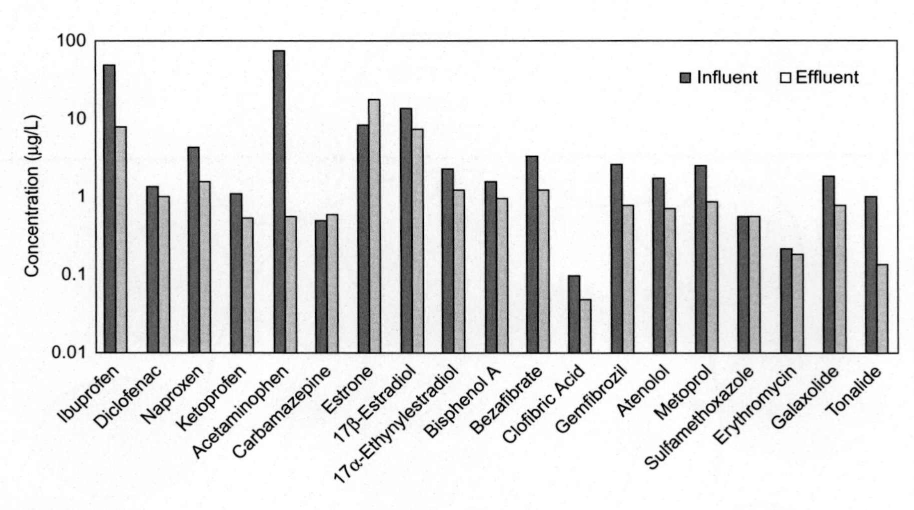

FIGURE 1.3

Average concentrations (logarithmic y-axis) for selected EMPs of influents and effluents of WWTPs after MBR and conventional ASP (Besha et al., 2017). *ASP*, Activated sludge process; *EMPs*, emerging micropollutants; *MBR*, membrane bioreactor; *WWTPs*, wastewater treatment plants.

With permission.

local common diseases led to especial pharmaceutical consumption, different climate change led to use of pesticide in agricultural actions, rainfall led to change of influent flow of wastewater, and weather conditions led to effect of temperature and sunlight on EMPs degradation (Luo et al., 2014).

Fig. 1.3 presents the mean concentrations of some selected EMPs in the influents and effluents after wastewater treatment. The treatment technologies applied in water and wastewater plants remove the EMPs with very high to low or negative efficiency. However, complete removal of the EMPs is impossible. This figure confirmed that conventional water treatment is not enough to eliminate the EMPs from wastewater.

1.4 Occurrence of emerging micropollutants in surface water

The discharge of WWTP effluent into surface water has been considered as a main reason for the EMPs occurrence in surface water resources. Due to water dilution in river, EMPs concentration may occur at lower levels than wastewater effluent. It was found that the natural attenuation of EMPs is more likely affected by dilution or sorption processes than degradation. Surface water dilution can be changed by rainfall. This subject has been confirmed through increase in EMPs concentration during dry weather conditions and reduction during wet weather conditions. It was found that pharmaceuticals in water samples revealed lower concentration in summer than those in winter. This could be due to two reasons: (1) stimulated biodegradation of pharmaceuticals in higher temperature in summer and (2) increased dilution during wetter summer. However, rainfall did not always decrease the EMPs. In some cases, rainfall was identified as a contributor to the emission of EMPs to surface water. In addition, rainfall might increase combined sewer overflows, resulting in a higher concentration of EMPs released in the surface water. Also the occurrence of pesticides as an EMPs in surface water depends on crop type, soil properties, characteristics of the water bodies (depth and flow rate), features of the land close to the water bodies (soil use, slope, and distance from water bodies), and climatic conditions (temperature, rainfall, moisture, and wind) (Luo et al., 2014).

According to Table 1.1, nonsteroidal antiinflammatory drugs (NSAIDs), carbamazepine, sulfamethoxazole (SMX), and triclosan were the most EMPs detected in surface water resources from different countries. Due to the release of high hospital wastewater effluents and other polluted waters to Costa Rica surface waters, the EMPs concentrations were higher in this country than other countries (Luo et al., 2014).

1.5 Occurrence of emerging micropollutants in groundwater

In comparison to surface water, ground water was found to be less contaminated with EMPs. Hence the presence of EMPs in groundwater mainly results from landfill leachate, groundwater–surface water interaction, infiltration of contaminated water from agricultural land, or seepage of septic tanks and sewer systems. Concentrations of EMPs in landfill leachate and septic tank leakage generally range from 10 to 104 ng/L and 10 to 103 ng/L, respectively. Soil is the major pathway for groundwater pollution by some EMPs (e.g., pesticides). EMPs can also be introduced in groundwater via bank filtration or artificial recharge using reclaimed water. The physicochemical properties of EMPs are important for the transfer of these compounds to groundwater. For example, octanol–water partition coefficient (K_{OW}) indicates contaminant mobility in the subsurface, where the compounds (e.g., trimethoprim and TCEP) with $K_{OW} < 1.5$ tend to stay in the dissolved phase (more mobility) and are more likely to occur in groundwater (Luo et al., 2014). Some EMPs were most commonly detected in both surface water and wastewater, evidencing a correlation of the presence of EMPs in different aquatic environments. Table 1.2 showed the occurrence of some common EMPs in groundwater in different countries.

Table 1.1 Occurrence of some common EMPs in surface waters in different countries (Luo et al., 2014).

Compound	Concentration (ng/L)											
	Canada	China	Costa Rica	France	Germany	Greece	Korea	Spain	Taiwan	United Kingdom	United States	PNEC
Ibuprofen	0.98 (79)	ND–1417	5 (36,788)	ND–8	–	1–67	<15–414	–	5–280	0.3–100	ND–77	5000
Naproxen	1 (87)	ND–328	–	ND–6.4	3–322	–	–	–	–	0.3–149	–	37,000
Ketoprofen	–	–	7 (9808)	ND–22.0	–	0.4–39.5	–	–	10–190	0.5–14	–	15.6×10^6
Diclofenac	–	–	14 (266)	ND–35.0	–	0.8–1043	–	–	–	0.5–261	–	10,000
Mefenamic acid	–	–	–	–	–	–	<30–326	–	–	0.3–169	–	–
Carbamazepine	3 (749)	–	1 (82)	ND–31.6	102–1194	–	<4–595		–	0.5–684	ND–9.6	5,000
Gemfibrozil	–	–	41 (17,036)	–	–	–	–	–	1.9–3.5	–	–	100,000
Atenolol	–	–	–	ND–34.0	–	–	<100–690	–	–	1–560	–	10×10^6
Sulfamethoxazol	0.2 (284)	–	11 (56)	ND–5.1	–	–	–	–	0.3–60	.5–4	ND–38	20,000
Trimethoprim	–	–	–	–	–	–	–	–	1–2.1	7–122	ND–9.1	1000
Triclosan	0.4 (25)	35–1023	11 (263)	–	124–220	3–39	ND	–	–	ND	ND–9.8	–
Galaxolide	–	–	–	–	35–1814	–	–	–	–	–	–	
Tonalide	–	–	–	–	5–273	–	–	–	–	–	–	–
Estrone	–	ND–65					3.6–69.1	–	–	–	–	18
Estradiol	–	ND–2					1.1–10.1	–	–	–	–	–
Ethinylestradiol	–	ND–1	–	–	–	–	ND–1.9	–	–	–	–	0.02
Estriol	–	ND–1	–	–	–	–	–	–	–	–	–	149
Caffeine	–	–	24 (1,121,446)	–	–	–	–	–	1–1813	–	ND–225	10×10^5
Nonylphenol	–	36–33,231	–	–	–	558–2704	15–336	–	–	–	–	330
Bisphenol A	2.1 (87)	6–881	–	–	192–215	55–162	.5–33	–	–	6–68	–	1000
TCEP	–	––	–	–	<3–184	–	–	–	–	–	–	–
TCPP	–	–	––	–	<4–379	–	–	–	–	–	–	–
Atrazine												
Atrazine	–	–	–	–	–	–	–	11 (39)	–	–	–	2000
Diazinon	–	–	–	–	–	–	–	10 (216)	–	–	–	–
Diuron	–	–	–	–	–	–	–	72 (408)	–	–	–	1800

EMPs, *Emerging micropollutants.*
With permission.

Table 1.2 Occurrence of some common EMPs in groundwater in different countries (Luo et al., 2014).

Compound	Europe	France	Germany	Spain	United States	PNEC
Ibuprofen	3 (395)	0	–	185 (185)	0, 3110	5000
Naproxen	–	1.2	–	204 (145–263)	–	37,000
Ketoprofen	26 (2886)	2.8				15.6×10^5
						10,000
Estrone	0 (4)	0.7	–	–	79	18
Estradiol	–	0.4	–	–	147	–
Ethinylestradiol	–	1.2	–	–	230	0.02
Estriol	–	–	–	–	1661	149
TCEP			4–51			
TCPP	–	–	14–355	–	–	–
Atrazine	–	–	–	36 (756)	–	2000
Diazinon	–	–	–	5.3 (30.8)		–
Diuron	–	–	–	8.8 (178)		1800

EMPs, *Emerging micropollutants.*
With permission.

1.6 Occurrence of emerging micropollutants in drinking water

The concentration of EMPs in drinking water is dependent on water resources and seasons, for example water samples in winter showing higher concentrations in comparison with water samples in summer. Furthermore, water treatment processes play a significant role in removal of EMPs from drinking water. Monitoring of EMPs in water treatment plants (WTPs) has shown the presence of EMPs in WTPs (Tröger et al., 2018) at concentration between parts per billion (ppb) or parts per trillion (ppt).

Based on review study of Luo et al. (2014) the maximum concentration of the most EMPs in drinking water was reported to be lower than 100 ng/L, with the exception of carbamazepine and caffeine. It is interesting that carbamazepine concentration was detected at a concentration higher than 600 ng/L. The high concentration of carbamazepine could be described by its high persistency. Table 1.3 summarized the removal efficiencies of detected 12 EMPs in each treatment stage and total removals in WTP (Nam et al., 2014).

Table 1.3 Average concentrations of 12 EMPs during each treatment and final removal efficiencies in the WTP (coagulation–sedimentation, sand filtration, and postchlorination) during the overall period (Nam et al., 2014).

	Average concentration (ng/L)				
Compounds	Influent	Coagulation effluent	Filtration effluent	Disinfection effluent	Removal (%)
SMZ	7.8 ± 5.4	1.9 ± 1.7	2.1 ± 1.6	2.1 ± 1.9	73
CBM	10.3 ± 9.5	5.1 ± 10.1	1.2 ± 3.7	1.7 ± 4.2	84.7
CFF	36.1 ± 20.8	6.7 ± 10.5	14.3 ± 5.3	5.8 ± 7.4	84
MPT	37.4 ± 34.1	34.2 ± 21.2	28.4 ± 19.7	35.18 ± 22.4	6
SMA	3.7 ± 2.4	2.5 ± 1.8	2.2 ± 1.7	2.31 ± 1.7	38
IBU	19.6 ± 21	6.9 ± 9.2	5.5 ± 11.5	4.3 ± 6	78
ACT	106.1 ± 147.1	24.7 ± 43.3	12.2 ± 22.6	5 ± 9.1	95.3
NPX	11.8 ± 12.4	4.3 ± 7.2	3.5 ± 7.8	2.7 ± 3.8	78
BPA	88 ± 102.8	17.5 ± 8.6	21.4 ± 12.9	19 ± 8.7	78.4
2,4-D	33.6 ± 25.1	15.6 ± 14	6.6 ± 1.8	6.5 ± 1.5	81
DCF	7.8 ± 2.45	<MDL	<MDL	<MDL	100
NP	41.5 ± 52.1	35 ± 75.54	11.4 ± 5.12	12.6 ± 5.9	70

BPA, *Bisphenol A;* EMPs, *emerging micropollutants;* NPX, *naproxen;* WTP, *wastewater treatment plants.*
With permission.

1.7 Processes controlling the fate of emerging micropollutants during wastewater treatment

According to the different of physical/chemical properties of EPMs, such as persistent organic pollutants (POPs), more polar substances (pesticides, pharmaceuticals, industrial chemicals), inorganic compounds (trace metals), and particulate contaminants (nanoparticles and microplastics), removal and detection methods of EMPs are very different.

Different factors affect the removal efficiency of EMPs from wastewater. The most important factors are: the EMPs physicochemical properties (MW, molecular diameter, pK_a, K_{ow}, K_d, and K_{bio}), the operating conditions [sludge retention time (SRT), hydraulic retention time (HRT), temperature, and redox conditions], and the wastewater characteristics (pH, organic matter concentration, and ionic strength). In the following section, the main factors have been discussed with respect to removal potential of the EMPs in WWTPs.

1.7.1 Effective factors related to emerging micropollutants properties

Sorption

The term sorption comprises two mechanisms: absorption (EMPs move from the aqueous phase and enter into the lipophilic cell membrane of biomass or into the lipid fraction of the sludge due to their hydrophobicity) and adsorption (EMPs are retained onto solids surface due to electrostatic interactions between positively charged compounds and the negatively charged surface of biomass cells) (Alvarino et al., 2018).

Sorption of EMPs is a function of their physicochemical properties, as well as of the characteristics of the sorbent agent. In the case of sludges, the octanol–water coefficient (K_{ow}) and the acid dissociation constant of OMPs (organic micropollutants) (K_a) determine their sorption trend (Suárez et al., 2008; Ternes et al., 2004).

Reported results indicate that those compounds with a medium or high lipophilic behavior (high K_{ow}), such as musk fragrances or hormones, are preferentially removed by absorption onto the sludge (Suárez et al., 2008; Joss et al., 2005), independently from the biomass conformation (granular or flocculent biomass) (Alvarino et al., 2014). On the other hand, OMPs which are ionized or dissociated in the aqueous phase can be removed by electrostatic interactions with the negatively charged surfaces of the biomass, as in the case of the cationic species of the antibiotic trimethoprim (Suárez et al., 2008). The use of the solid–water distribution coefficient (K_d, in L/kg), defined as the ratio between the concentrations in the solid and liquid phases at equilibrium conditions, is commonly used to determine the fraction sorbed onto sludge (Hörsing et al., 2011).

Adsorption of EMPs to solids mostly depends on its hydrophobicity. The octanol–water partition coefficient (K_{ow}) of a compound can be used to predict its sorption potential. As a general rule compounds with logK_{ow} <2.5 will have low sorption potential, logK_{ow} values between 2.5 and 4.0 will indicate moderate sorption potential, and compounds with logK_{ow} >4.0 will have strong sorption potential. Thus hydrophobic compounds, such as galaxolide and tonalide (logK_{ow} >5), have been reported to be sorbed to the biomass (i.e., algae) in HRA. On the other hand compounds that are highly or moderately hydrophilic are unlikely to significantly sorbed to the organic matter in pond sediments and will remain in the dissolved phase of the pond where the likely mechanisms of elimination will be biodegradation and/or photodegradation (Gruchlik et al., 2018).

The amount of EMPs present in the solid phase can be divided into different fractions according to the strength of the binding. The first fraction relates to the amount of substance more weakly sorbed (extractable fraction), while the second is referred to that portion more strongly retained (nonextractable fraction). The ultrasonic solvent extraction (USE) methodology is the most common technology used for measuring the EMPs extractable fraction, whereas the nonextractable fraction can only be measured by using more advanced techniques such as radiolabeled isotope based methods. These advanced methods can be used to assess the changes over time in the distribution of the radioactivity of the radiolabeled parent compound in both the liquid and solid phases, as well as in the gases produced. The 14C technique is used to determine the sorption of the antibiotic SMX in bioreactors using different biomass conformations (granular and flocculent biomass) and applying different redox conditions (anaerobic, anoxic, and aerobic). Except in the case of the anaerobic granules, the contribution of the nonextractable fraction to the total sorption was always above 50%, with an increasing trend along the duration of the assays, and the overall removal by sorption 26% of the total radioactivity (Alvarino et al., 2018).

Photolysis

In the case of direct photolysis, sunlight absorption occurs when light is absorbed by the EMP, while indirect photolysis refers to processes initiated through the absorption of sunlight by intermediary compounds called photosensitizers (Wang et al., 2017). Photolysis is affected by the amount of light absorbance by the EMP and the suspended solids concentration which limits the penetration of light aquatic environments. The manual GCSOLAR program which contains a set of routines can compute direct photolysis rates and half-lives of pollutants in the aquatic environment. Based on this program, photolysis is a function of the season, latitude, time of day, depth of water, and ozone layer thickness (Zepp and Cline, 1977).

Volatilization

Vaporization of an element or compound is a phase transition from the liquid phase to vapor until equilibrium between the two phases is occurred, and this phenomenon can be described by Henry's Law.

Because of the very low concentration of EMPs in the atmosphere, the transfer can be assumed to only occur from the wastewater to the atmosphere. Henry's Law constants can be used to predict the behavior of a compound at the interface between air and water. Volatilization of EMPs is almost negligible since most EMPs have Henry constant less than 0.005. Although most EMPs have Henry's Law constants, volatilization can be a remarkable process for some EMPs in treatment unites with large surface areas (such as waste stabilization ponds) or with high aeration rate [such as membrane bioreactors (MBR)].

Biotransformation

Biotransformation or biodegradation process is the uptake of EMPs by cell and biodegradation of them by the bacterial enzymes. EMPs structure also has a key role in their resistance against to biodegradation process. Cyclic structures, such as monocyclic or polycyclic, and functional groups, such as halogen groups, are important in the biodegradability of EMPs.

In general the easily degraded substances include (1) linear compounds with short branched chain, (2) unsaturated aliphatic compounds, and (3) compounds with electron donating functional groups. On the other hand the persistent EMPs contain (1) compounds with long, branched side chains, (2) saturated or polycyclic compounds, and (3) compounds with sulfate, halogen, or electron withdrawing functional groups (Luo et al., 2014).

Two biodegradation pathways are described for EMPs removal (Pomies et al., 2013). First, the direct biodegradation is carried out by a fraction of the total biomass which is able to degrade EMPs. Secondly, the biodegradation reaction of micropollutants can occur during the conversion of macropollutants. Micropollutants are not a source of carbon or energy for the biomass; a co-substrate(like readily biodegradable COD or ammonium) is necessarily present and serves as growth substrate.The simultaneous degradation of the co-substrate and the micropollutant is linked to the capacity of the enzymes to degrade many substrates. For example, methanotroph bacteria are known to co-oxidize PAHs, alkanes and aromatic com-pounds. When a co-substrate is present, direct biodegradation and cometabolism should occur simultaneously (Pomies et al., 2013).

1.7.2 Effective factors related to the operation condition of wastewater treatment plants

Sludge Retention Time

SRT that controls the size and variety of a microbial community is very important for EMPs removal. Enhanced EMPs removal is possible through extended SRTs. Suitable SRT facilitates some bacteria growth, such as nitrifying bacteria. In nitrifying situations, cometabolism process performed by ammonium monooxygenase enzyme is a possible pathway for EMPs degradation. Nitrifying biomass has been found to have positive effects on the removal of a range of EMPs such as ibuprofen, naproxen, trimethoprim, erythromycin, galaxolide, tonalide, ethinylestradiol, bisphenol A, and nonylphenol (Luo et al., 2014).

It is suggested that the SRTs allowing nitrogen removal (nitrification and denitrification) above 10 days can enhance the elimination of some biodegradable compounds (e.g., ibuprofen, bezafibrate, natural estrogens, and bisphenol A). In a study the activated sludge treatment with an increased SRT of 18 days could achieve considerably higher removal of beta blockers and psychoactive drugs in comparison with the same treatment with shorter SRT of 0.5 day. It was identified as 10% higher removal efficiency for fluoxetine, citalopram, and ethinylestradiol when prolonged SRT was applied. Enhanced biodegradation was found for 4-*n*-nonylphenol and triclosan at SRT of 20 days (compared with 3 and 10 days). However, high SRT does not necessarily mean better removal performance. The variation of the sludge age between 10 and 60–80 days showed no noticeable effects on removal efficiency of the investigated pharmaceuticals. High SRT (20 days) also seemed not to appreciably affect the biodegradation of bisphenol. Also application of low SRTs (1.5–5.1 days) had minor effects on the removal of some pharmaceutical compounds (e.g., ibuprofen, diclofenac, naproxen, and carbamazepine) (Luo et al., 2014).

The study of effect of sludge retention time on endocrine disrupter sorption and their biodegradation during activated sludge treatment showed that SRT did not affect the sorption potential of triclosan (TCS) and bisphenol A (BPA), while higher sorption constants were observed for 4-n-nonylphenol (4-n-NP) at SRT of 20 days. The use of lab-scale continuous-flow systems showed that for an HRT of 10 h and SRT ranging between 3 and 20 days, the major part of EDCs ($>90\%$) can be removed during activated sludge process mainly via biodegradation. The calculation of the mean pseudo-first-order biotransformation rates at different SRTs showed that EDC values were ranged between 178–507, 30–288, and 17–113 Lg/VSS day for 4-n-NP, TCS, and BPA, respectively (Stasinakis et al., 2010).

Hydraulic Retention Time

HRT is the amount of time that allows for biodegradation and sorption. The micropollutants having slow/intermediate kinetics such as fluoxetine or some antibiotics will experience less effective biodegradation at shorter HRTs or increasing loading rates. It has been indicated that HRT in the range of 5–14 h actives minor removal of DEHP, while higher HRT increases DEHP accumulation in the system and DEHP retention in the waste sludge. Redox conditions may cause the observed differences by having an effect on certain wastewater or sludge characteristics as well as on the biodiversity of the microbial flora present (Luo et al., 2014).

The study of the effect of HRT on wastewater treatment and removal of pharmaceuticals, hormones, and phenolic utility substances showed that the strength of the effect of HRT on the removal efficiency of single substances varied, but all substances except for carbamazepine were removed more efficiently with increasing HRT. Carbamazepine showed negative removal values, and it was consequently not possible to evaluate removal efficiency for this substance. The strongest dependency of the removal efficiency to HRT is equal to the steepest slope (highest calculated slope) applying a first-order linear relationship. The removal of propranolol was most strongly dependent on HRT (slope = 31, $R^2 = 0.58$), followed by estrone (slope = 14, $R^2 = 0.37$), estradiol (slope = 12, $R^2 = 0.71$), metoprolol (slope = 12, $R^2 = 0.81$), and atenolol (slope = 11, $R^2 = 0.80$). These results showed that HRT had effects on the structures of bacteria communities and the changes of bacterial communities. It was observed that a hydraulic retention time of 7 days was critical for increased removal efficiencies of Ketoprofen and Naproxen in large active sludge WWTPs (Ejhed et al., 2018).

The overall average removal efficiency of all pharmaceuticals and hormones showed a strong first-order linear relationship with HRT. Results showed that onsite wastewater treatment facilities would remove micropollutants more efficiently if the HRT was increased from 2–3 to 4–6 days. However, the relationship between removal efficiency and HRT can be a seen as a pseudo-first-order relationship. Increased HRT may affect several processes in wastewater treatment, that is, increasing possible nitrification and denitrification processes by reducing ammonium and total nitrogen in the effluent or enhancing sludge and particle removal that would result in reduced turbidity in the effluent. The average removal of total nitrogen (Ntot) and turbidity (%) was increased with the longer hydraulic retention time. There was an overall weak, but positive correlation between the removal of Ntot and turbidity with HRT. The removal of ammonium and total phosphorous was dependent on the type of technique (activated sludge or phosphorous filter), and a there was no obvious relationship between removal of these substances and the hydraulic retention time (Ejhed et al., 2018).

Redox condition

The unfavorable redox conditions (anaerobic conditions) could result in inefficient biodegradation of some EMPs. In another study, naproxen, ethinylestradiol, roxithromycin, and erythromycinwere found only considerably eliminated under aerobic condition, and anoxic removal was much less effective when compared to short term biodegradation of clofibric acid, ibuprofen, and diclofenac in oxic and anoxic (denitrification conditions, absence of oxygen while presence of nitrate) biofilm reactor. In the oxic biofilm reactor, clofibric acid and diclofenac were not eliminated, with only 1%−4% loss of their initial concentration being observed. Ibuprofen was reduced by 64%−70%. By contrast the anoxic biofilm reactor achieved much lower removal of ibuprofen (17%−21%) and higher removal of diclofenac (34%−38%) and clofibric acid (26%−30%). It was reported that removal of the nonylphenol ethoxylate surfactant was higher in the oxic reactors (50% to 70%) compared to the anoxic reactors (30% to 50%). Similarly DEHP were removed by 15%, 19%, and 62% in anaerobic, anoxic, and aerobic reactors. Anoxic redox conditions were not necessarily less favorable environments for micropollutant removal. For instance anoxic conditions could lead to improved elimination of iodinated X-ray contrast media, while aerobic environments witnessed minor removal. Some persistent substances such as diclofenac, SMX, trimethoprim, and carbamazepine showed minor removals (<25%) by the biological treatment with either nitrifying (oxic) or denitrifying bacteria (anoxic) (Luo et al., 2014).

PH

pH can be a contributing factor to the dissolution of micropollutants. In addition, optimum pH is essential to maintain the enzymatic activity of microbes, which is a dark horse for EMPs removal in aquatic environments. The change in pH of sewage may have a negative effect on EMPs removal. Nevertheless, pH has different impacts on the sorption mechanism. Depending on the value of dissociation constant pK_a of micropollutants, cation or anion formation can occur. Cations can easily interact with the negatively charged sludge surface through van der Waals-type attractions. On the contrary, anions show poor interaction with the sludge surface. Certain ionization EMPs such as ibuprofen, ketoprofen, and sulfamethoxazole were reported as being strongly pH dependent (Varjani et al., 2020).

The acid dissociation constants (pK_a) of micropollutants vary according to their molecular structure. Ionizable chemicals are converted to either cations or anions depending on their pK_a and pH in solution. Hydrophilic compounds easily interact with water by polar–polar interaction. This phenomenon can cause adsorption of the ionized micropollutants on the activated carbon surface due to electrostatic interactions. It was examined the effect of pH on adsorption removal of selected micropollutants. The adsorption of hydrophobic compounds was not affected by pH. Among the hydrophobic micropollutants, cetaminophen, exhibited the highest removal rate (77.2%–79.3%). In contrast, adsorption of hydrophilic compounds (with the exception of caffeine) was significantly affected by pH. Electrostatic and specific sorbate–sorbent interactions between micropollutants and activated carbon surfaces may affect the adsorption of hydrophilic compounds (Nam et al., 2014).

The modest pH variation had significant effects on the removal of acidic pharmaceuticals (clofibric acid, ibuprofen, ketoprofen, naproxen, and mefenamic acid) by the biosolids, which was presumably ascribed to activation of enzymes involved or enhancement of affinity between the biosolids and pharmaceuticals due to protonation of acidic pharmaceuticals. Seasonal variation of temperature may have impact on EMPs removal in WWTPs.

Temperature

Temperature also plays an important role. An average of 15°C −20°C is the most favorable temperature for wastewater treatment processes. In cold countries, temperatures often reach below 10°C. However, temperature variation has been observed in winter and summer, which also affects biodegradation. For example, some compounds such as sulfamethoxazole exhibited almost similar removal characteristics in both summer and winter seasons, and specific compounds such as trimethoprim and roxithromycin show higher removal in summer. Actually, except for a few pharmaceuticals, overall removal efficiency for most pharmaceuticals was not affected by seasonal variations. For example, minor changes in climate of North Island of New Zealand did not produce noticeable differences in temperature throughout the year. Another study also reported about 65 percent antibiotic removal in summer. Adsorption and biodegradation were the predominate mechanisms in MBRs and higher temperatures in summer supported biodegradation. However, the effect of temperature variation and EMPs removal in MBRs are methodically reported. It was reported that by changing temperature in range 10°C–35°C did not have any significant effect on EMPs removal. However, increasing SRT to 90 days, even at a lower temperature of 8°C, enhanced EMPs removal (Varjani et al., 2020).

Temperature variation can affect biodegradation and partition (sorption and volatilization) of EMPs. To eliminate the seasonal effect, alteration of operation parameters can be taken into consideration. For example, a possible strategy to improve EDC removal in the cold temperature is to increase the mixed liquor suspended solids (MLSS) concentration by raising the SRT. Generally enhanced EMPs removal can be achieved at warmer temperature due to promoted microbial activities. Nevertheless, it was found that operation at high temperature levels (45°C) could lead to lower EMPs removal. Some other studies showed that EMPs elimination was independent of temperature fluctuation (Luo et al., 2014).

1.8 Emerging micropollutants and sustainable treatment options

Water and wastewater treatment process for potable reuse is an important policy globally applied to enhancement water supply, especially in semiarid and arid areas, seaside communities faced with saltwater resources and regions where the quantity and/or quality of the water supply may be not potable (Etchepare and Van Der hoek, 2015). There is no specific treatment technology able to remove all EMPs types in WTPs and WWTPs due to their different behavior during treatments and climate change (hot seasons to cold seasons) (Verlicchi et al., 2010).

Environmental monitoring in water pollution control and management has been generally focused on conventional primary pollutants (Cruz and Barceló, 2015), nutrients, microbial pollutants, and heavy metals (Rodriguez-Narvaez et al., 2017).

The efficiency of a conventional WWTP varies depending on the characteristics of the pollutant and on the treatment process employed. The main processes for EMPs removal during the secondary treatment at WWTPs are Biological and/or chemical transformation and sorption mechanisms (Radjenović et al., 2009; Verlicchi et al., 2012).

The upgrading of the treatment processes for effluents generated by conventional WWTPs might minimize the discharge of micropollutants into the receiving waters and can even improve the overall quality status of effluents for possible reuse (Hollender et al., 2009; Eggen et al., 2014). Transformation of EMPs into less harmful or mineral compounds is one of the promising strategies to achieve the upgrading of the treatment processes. An overview of the current treatment options is present in the following sections to reveal the performance of each technique for EMPs removal and to identify the need for improvement.

1.8.1 Coagulation–flocculation

The EMPs removal from the treated secondary effluent in a coagulation/flocculation clarifier was evaluated. Based on reported data, around 80% reduction of musk (e.g., tonalide and galaxolide), 46% reduction of diclofenac, 42% reduction of naproxen and 23% reduction of ibuprofen were observed during coagulation–flocculation process in hospital wastewater treatment. It was cleared that during landfill leachate treatment by coagulation–flocculation process, some EDCs such as DHHP and nonylphenol were reduced around 70% and 90% respectively, but bisphenol A was not removed. Totally most EMPs is removed during coagulation–flocculation processes, except some musks, a few pharmaceuticals (e.g., diclofenac), and nonylphenol due to their high K_{ow} (4–6). In addition, temperature and coagulant dose do not have significantly effect on amount of EMPs removal during coagulation–flocculation processes. For example, hydrophobic compounds are better removed in water sources with high fat content. The elimination of some pharmaceutical (diclofenac, ibuprofen, and bezafibrat) it better done by the presence of dissolved humic acid. On the contrary, the presence of dissolved organic matters (DOM) with low molecular weight can possibly inhibit the micropollutant removal due to the preferential removal of DOM through coagulation. Negative charge DOM might react with positive charge aluminum hydrolysis species, resulting in a less dose coagulant available for EMPs removal (Luo et al., 2014).

1.8.2 Activated carbon adsorption

Adsorption by activated carbons (ACs) is commonly employed for controlling taste and odor in drinking water. Both powdered activated carbon (PAC) and granular activated carbon (GAC) have been widely used in adsorption processes (Luo et al., 2014).

Powdered activated carbon

PAC has been considered as an effective adsorbent for treating persistent/nonbiodegradable organic compounds. Also removal efficiency of >94% is confirmed for 100–1600 μg/L of EMPs under a carbon dose of 1.25 g/L and with a contact time of 5 min. PAC addition in activated sludge tank or posttreatment configurations is a major application of PAC in the full-scale municipal WWTPs. PAC addition in WWTPs was shown to be able to reduce EMPs levels by more than

80%. The performance of PAC in EMPs removal depends on PAC dose, contact time, the molecular structure, and behavior of the targeted compound, as well as the water/wastewater properties. Water/wastewater properties also affect the EMPs adsorption. Either higher dose or longer contact time can probably result in greater removal of EMPs. PAC addition appears an attractive method for upgrading municipal WWTPs for improved EMPs removal (Luo et al., 2014).

Granular activated carbon

Granular activated carbon (GAC) dosage typically applied to taste and odor control in drinking water (<10 mg/L) is sufficient to provide a 2-log removal for most of various compounds in a lake water. Specifically the removal efficiency under GAC can be ranged from 50% (tonalide and nonylphenol) to more than 90% (galaxolide). Considerable removals of steroidal estrogens from sewage effluent have been observed during the GAC tertiary treatment. By comparison, the reduction of pharmaceutical concentrations under granular activated carbon more variable. For example, higher removals (84%–99%) have been observed for mebeverine, indomethacine, and diclofenac, while some compounds (e.g., carbamazepine and propranolol) displayed much less removals (17%–23%). Similar to PAC, the contact time is a major factor that affects the degree of adsorption. Short contact time is likely to lead to significantly lowered adsorption efficiency. Properties such as pore shape/size and volumes of ACs, carbon type, surface charge of compounds, and operation year are noted to have influence on the removal performances. Broader micropore size distribution of the GAC leads to more efficient adsorption of EMPs with different shapes and sizes. Pore volume is more important to adsorption capacity than specific area, and larger pore volume was commonly associated with greater removal efficiency. Negatively charged EMPs are likely to be poorly adsorbed by the negatively charged carbon and well adsorbed by the positively charged carbon, and adsorption capacity reduced with operation year. GAC and PAC appear to be attractive methods for EMPs removal. In general efficient removal is potentially achievable when the compounds have nonpolar characteristics ($K_{ow} > 2$) as well as matching pore size/shape requirements. However, AC efficacy might be significantly lowered by the presence of natural organic matter (NOM) which competes for binding sides, thereby resulting in blocked pores. Besides, GAC dose, GAC regeneration and contact time play important roles in efficient removal of EMPs (Luo et al., 2014).

1.8.3 Ozonation and advanced oxidation processes

The refractory feature of some EMPs is led to conventional treatment processes such as physicochemical and biological processes could not provide adequate removal efficiency for these chemicals. To solve this problem, ozonation and advanced oxidation processes (AOPs) can be favorable options. Ozonation and AOPs are effective redox mechanisms which reveal more advantages (high degradation rates and nonselectivity) than conventional treatments. By direct or indirect production of stronger and less oxidizer agents such as $\cdot OH$, ozone can reduce

EMPs concentration. Some EMPs are vulnerable to both ozone and AOPs (e.g., naproxen and carbamazepine), whereas some are only subject to ˙OH (e.g., atrazine and meprobamate), and some are resistant to both forms of oxidation (e.g., TCEP and TCPP). The generation of ˙OH can be promoted with the presence of H_2O_2, Fenton reagent, and ultraviolet (Luo et al., 2014).

Ozonation is an acceptable process to reduce the EMPs load from full-scale WWTPs. In general under ozone dose of 15 mg/L, the concentrations of carbamazepine, diclofenac, indomethacin, sulpiride, and trimethoprim were considerably reduced by more than 95%. The reductions of DEET and metoprolol were modest. By contrast, bezafibrate was very resistant to ozonation and was removed by only 14% (Luo et al., 2014).

The role of O_3/H_2O_2 process in removal of some EMPs such as PPCPs and steroid hormones during water treatment has been confirmed. The process showed considerable removal efficiency (>90%) for almost all of the target contaminants, except TCEP (13%), TCPP (26%), atrazine (69%), meprobamate (80%), and ibuprofen (83%). They indicated that EMPs with the highest oxidation level were affected by high ozone and ˙OH rate constants. This feature is associated with their electron-rich groups such as olefins and activated aromatic, olefins, phenols, and anilines (Gerrity et al., 2011).

Although the ˙OH production is greater under alkaline conditions, it is reported that lower pH is favorable for EDCs removal by ozone from synthetic secondary effluent. It is notable that ozone was less reactive to the inorganic and organic matters in the synthetic secondary effluent in comparison with ˙OH (generated at high pH). For the more reactions with target compounds, a greater amount of O_3 can be used. Although suspended sludge particles are able to consume higher O_3 dose, the EMPs removal efficiency under O_3 process is not significantly affected by their presence (Hernández-Leal et al., 2011; Huber et al., 2003).

The removal efficiency of 41 pharmaceutical compounds under UV and UV/H_2O_2 processes was examined (Kim et al., 2009). UV alone could significantly remove (>90%) only some EMPs such ketoprofen, diclofenac, and antipyrine, while macrolides removal efficiency was 24%–34%. But addition of H_2O_2 (7.8 mg/L) improved removal efficiency up to 90% for 39 out of 41 compounds. Also effect of UV (wavelength: 254 nm), UV/H_2O_2, Fenton (Fe^{2+},$^{3+}$/H_2O_2), and Photo-Fenton ($Fe^{2+,3+}/H_2O_2$/UV and $Fe^{2+,3+}/H_2O_2$/simulated sunlight) on removal efficiency of 32 selected EMPs in municipal wastewater effluents has been investigated.

UV and H_2O_2 (50 mg/L) revealed the increased removal efficiency of EMPs (a total degradation of 81%). But after 30 min under UV/H_2O_2 process, the removal efficiency increased up to 97%. Also 31% removal efficiency for EMPs has been reported under the Fenton process (5 mg/L $Fe^{2+,3+}$/50 mg/L H_2O_2). Increasing H_2O_2 dosage and reaction time under the Photo-Fenton process had positive effect on the removal efficiency.

Fenton/UV254 (100% degradation after 90 min) displayed much higher degradation efficiency compared with Fenton/sunlight (47% degradation after 90 min).

In addition, the presence of DOM in the wastewater seemed to enhance the EMPs removal during all the processes. The higher removal efficiency of Photo-Fenton with solar light for removal of 52 EMPs in a WWTP effluent has been reported. This process is able to reduce 48 out of 52 compounds than its limitation of detection. Because under oxidation processes, EMPs mineralization do not completely occur, the formation of by-products is the major concern associated with these processes (Klamerth et al., 2010). In comparison with parent compounds, by-products compounds show the lower estrogenic and antimicrobial activities (Hollender et al., 2009; Reungoat et al., 2010). To overcome this problem, biological postfiltration (sand filtration or AC filtration) can be suggested (Luo et al., 2014). Rodriguez-Narvaez et al. (2017) compared removal efficiency of some EMPs under several AOPs processes (Table 1.4).

1.8.4 Membrane processes

Membrane filtration is a promising technique to remove micropollutants from water. The retention of compounds depends on their nature and is strongly influenced by the applied membrane material, the overall compounds of the raw water, and the process conditions (Hofman et al., 1993).

Although microfiltration (MF) and ultrafiltration (UF) processes are proved to efficiently remove turbidity, EMPs are generally poorly removed during UF and MF as the membrane pore sizes are much larger than the molecular sizes of EMPs. However, due to interaction with NOM in wastewater and adsorption on to membrane polymers, EMPs can be removed. Without NOM, removal efficiency of EMPs by UF with hydrophilic membrane is lower or insignificant than UF with hydrophobic membrane. NOM substances with high molecular weight such as humic acid shows greater effects on removal efficiency of EMPs than the lower molecular weight compounds (Jermann et al., 2009). The combination of MF or UF with other processes such as NF or RO is essential for enhanced removal of different EMPs. Removal efficiency under MF and MF/RO processes has been found to be 50% for DEHP and 65% to 90% for most EMPs, respectively (Garcia et al., 2013). The removal efficiency of NF membranes is very similar to that conducted by RO membranes. The average removal efficiency by tight NF membrane has been measured 82% and 97% for neutral and ionic contaminants, respectively, while RO process is able to show 85% and 99% removal for neutral and ionic contaminants, respectively (Luo et al., 2014).

1.8.5 Membrane bioreactor

MBR process combines activated sludge biological treatment and membrane filtration (MF and UF). MBR has many advantages including the high effluent quality, excellent microbial separation ability, absolute control of SRTs and HRTs, high biomass content and less sludge bulking problem, low-rate sludge production, small footprint and limited space requirement, and possibilities for a flexible and phased extension of existing WWTPs. MBRs are able to effectively remove a wide spectrum of EMPs including compounds that are resistant to activate sludge processes. This is because: (1) they are able to retain sludge to which many

Table 1.4 Removal of EMPs using AOPs (Rodriguez-Narvaez et al., 2017).

System	EMPs	Removal efficiency	Notes	References
UV	Estrone	90	$[C]_0 = 5$ mg/L pH = 6.5; 13 W low-pressure Hg lamp (254 nm; 18 mW/cm^2) $T = 20°C$; 30 min	Sarkar et al. (2014)
UV/H_2O_2	Doxycycline	100	pH = 3; $[Dox]_0 = 10$ mg/L; $[H_2O_2]_0 = 100$ mmol/L UV-C radiation; 5.03×10^{-5} Es^{-1}; 20 min	Bolobajev et al. (2016)
UV/ ozone	Caffeine	>95	$[C]_0 = 40$ mg/L; pH = 7; UV 32 W; 22.5 min	Souza and Feris (2015)
Ozone	E2	>99	$[O_3]_0$ E2 = 2.4 mol/L; 1 s	Vallejo-Rodríguez et al. (2014)
	EE2	80	$[O_3]_0$EE2 = 3.7 mol/L; 3 s	
	NPX	80	$[O_3]_0$ NPX = 4.75 mol/L; 30 s	
	Ibuprofen	90	$[O_3]_0$ IBP = 100 mmol/L; 2500 s	
	Ketoprofen	90–96	Lab water; $T = 24°C$;$[O_3]$ = 2.4 mg/L	
Ozone/ H_2O_2	Naproxen	96–98		Feng et al. (2015)
Ozone/ H_2O_2/UV	Estrone	>99	$[C]_0 = 5$ mg/L; pH = 6.5; low-pressure Hg lamp (254 nm; 18 mW/cm^2); $T = 20°C$; 30 min	Sarkar et al. (2014)
Fenton process	Doxycycline	100	$[C]_0 = 100$ mg/L; $[Fe^{+2}]_0 = 25$ mg/L; $[H_2O_2]$ 0 = 611 mg/L; $T = 35°C$	Borghi et al. (2015)
Photo-Fenton	Acetamiprid	70–90	[Fe] = 1, 2, 3 mg/L; H_2O_2/Fe ratio: 2:1; 4:1; pH = 2.8	Carra et al. (2015)
		90–100 100–100	Synthetic secondary effluent; 15 min; low-pressure UV lamp (30 W/m^2; 254 nm)	
Sono chemical	Dicloxacillin	>99	pH = 5.5; 600 kHz; [C] 0 = 0.21 mM; 180 min	Villegas-Guzman et al. (2015)

AOP, *Advanced oxidation processes;* E2, *estradiol;* EE2, *ethynilesradiol;* EMPs, *emerging micropollutants;* NPX, *naproxen.*
With permission.

compounds are adhered; (2) the membrane surface can also intercept the compounds; and (3) the longer SRT in MBRs may promote microbial degradation of the compounds. The removal of EMPs in MBR can be affected by a number of factors, such as sludge age and concentration, existence of anoxic and anaerobic compartments, composition of the wastewater, operating temperatures, pH, and conductivity (Luo et al., 2014).

Because partial removal efficiency (24%–68%) of EMPs occurs during conventional wastewater treatment, using technical removal methods such as MBRs with removal efficiency >90 percent is necessary.

During MBR processes, several operational parameters (e.g., SRT, HRT, and temperature) can influence the removal of EMPs. In general, MBRs have high SRTs, thus diverse microorganisms, including some slow growing bacterial, can reside in the reactors. When biomass is rich in nitrifying bacteria, higher biodegradation efficiency for certain EMPs can be achieved. It has been reported a high elimination (99%) of 17α-ethinylestradiol (at initial concentration of 83 ng/L) when a nitrifier enrichment culture was applied in a MBR. The degradation of EMPs by nitrifying bacteria has also been evaluated in other types of systems (e.g., activated sludge and fixed bed reactor). A general conclusion drawn from these studies is that nitrifying conditions have positive effects on EMPs removal. Temperature variability has been linked to decrease in bulk water quality parameters and unreliability of system, as microbial growth and activity as well as solubility and other physicochemical properties of organics are significantly affected by temperature. Effects of temperature variation were explored in a lab-scale MBR treating wastewater containing selected EMPs. Both hydrophobic compounds (logD > 3.2) and less hydrophobic compounds (logD < 3.2) showed reduced elimination at 45°C, which was ascribed to disrupted metabolic activity typically linked to such elevated temperature. The removal of hydrophobic compounds was unaffected in the temperature range of 10°C–35°C, while a relatively more obvious variation was found in the removals of less hydrophobic compounds. The advantages and challenges of different processes for the removal of EMPs are shown in Table 1.5.

1.9 Future challenges and perspectives

In order to advance the potential applications of MBR in the treatment of wastewater rich in EMPs, it is highly imperative to address the following issues.

- The efficiency removal of EMPs under MBR is not yet significant. The integrated system (MBR-NF and MBR-RO, MBR-PAC, and MBR-GAC) is effective to significantly remove refractory EMPs. But these integrated systems are not cost effective and energy saving. However, if the EMPs environmental impact is considered, this cost might be acceptable.
- The application of specific enzymes by using MBR could be an alternative to remove some refractory compounds such as carbamazepine, diclofenac, clofibric acid, SMX, and erythromycin. In addition, the use of enzymes instead of microorganisms plays a crucial role to reduce developing bacteria resistant to EMPs.

Table 1.5 Advantages, drawbacks, and recommendations for each advanced treatment (Rizzo et al., 2019).

Advanced treatment	Advantages	Drawbacks	Recommendations
UV/H_2O_2	• Moderate-good CEC removal at lab/pilot scale • Effective as disinfection process too	• Formation of oxidation transformation products • No full-scale evidences on CEC removal • Higher energy consumption compared to ozonation, specifically when high organic matter concentration acts as inner filter for UV radiation	Toxicity tests recommended
Photo-Fenton	• High CEC removal • Use of solar irradiation • Effective as disinfection process too	• Formation of oxidation transformation products • No full-scale evidences on CEC removal • At neutral pH 7 addition of chelating agents necessary. • Large space requirements for solar collectors	Toxicity tests recommended
UV/TiO_2	• High CEC removal • Use of solar irradiation • Effective as disinfection process too	• Low kinetics • Formation of oxidation transformation products • Catalyst removal	Not possible to apply until more efficient photocatalysts (at least one order of magnitude) will be developed
Ozonation	• High CEC removal • Full-scale evidence on practicability • Partial disinfection • Lower energy demand compared to UV/H_2O_2 and membranes	• Formation of by-products (NDMA, bromate) and other unknown oxidation transformation products • Need for a subsequent biological treatment (e.g., slow sand filtration) to remove organic by-products	• Toxicity tests recommended • NDMA and bromate should be monitored

(*Continued*)

Table 1.5 Advantages, drawbacks, and recommendations for each advanced treatment (Rizzo et al., 2019). *Continued*

Advanced treatment	Advantages	Drawbacks	Recommendations
PAC	• High CEC removal • Full-scale evidence on practicability • Additional DOC removal • No formation of by-products • Partial disinfection possible by the combination with membrane filtration (UF)	• PAC must be disposed • Posttreatment required (membrane, textile or sand filter) to prevent discharge of PAC • Production of PAC needs high energy • Adsorption capacity may fluctuate with each batch	Test with different products/process configurations recommended
GAC	• High CEC removal • Full-scale evidence on practicability • Additional DOC removal • No formation of by-products • An existing sand filtration can relative easily be replaced by GAC • GAC can be regenerated	• Production of GAC needs high energy • Still under investigation if more activated carbon is needed compared to PAC • Less flexible in operation than PAC and ozonation to react to changes in wastewater composition • Adsorption capacity may fluctuate with each batch	Test with different products recommended
NF and RO	• High CEC removal • RO can reduce salinity • Effective disinfection • Full rejection of particles and particle-bound substances	• High energy requirements • High investment and reinvestment costs • Disposal of concentrated waste stream • Need for pretreatment to remove solids	

GAC, *Granular activated carbon;* PAC, *powdered activated carbon;* UF, *ultrafiltration.*
With permission.

- The production of toxic transformation is the major limitation in enzymatic treatment. Posttreatment could be used to reduce the toxicity generated by the enzyme and the mediator.
- Due to enzyme and mediator cost, the use of enzymatic membrane reactor (EMR) at large scale is limited. To overcome this problem, the use of immobilized enzyme in a biocatalytic membrane reactor (BMR) is suggested.
- Membrane fouling due to extracellular polymeric substances (EPS) is the major challenge in MBR operation. Because of seasonal difference and unsuitable disposal of EMPs from homes and hospitals, the EPS level in the wastewater may be increased. To reduce the EMPs entry into water bodies, some changes are essential. The return of expired or unused EMPs back to the pharmacies can be accepted as regulation. Moreover an acceptable household strategy for disposal of drugs based on their labels may help to safe disposal of these compounds. Further studies are required in this field.
- Due to more complication of ESP fouling than other fouling, control MBR fouling is not effective by methodologies such as aeration, low flux operations, and aggressive cleaning. To overcome this problem, influent characteristics, pretreatments, operating conditions, membrane type (morphology, charge, roughness, hydrophobicity, and electrostatic interactions) should be considered. Besides, the development of new method may reduce energy consumption and cost of fouling control.
- Novel MBR designs such as electrochemical membrane bioreactor (EMBR) can save energy in wastewater treatment. Also integrated application of MBR with reverse electrodialysis (RED) and pressure retarded osmosis (PRO) can be suggested for saving energy and cost.
- Determination of biofilm thickness, concentration gradient of nutrients, and dissolved oxygen with modeling and/or simulation may save energy, time, and money by predicting long-term fouling.
- EMPs alone or in combination with biological systems can affect microbial activities. Exposure to especially high concentrations of EMPs that come from special locations like hospital sewage lead to a greater shift in the structure of the sludge communities (Besha et al., 2017). A shift in microbial community structure might be characterized by appearance or disappearance of a certain group of bacteria from the wastewater when exposed to chemical stress. Because of the presence of EMPs, microbial community structural shifts are already reported in several literatures (Kraigher et al., 2008; Harb et al., 2016; Lay et al., 2012; Lozada et al., 2004; Rosenkranz et al., 2013). In the presence of EMPs, microorganism may be clustered together or dispersed, and hence microbial community structural shifts (Besha et al., 2017).
- Microbial community in the activated sludge responds to new environment conditions to adapt with it. It is found that EMPs (ibuprofen, naproxen, ketoprofen, diclofenac, and clofbric acid) had a stronger effect on the microbial community structure. Also analysis of 16S rRNA gene sequencing indicats that the addition of EMPs leads to the absence of the genus *Nitrospira* that

plays a key role in the second stage of nitrification in wastewater treatment. Respirometric analysis helps to study the microbial activity of the sludge in the presence of EMPs. The presence of EMPs may affect on the respiration rate (as an index for microbial activity) and the activated sludge properties (Alvarino et al., 2018).

- The few monitoring efforts conducted at a larger scale, such as the river basin wide international exercises in the Danube or Elbe Rivers for water, the EMEP network for ambient air, or the LUCAS soil survey are related to the policies made for a single environmental compartment. But EMPs as many other toxicants are detected in the multi environments. Mono compartmental policies and the noncomprehensive monitoring schemes cannot provide the full picture on the exchange and transformation dynamics between all environments and EMPs (Geissen et al., 2015).
- Detection, identification, and quantification of EMPs and their by-products in the various environmental compartments are essential for obtaining knowledge on their occurrence and fate. This is highly challenging for several reasons: the number of currently known potential EMPs is very high. The variety in concentration of EMPs and changes in their characteristics with time in aquatic environments can be a challenge for quality control of water bodies. In recent years generic chemical screening approaches based on chromatography with full scan mass spectrometry have become increasingly popular to complement or replace existing targeted methods (Geissen et al., 2015).
- The 80 EMPs fate in WWPTs was predicted by development methods. For example, fate of volatile organic carbones (VOCs), surfactants, and metals (particularly priority metals Cd, Pb, and Ni), poly aromatic hydrocarbon (PAHs), bisphenol A (BPA) and some pesticides (like DDT, dieldrin, and lindane) were predicted by models. Based on these models and depending on their properties, fate of EMPs can be predicted under different processes such as runoff, erosion, or leaching and entering to groundwater or surface water. It is possible that some EMPs is captured in soil by adsorption process and never reached to water bodies.
- Although modeling EMPs fate in watershed scale has been conducted by some studies, it is necessary to predict EMPs fate at the interface with different media environments including water, particles and chemicals compounds in soil, groundwater, and surface water. Because environmental properties have impact on EMPs concentration in an environment, the calculation of EMPs concentration is different for the unsaturated soil, saturated groundwater, and surface water (Geissen et al., 2015). Also considering the EMPs fate without their parent compounds and/or their by-products in media water is an incomplete study. It can hide some more toxic by-products than their parent compounds (Pomies et al., 2013). Amount of EMPs transport and accumulation in watershed under different situation can be assessed by parameterized models. A model should give information on how the discharge and chemical fluxes at the outlet are related to EMPs emissions under certain meteorological

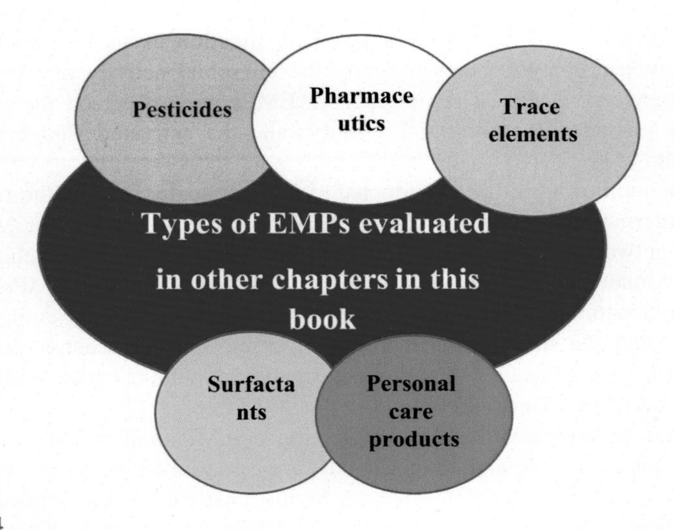

FIGURE 1.4

Type of EMPs evaluated by authors in the other chapters in this book. *EMPs*, Emerging micropollutants.

conditions, for watershed. In other words, model simulations should provide the discharge and chemical fluxes at the outlets, which can then be postprocessed to infer the distribution so travel times for EMPs in the aquatic bodies (Geissen et al., 2015).

In other chapters, the types of EMPs along with their challenges in aquatic environments will be described (Fig. 1.4).

1.10 Conclusion

Results of this chapter indicate that EMPs continue to cause new and serious challenges in aquatic environments and human health. Although production of new EMPs are shifting to growing economies worldwide, integrated safety management of EMPs by new technology, risk assessment methods, in situ and ex situ conservation strategies, and remediation technologies are necessary. For this reason, several factors must be taken into account: EMPs category and characteristics, EMPs concentration, EMPs risk assessment for environment and living organisms, and available resources and using developed methods to predict the fate of EMPs in aquatic environments. Also consumer education, minimize emerging micropollutants waste in places of production and consumption, government regulation, none disposal of EMPs in drains or toilets, medical innovation proper disposal after use and effective filtration in wastewater facilities could be employed to management of EMPs entering into water bodies. The results of this chapter along with future results will lead to future regulations and standards for management and control of risk effects of EMPs. In practical terms, obtained results will help to the knowledge of EMPs in aquatic environment and develop significant solutions to fill the gap faced in aquatic environments.

References

Alvarino, T., Suarez, S., Lema, J., Omil, F., 2014. Understanding the removal mechanisms of PPCPs and the influence of main technological parameters in anaerobic UASB and aerobic CAS reactors. J. Hazard. Mater. 278, 506–513.

Alvarino, T., Suarez, S., Lema, J., Omil, F., 2018. Understanding the sorption and biotransformation of organic micropollutants in innovative biological wastewater treatment technologies. Sci. Total. Environ. 615, 297–306.

Antakyali, D., Morgenschweis, C., De Kort, T., Herbst, H., German, G.G. & Grontmij, N. Micropollutants in the aquatic environment and their removal in wastewater treatment works. *Aqua Environ*.

Barbosa, M.O., Moreira, N.F., Ribeiro, A.R., Pereira, M.F., Silva, A.M., 2016. Occurrence and removal of organic micropollutants: an overview of the watch list of EU Decision 2015/495. Water Res. 94, 257–279.

Benotti, M.J., Brownawell, B.J., 2009. Microbial degradation of pharmaceuticals in estuarine and coastal seawater. Environ. Pollut. 157, 994–1002.

Besha, A.T., Gebreyohannes, A.Y., Tufa, R.A., Bekele, D.N., Curcio, E., Giorno, L., 2017. Removal of emerging micropollutants by activated sludge process and membrane bioreactors and the effects of micropollutants on membrane fouling: a review. J. Environ. Chem. Eng. 5, 2395–2414.

Bolobajev, J., Trapido, M., Goi, A., 2016. Effect of iron ion on doxycycline photocatalytic and Fenton-based autocatatalytic decomposition. Chemosphere 153, 220–226.

Borghi, A.A., Silva, M.F., Al arni, S., Converti, A., Palma, M.S., 2015. Doxycycline degradation by the oxidative fenton process. J. Chem. 2015.

Boxall, A.B.A. 2012. New and emerging water pollutants arising from agriculture.

Brunsch, A.F., Langenhoff, A.A., Rijnaarts, H.H., Ahring, A., Ter laak, T.L., 2019. In situ removal of four organic micropollutants in a small river determined by monitoring and modelling. Environ. Pollution. .

Bunke, D., Moritz, S., Brack, W., HERRáEz, D.L., Posthuma, L., Nuss, M., 2019. Developments in society and implications for emerging pollutants in the aquatic environment. Environ. Sci. Europe 31, 32.

Carra, I., Sirtori, C., Ponce-Robles, L., Perez, J.A.S., Malato, S., Aguera, A., 2015. Degradation and monitoring of acetamiprid, thiabendazole and their transformation products in an agro-food industry effluent during solar photo-Fenton treatment in a raceway pond reactor. Chemosphere 130, 73–81.

Chau, H., Kadokami, K., Duong, H., Kong, L., Nguyen, T., Nguyen, T., et al., 2018. Occurrence of 1153 organic micropollutants in the aquatic environment of Vietnam. Environ. Sci. Pollut. Res. 25, 7147–7156.

Cruz, S., Barceló, D., 2015. Personal Care Products in the Aquatic Environment. Springer.

Dalton, H., Stirling, D., 1982. Co-metabolism. Philos. Trans. R. Soc. London. B, Biol. Sci. 297, 481–496.

Ebrahimi, P., Barbieri, M., 2019. Gadolinium as an emerging microcontaminant in water resources: threats and opportunities. Geosciences 9, 93.
Eggen, R.I., Hollender, J., Joss, A., Schärer, M., Stamm, C., 2014. Reducing the Discharge of Micropollutants in the Aquatic Environment: the Benefits of Upgrading Wastewater Treatment Plants. ACS Publications.
Ejhed, H., Fang, J., Hansen, K., Graae, L., Ramberg, M., Magner, J., et al., 2018. The effect of hydraulic retention time in onsite wastewater treatment and removal of pharmaceuticals, hormones and phenolic utility substances. Sci. Total Environ. 618, 250–261.
Etchepare, R., Van Der hoek, J.P., 2015. Health risk assessment of organic micropollutants in greywater for potable reuse. Water Res. 72, 186–198.
Feng, L., Watts, M.J., Yeh, D., Esposito, G., Van hullebusch, E.D., 2015. The efficacy of ozone/BAC treatment on non-steroidal anti-inflammatory drug removal from drinking water and surface water. Ozone: Sci. & Eng. 37, 343–356.
Garcia, N., Moreno, J., Cartmell, E., Rodriguez-Roda, I., Judd, S., 2013. The application of microfiltration-reverse osmosis/nanofiltration to trace organics removal for municipal wastewater reuse. Environ. Technol. 34, 3183–3189.
Gavrilescu, M., Demnerová, K., Aamand, J., Agathos, S., Fava, F., 2015. Emerging pollutants in the environment: present and future challenges in biomonitoring, ecological risks and bioremediation. N. Biotechnol. 32, 147–156.
Geissen, V., Mol, H., Klumpp, E., Umlauf, G., Nadal, M., Van Der ploeg, M., et al., 2015. Emerging pollutants in the environment: a challenge for water resource management. Int. Soil. Water Conserv. Res. 3, 57–65.
Gerrity, D., Gamage, S., Holady, J.C., Mawhinney, D.B., Quiñones, O., Trenholm, R.A., et al., 2011. Pilot-scale evaluation of ozone and biological activated carbon for trace organic contaminant mitigation and disinfection. Water Res. 45, 2155–2165.
Grandclement, C., Seyssiecq, I., Piram, A., Wong-Wah-Chung, P., Vanot, G., Tiliacos, N., et al., 2017. From the conventional biological wastewater treatment to hybrid processes, the evaluation of organic micropollutant removal: a review. Water Res. 111, 297–317.
Gruchlik, Y., Linge, K., Joll, C., 2018. Removal of organic micropollutants in waste stabilisation ponds: a review. J. Environ. Manag. 206, 202–214.
Harb, M., Wei, C.-H., Wang, N., Amy, G., Hong, P.-Y., 2016. Organic micropollutants in aerobic and anaerobic membrane bioreactors: changes in microbial communities and gene expression. Bioresour. Technol. 218, 882–891.
Heberer, T., 2002. Occurrence, fate, and removal of pharmaceutical residues in the aquatic environment: a review of recent research data. Toxicol. Lett. 131, 5–17.
Hecker, M., Hollert, H., 2011. Endocrine disruptor screening: regulatory perspectives and needs. Environ. Sci. Europe 23, 15.
Hernández-Leal, L., Temmink, H., Zeeman, G., Buisman, C., 2011. Removal of micropollutants from aerobically treated grey water via ozone and activated carbon. Water Res. 45, 2887–2896.
Hofman, J., Noij, T., Schippers, C., 1993. Removal of pesticides and other organic micropollutants with membrane filtration. Water Supply 11, 129–139.
Hollender, J., Zimmermann, S.G., Koepke, S., Krauss, M., Mcardell, C.S., Ort, C., et al., 2009. Elimination of organic micropollutants in a municipal wastewater treatment plant upgraded with a full-scale post-ozonation followed by sand filtration. Environ. Sci. & Technol. 43, 7862–7869.

Hörsing, M., Ledin, A., Grabic, R., Fick, J., Tysklind, M., La Cour Jansen, J., et al., 2011. Determination of sorption of seventy-five pharmaceuticals in sewage sludge. Water Res. 45, 4470–4482.

Huber, M.M., Canonica, S., Park, G.-Y., Von Gunten, U., 2003. Oxidation of pharmaceuticals during ozonation and advanced oxidation processes. Environ. Sci. & Technol. 37, 1016–1024.

Jermann, D., Pronk, W., Boller, M., Schäfer, A.I., 2009. The role of NOM fouling for the retention of estradiol and ibuprofen during ultrafiltration. J. Membr. Sci. 329, 75–84.

Jiang, J.-Q., Zhou, Z., Sharma, V., 2013. Occurrence, transportation, monitoring and treatment of emerging micro-pollutants in waste water—a review from global views. Microchemical J. 110, 292–300.

Joss, A., Keller, E., Alder, A.C., Göbel, A., Mcardell, C.S., Ternes, T., et al., 2005. Removal of pharmaceuticals and fragrances in biological wastewater treatment. Water Res. 39, 3139–3152.

Kim, M.-K., Zoh, K.-D., 2016. Occurrence and removals of micropollutants in water environment. Environ. Eng. Res. 21, 319–332.

Kim, I., Yamashita, N., Tanaka, H., 2009. Performance of UV and UV/H_2O_2 processes for the removal of pharmaceuticals detected in secondary effluent of a sewage treatment plant in Japan. J. Hazard. Mater. 166, 1134–1140.

Klamerth, N., Malato, S., Maldonado, M., Aguera, A., Fernández-Alba, A., 2010. Application of photo-fenton as a tertiary treatment of emerging contaminants in municipal wastewater. Environ. Sci. & Technol. 44, 1792–1798.

Kraigher, B., Kosjek, T., Heath, E., Kompare, B., Mandic-Mulec, I., 2008. Influence of pharmaceutical residues on the structure of activated sludge bacterial communities in wastewater treatment bioreactors. Water Res. 42, 4578–4588.

Kwon, M., Kim, S., Yoon, Y., Jung, Y., Hwang, T.-M., Lee, J., et al., 2015. Comparative evaluation of ibuprofen removal by UV/H_2O_2 and UV/S2O82 − processes for wastewater treatment. Chem. Eng. J. 269, 379–390.

Lay, W.C., Zhang, Q., Zhang, J., Mcdougald, D., Tang, C., Wang, R., et al., 2012. Effect of pharmaceuticals on the performance of a novel osmotic membrane bioreactor (OMBR). Sep. Sci. Technol. 47, 543–554.

Lozada, M., Itria, R.F., Figuerola, E.L., Babay, P.A., Gettar, R.T., De Tullio, L.A., et al., 2004. Bacterial community shifts in nonylphenol polyethoxylates-enriched activated sludge. Water Res. 38, 2077–2086.

Luo, Y., Guo, W., Ngo, H.H., Nghiem, L.D., Hai, F.I., Zhang, J., et al., 2014. A review on the occurrence of micropollutants in the aquatic environment and their fate and removal during wastewater treatment. Sci. Total. Environ. 473, 619–641.

Nam, S.-W., Jo, B.-I., Yoon, Y., Zoh, K.-D., 2014. Occurrence and removal of selected micropollutants in a water treatment plant. Chemosphere 95, 156–165.

Nam S.W., Choi D. J., Kim S. K., Her N., Zoh K. D., 2014. Adsorption characteristics of selected hydrophilic and hydrophobic micropollutants in water using activated carbon, *J. Hazard. Mater.* **270,** 144–152.

Pablos, M.V., García-Hortiguela, P., Fernández, C., 2015. Acute and chronic toxicity of emerging contaminants, alone or in combination, in Chlorella vulgaris and Daphnia magna. Environ. Sci. Pollut. Res. 22, 5417–5424.

Pomies, M., Choubert, J.-M., Wisniewski, C., Coquery, M., 2013. Modelling of micropollutant removal in biological wastewater treatments: a review. Sci. Total. Environ. 443, 733–748.

Radjenović, J., Petrović, M., Barceló, D., 2009. Fate and distribution of pharmaceuticals in wastewater and sewage sludge of the conventional activated sludge (CAS) and advanced membrane bioreactor (MBR) treatment. Water Res. 43, 831–841.

Rasheed, T., Bilal, M., Nabeel, F., Adeel, M., Iqbal, H.M., 2018. Environmentally-related contaminants of high concern: potential sources and analytical modalities for detection, quantification, and treatment. Environ. international. .

Reungoat, J., Macova, M., Escher, B., Carswell, S., Mueller, J., Keller, J., 2010. Removal of micropollutants and reduction of biological activity in a full scale reclamation plant using ozonation and activated carbon filtration. Water Res. 44, 625–637.

Rizzo, L., Malato, S., Antakyali, D., Beretsou, V.G., Đolić, M.B., Gernjak, W., et al., 2018. Consolidated vs new advanced treatment methods for the removal of contaminants of emerging concern from urban wastewater. Sci. Total. Environ. .

Rizzo, L., Malato, S., Antakyali, D., Beretsou, V.G., Đolić, M.B., Gernjak, W., et al., 2019. Consolidated vs new advanced treatment methods for the removal of contaminants of emerging concern from urban wastewater. Sci. Total. Environ. 655, 986–1008.

Rodriguez-Narvaez, O.M., Peralta-Hernandez, J.M., Goonetilleke, A., Bandala, E.R., 2017. Treatment technologies for emerging contaminants in water: a review. Chem. Eng. J. 323, 361–380.

Rosenkranz, F., Cabrol, L., Carballa, M., Donoso-Bravo, A., Cruz, L., Ruiz-Filippi, G., et al., 2013. Relationship between phenol degradation efficiency and microbial community structure in an anaerobic SBR. Water Res. 47, 6739–6749.

Sarkar, S., Ali, S., Rehmann, L., Nakhla, G., Ray, M.B., 2014. Degradation of estrone in water and wastewater by various advanced oxidation processes. J. Hazard. Mater. 278, 16–24.

Souza, F.S., Feris, L.A., 2015. Degradation of caffeine by advanced oxidative processes: O_3 and O_3/UV. Ozone: Sci. & Eng. 37, 379–384.

Stasinakis, A.S., Kordoutis, C.I., Tsiouma, V.C., Gatidou, G., Thomaidis, N.S., 2010. Removal of selected endocrine disrupters in activated sludge systems: effect of sludge retention time on their sorption and biodegradation. Bio. Technol. 101, 2090–2095.

Suárez, S., Carballa, M., Omil, F., Lema, J.M., 2008. How are pharmaceutical and personal care products (PPCPs) removed from urban wastewaters? Rev. Environ. Sci. Bio/Technology 7, 125–138.

Talib, A., Randhir, T.O., 2017. Managing emerging contaminants in watersheds: need for comprehensive, systems-based strategies. Sustainability Water Qual. Ecol. 9, 1–8.

Ternes, T.A., Herrmann, N., Bonerz, M., Knacker, T., Siegrist, H., Joss, A., 2004. A rapid method to measure the solid–water distribution coefficient (Kd) for pharmaceuticals and musk fragrances in sewage sludge. Water Res. 38, 4075–4084.

Tröger, R., Klöckner, P., Ahrens, L., Wiberg, K., 2018. Micropollutants in drinking water from source to tap-method development and application of a multiresidue screening method. Sci. Total. Environ. 627, 1404–1432.

Varjani, S., Pandey, A., Tyagi, R., Larroche, C., Ngo, H.H., 2020. Current Developments in Biotechnology and Bioengineering: Emerging Organic Micro-Pollutants. Elsevier.

Vallejo-Rodríguez, R., Murillo-Tovar, M., Navarro-Laboulais, J., León-Becerril, E., López-López, A., 2014. Assessment of the kinetics of oxidation of some steroids and pharmaceutical compounds in water using ozone. J. Environ. Chem. Eng. 2, 316–323.
Verlicchi, P., Galletti, A., Petrovic, M., Barceló, D., 2010. Hospital effluents as a source of emerging pollutants: an overview of micropollutants and sustainable treatment options. J. Hydrol. 389, 416–428.
Verlicchi, P., Al Aukidy, M., Zambello, E., 2012. Occurrence of pharmaceutical compounds in urban wastewater: removal, mass load and environmental risk after a secondary treatment—a review. Sci. Total. environment 429, 123–155.
Villegas-Guzman, P., Silva-Agredo, J., González-Gómez, D., Giraldo-Aguirre, A.L., Flórez-Acosta, O., Torres-Palma, R.A., 2015. Evaluation of water matrix effects, experimental parameters, and the degradation pathway during the TiO_2 photocatalytical treatment of the antibiotic dicloxacillin. J. Environ. Sci. Health, Part. A 50, 40–48.
Virkutyte, J., Varma, R.S., Jegatheesan, V., 2010. Treatment of Micropollutants in Water and Wastewater. IWA Publishing.
Wanda, E., Nyoni, H., Mamba, B., Msagati, T., 2017. Occurrence of emerging micropollutants in water systems in Gauteng, Mpumalanga, and North West Provinces, South Africa. Int. J. Environ. Res. public. health 14, 79.
Wanda, E.M., Nyoni, H., Mamba, B.B., Msagati, T.A., 2018. Application of silica and germanium dioxide nanoparticles/polyethersulfone blend membranes for removal of emerging micropollutants from water. Phys. Chem. Earth, Parts A/B/C 108, 28–47.
Wang, Y., Roddick, F.A., Fan, L., 2017. Direct and indirect photolysis of seven micropollutants in secondary effluent from a wastewater lagoon. Chemosphere 185, 297–308.
Williamson, R.L., Burton, D.T., Clarke, J.H., Fleming, L.E., 1993. Gathering danger: the urgent need to regulate toxic substances that can bioaccumulate. Ecol. LQ 20, 605.
Zepp, R.G., Cline, D.M., 1977. Rates of direct photolysis in aquatic environment. Environ. Sci. & Technol. 11, 359–366.

CHAPTER

2 Pharmaceuticals as emerging micropollutants in aquatic environments

Afsane Chavoshani[1], Majid Hashemi[2,3], Mohammad Mehdi Amin[1,4] and Suresh C. Ameta[5]

[1]*Department of Environmental Health Engineering, School of Health, Isfahan University of Medical Sciences, Isfahan, Iran*

[2]*Environmental Health Engineering, School of Public Health, Kerman University of Medical Sciences, Kerman, Iran*

[3]*Environmental Health Engineering Research Center, Kerman University of Medical Sciences, Kerman, Iran*

[4]*Environment Research Center, Research Institute for Primordial Prevention of Non-Communicable Disease, Isfahan University of Medical Sciences, Isfahan, Iran*

[5]*Department of Chemistry, PAHER University, Udaipur, India*

2.1 Introduction

The progress of medical science and pharmaceuticals during the last century have allowed to better prevention of diseases, mortality, improve of life expectancy, health quality, and quality of daily life (Ebele et al., 2017; Sangion and Gramatica, 2016). It is estimated that the use of pharmaceuticals will reach 4500 billion doses by 2020 (Aitken and Kleinrock, 2015).

The largest use of medicines occurs China, Brasil, India, Russia, and Indonesia. Compared with medicines consumption, global spending for pharmaceuticals is mostly attributed to the USA, EU (Germany, UK, Italy, France, Spain), Japan, Canada, South Korea, and Australia (Mezzelani et al., 2018b). Unfortunately the pharmaceuticals overuse is very common that not only leads to health care costs and negative consequences on societies and individuals but also leads to the increased resistance of infectious microorganisms to many antibiotics in human and veterinary (Sangion and Gramatica, 2016). It was approved that humans apply around 10,000 pharmaceuticals with up to 3000 ingredients. Between 2000 and 2015, the rate of pharmaceuticals consumption has been increased, and the most consumption has been attributed to cholesterol-lowering drugs (Stadlmair et al., 2018).

The international market has showed substantial regional differences in the use of pharmaceuticals, influenced by economic status, health requirements, capacity for local manufacture, and legal restrictions. Leading developed countries dominate global pharmaceutical sales, with North America accounting for 45% (US$248 billion

Micropollutants and Challenges. DOI: https://doi.org/10.1016/B978-0-12-818612-1.00002-7

in 2004), Europe 13%, Japan 10%, and Australia 1% of recorded sales. In accordance with this pattern, the detection of pharmaceuticals in the aquatic environment has predominantly been reported in the developed world (USA, EU, Japan, and Australia) (Corcoran et al., 2010). Also some reports from India and China show high pharmaceutical concentrations discharged into aquatic environments (Balakrishna et al., 2017; Corcoran et al., 2010; Chen et al., 2012; Peng et al., 2008; Wu et al., 2015).

Generally there is a positive correlation between the most frequently used classes of pharmaceutical and their detection in the aquatic environment. Many of the top pharmaceuticals sold in the USA and the UK are in drug classes used to treat diseases associated with westernized society: for example, maintaining cholesterol balance, combating mental illness, treating stress, ulcers, asthma, etc., and they include specific beta blockers, lipid regulators, antidiabetic, antianginal drugs, analgesics, and antibiotics. Paralleling this usage, these therapeutic classes are the most commonly detected pharmaceuticals in both WWTPs effluents and surface water (Corcoran et al., 2010).

According to German consumption data in 2012, many existing substances are produced in high amounts, such as metformin (1.200 t), ibuprofen (IBU) (975 t), metamizole (615 t), acetaminophen (AMP) (458 t), iomeprol (255 t), and metoprolol (157 t). Also German consumption data of the past 10 years additionally show that the volume of these compounds may still heavily increase. As an example between 2002 and 2012 the consumption of metformin and IBU in Germany increased from 390 to 1200 t and from 250 to 975 t, respectively. Many of the existing substances have been detected in the environment (Kuster and Adler, 2014). Based on one report, worldwide pharmaceuticals consumption will reach 4.5 trillion doses until 2020. Also antibiotics consumption alone reached in 34.8 billion in 2015 (Patel et al., 2019).

Although the unique global increase of pharmaceuticals is referred to one of the greatest benefits of health in society, it has also been paralleled by discharge of these compounds in natural environment and ecosystem (Garrison, 1976). Nowadays pharmaceuticals are known as emerging micropollutants (EMPs) in environment (Snyder et al., 2003). First during the Catchment Quality Control (CQC) in US, it was found that pharmaceutical compounds can enter the water bodies by two main routes: the industrial and the domestic routes (Garrison, 1976).

Pharmaceutical contaminants differ from most other contaminants based on: (a) having molecular masses $<$500 Da, although larger for some compounds, (b) containing chemically complex molecules with a large variety of structures, shapes, molecular masses, and functionalities, and (c) consisting of polar compounds having more than one ionizable group. Pharmaceuticals can also (d) exhibit properties and a degree of ionization that depends on the medium's pH, and (e) have lipophilic properties, while some may also have moderate water solubility. They also share (f) the ability to persist in nature, accumulate in life forms, and remain biologically active. For example, naproxen, sulfamethoxazole, and erythromycin can persist for almost one-year while clofibric acid can remain unchanged for multiple years. Finally (g) these molecules tend to adsorb and be distributed in a living body, which metabolically modifies their chemical structure (Patel et al., 2019).

Not long after, the discharges of different pharmaceutical compounds were reported in British rivers and samples of Waste Water Treatment Plants (WWTPs) (Aherne and Briggs, 1989; Jones et al., 2002; Ashton et al., 2004; Bound and Voulvoulis, 2005, 2006). Technical advancements in detection methods have led to accurate measurements of low levels (ng/L and μg/L range) of pharmaceuticals in aquatic environments (Öllers et al., 2001; Gros et al., 2006; Trenholm et al., 2006).

Unlike traditional chemical pollutants, medicines are bioactive compounds, designed to be effective at very low concentrations, and they are continuously discharged in natural ecosystems, potentially affecting aquatic species. Scientific investigations suggested that pharmaceuticals have the capability to interfere with biota at very low concentrations, but these substances are not routinely monitored in environments (Mezzelani et al., 2018b).

Pharmaceutically active compounds (PhACs) found in the both human and veterinary drugs include a mixture of parent and metabolite compounds. These active ingredients are usually more polar and hydrophilic than the original compounds. They can reach to the aquatic environments by different routes such as human excretion, disposal of unused and expired drugs, agricultural, and livestock practices. Their continuous discharge into the aquatic environment may cause adverse effects on living organisms and the environment for instance, it can induce behavioral changes in fish, affecting fish aggression, reproduction, and feeding activity (Mandaric et al., 2019).

Therefore due to their environmental side effects in aquatic and terrestrial environments, in flora, biota, and human health, during the past three decades, the occurrence of pharmaceuticals in the environment has received increased scientific attention (Kalyva, 2017; Kuster and Adler, 2014).

Although, for safety and efficacy aspects, strict regulations and controls are derived by pharmaceutical companies and institutions such as Food and Drug Administration (FDA) or the EMA (Debaveye et al., 2016), there are many environmental challenges to face now and in the future (Celiz et al., 2009; Rahman et al., 2009; Balaram, 2016; Bu et al., 2016).

Based on the review of priority substances in surface water bodies, diclofenac (DIC), ethinylestradiol (EE2), and beta-estradiol had widespread use on the European monitoring list. (Zenker et al., 2014). Increasing scientific and technology knowledge on effects of environmental pharmaceuticals was led to that EMA and European Parliament approved guidelines and legislation on environmental risks assessment of pharmaceuticals. This concept emphasized on the detection, assessment, understanding, and prevention of adverse effects of the presence of pharmaceuticals in the environment, which affect both human and the other animal species (Mezzelani et al., 2018b). After new findings on the numerous adverse effects of these compounds, European Commission added 17a-ethinyl estradiol, 17b-estradiol, and DIC to the Watch List of the daughter Water Framework Directive (2013/39/EU), modifying Environmental Quality Standards Directive (2008/105/EC), and Member States will thus be obliged to monitor such substances at least annually, on a limited number of representative monitoring stations, for up to 4 years

(Mezzelani et al., 2018b). It is clear that these mentioned pharmaceuticals are only examples of the large variety of substances that can impact the quality of the water bodies and the environment (Zenker et al., 2014). Pharmaceuticals with basic features ($pK_a > 7$) prefer to bind to suspended solids. These compounds tend to accumulate in biota (bioaccumulation) (Zenker et al., 2014).

2.2 Persistence, bioaccumulation potential, and toxicity of pharmaceuticals

The absorption and interaction capacities of pharmaceuticals with living organism makes them a threat for the whole ecosystem. Biologically active compounds existed in pharmaceuticals may lead to their adverse effects on the wildlife (so-called nontarget organisms). Pharmaceutical occurrence in environment is a reason for scientific challenges to prevent the increase of negative impacts and provide the good quality standards (Sangion and Gramatica, 2016).

Consequently, the Environmental Risk Assessment (ERA) for human and veterinary pharmaceuticals was accepted by the European legislation with two EC Directives 2004/27/EC and 2004/28/EC laws. Today the ERA is conducted based on the EMA guidelines (Sangion and Gramatica, 2016).

An ERA refers to acute toxic risk occurred in the aquatic environment. This risk calculated based on the ratio between the predicted environmental concentration (PEC) of the compounds, and the highest predicted no-effect concentration (PNEC) of these compounds. A PEC: PNEC ratio <0.1 is considered insignificant risk; 0.1–1, low risk; 1–10, moderate risk; and >10, high risk. The environmental hazard of a substance is classified by the following characteristics (Deblonde and Hartemann, 2013):

- persistence-ability to resist degradation in the aquatic environment;
- bioaccumulation–accumulation in adipose tissue of aquatic organisms; and
- toxicity-potential to poison aquatic organism.

Each of these features is assigned a numerical value (0–3). The total of these numerical values shows the persistence, bioaccumulation, and toxicity index (PBT index) for the substance, which ranges from zero to nine. The higher PBT index of a pharmaceutical is related to its greater hazard in the environment. Pharmaceuticals with a PBT index of 9 are antifungals (ketoconazole, miconazole, terbinafine), antiinfectives (ofloxacin, efavirenz), antineoplastic agents (dasatinib, docetaxel, tamoxifen, and megestrol) and drugs for the nervous system (propofol, bromocriptine, clozapine, citalopram, etc.) (Deblonde and Hartemann, 2013).

Bioconcentration is the ability of an organism to accumulate a chemical from the ambient environment. The bioconcentration factor (BCF) is the ratio of the concentration of the chemical accumulated in the organism to its concentration in the ambient environment. A range BCF values for the bioaccumulation of pharmaceuticals have been reported. Table 2.1 show BCF ranges related to several pharmaceuticals in fish prey.

According to the OECD the main criteria for accumulation is a $\log K_{ow} > 3$, meaning there is a tendency for accumulation (Rahman et al., 2012). The authors Howard and Muir shown 92 out of 275 pharmaceuticals detected in the environment have the potential bioaccumulative properties (Howard and Muir, 2011). Some ionophore antibiotics used as veterinary drugs were detected in sediments at higher concentrations than in water (Kim and Carlson, 2006). Large amounts of gemfibrozil, IBU, and DIC were also found to be bound to sewage sludge (Yu and Wu, 2012).

2.3 Necessary data for ecological risk assessment of pharmaceuticals

Most available data are associated with ecological risk assessment of pharmaceuticals in different aquatic environments, but few data are available for the

Table 2.1 BCF ranges of environmentally relevant pharmaceuticals in fish prey (Zenker et al., 2014).

Compound	BCF	Compartment	References
Diclofenac	4.9	Blood plasma	Lahti et al. (2011)
Diclofenac	657, 320–950	Bile	Mehinto et al. (2010)
Diclofenac	12–2732	Liver	Schwaiger et al. (2004)
Diclofenac	5–971	Kidney	Schwaiger et al. (2004)
Diclofenac	3–763	Gills	Schwaiger et al. (2004)
Diclofenac	0.3–69	Muscle	Schwaiger et al. (2004)
Fluoxetine	185–900	*Gammarus sp.*	Meredith-Williams et al. (2012)
Fluoxetine	8.8–260a, 80	Body	Nakamura et al. (2008) and Metcalfe et al. (2010)
Norfluoxetine	80–650	Body	Nakamura et al. (2008)
Ibuprofen	0.08–1.4	Blood plasma	Nallani et al. (2011)
Ibuprofen	28	Body	Wang and Gardinali (2013)
Diphenhydramine	16	Body	Wang and Gardinali (2013)
Diltiazem	16	Body	Wang and Gardinali (2013)
Carbamazepine	1.4	Body	Wang and Gardinali (2013)
Carbamazepine	1.9	Muscle	Garcia et al. (2012)
Carbamazepine	4.6	Liver	Garcia et al. (2012)
Gemfibrozil	113	Blood plasma	Mimeault et al. (2005)
Sulfamethazine	0.61–1.19	Muscle	Hou et al. (2003)

BCF, *Bioconcentration factor.*
With permission.

pharmaceuticals fate in aquatic environments and other settings. Acute lethal toxicity of pharmaceuticals can be determined by testing algae, invertebrates, and fish in laboratory studies.

Some pharmaceutical compounds like estrogenic hormones can be toxic at low concentrations (ng/L) in several organisms such as fish, amphibians, or molluscs. In France, carbamazepine, DIC, and IBU are frequently isolated in surface water at concentrations of >10 ng/L. Several indices are used to evaluate toxic effects:

- biomarkers;
- toxicity bioassays;
- mesocosm studies (or multiple species bioassays); and
- field studies (population measurements, index of biotic integrity).

Table 2.2 shows main ecotoxicological effects of some pharmaceuticals.

Acute toxicity tests with bacteria, algae, invertebrates, and fish have been performed on more than 150 individual compounds. Crustacea is the most sensitive trophic group toward antibiotics during a 24 h EC_{50} ranging between 10.23 and 18.66 mg/L, while antiinflammatory compounds exhibited a 24 h EC_{50} between 13 and 140 mg/L for invertebrates (Mezzelani et al., 2018b).

2.4 Sources and pathways of pharmaceuticals in aquatic environments

The pharmaceutical resources can be classified into point source and nonpoint source. Point source pollution is a single identifiable source of pollutants (Lapworth et al., 2012). For instance, industrial, hospital, and sewage treatment effluents, as well as the septic tank are the main point sources for pharmaceuticals discharged to the water bodies. On the contrary, in nonpoint source, it is inherently difficult to be addressed the pollution source location based on geographical scales. Nonpoint source pollution usually results from runoff such as agricultural runoff and rainfall and snowmelt runoff. Due to higher chance for natural treatment in soil and other environmental media, pollution load of nonpoint source is lower than point source (Li, 2014).

Fig. 2.1 shows the pathways of pharmaceuticals discharged from the resources and end up to the receiving water bodies. Based on this figure, the main sources are including landfill, animal waste, freshwater aquaculture waste, hospital waste, industrial waste, and domestic waste. Also the pharmaceutical receptors are soil zone, groundwater, and surface water (Li, 2014).

The disposal of unwanted pharmaceuticals by consumers in domestic trash is a reason for detection of these compounds in landfill leachate. It has been found that several kinds of pharmaceuticals have been detected in the wells located near of the landfill (Li, 2014; Tischler et al., 2013). Entry into the environment via this way is dependent on the patient habits, and the efficiency of prescription practices

Table 2.2 Main ecotoxicological effects reported for marine species exposed to various molecules of different pharmaceutical classes at selected doses of exposure (Mezzelani et al., 2018b).

Therapeutical class	Species	Molecule	Exposure doses	Effects	References
Antiinflammatory drugs	*Mytilus galloprovincialis*	Acetaminophen Diclofenac Ibuprofen Ketoprofen Nimesulide	0.5, 2.5, 25 μg/L	Alteration of immunological parameters (lysosomal membrane stability, granulocyteshyalinocytes ratio, and phagocytosis activity), modulation of lipid metabolisms (ACOX, neutral lipids, and lipofuscin content), DNA fragmentation. Variation of transcriptional profile (genes involved in cellular turnover, immunological functions, and arachidonic acid and lipid metabolisms)	Mezzelani et al. (2016a,b, 2018a)
	M. galloprovincialis	Diclofenac	0.25 μg/L	Induction of antioxidant enzymes, increment of lipid peroxidation, enhanced levels of gonad vitellogenin-like proteins, and modulation of NF-kB pathway	Gonzalez-Rey and Bebianno (2011, 2012, 2014), Maria et al. (2016)
		Ibuprofen	1–100 μg/L	Oxidative stress, activation of immune responses	Ericson et al. (2010), Carvalho and Santos (2016)
	Mytilus spp.		1, 100, 1000, 5000, 10,000 μg/L	Decrement of byssus strength and energy available for growth and reproduction	Ericson et al. (2010)
	Ruditapes philippinarum	Ibuprofen	10, 100, 500, 1000 μg/L	Alteration of immunological parameters, transcriptional changes for genes involved in arachidonic acid metabolisms, apoptosis, peroxisomal proliferator-activated receptors, NF-kB and xenobiotic metabolisms, inhibition of superoxide dismutase, acetylcholinesterase, and lysozyme activities	Matozzo et al. (2012) and Milan et al. (2013)
	Carcinus maenas	Ibuprofen	0.1, 1, 5, 10, 15, 50 μg/L	Alteration of immunological parameters, oxidative stress, and genotoxic damage	Aguirre-Martínez et al. (2013, 2016)
	Crassostrea gigas	Acetaminophen Ibuprofen	1–100 μg/L	Transcriptional changes in genes involved in drug metabolism and biotransformation	Bebianno et al. (2017), Serrano et al. (2015), and

(Continued)

Table 2.2 Main ecotoxicological effects reported for marine species exposed to various molecules of different pharmaceutical classes at selected doses of exposure (Mezzelani et al., 2018b). *Continued*

Therapeutical class	Species	Molecule	Exposure doses	Effects	References
Psychiatric drugs	*Sepia officinalis*	Fluoxetine	1 ng/L	Inhibition of striking prey efficiency	Di Poi et al. (2013)
	M. galloprovincialis	Fluoxetine	0.03, 0.3, 3, 30, 300 ng/L	Lysosomal alterations, lipid peroxidation, and modulation of genes involved in detoxification	Franzellitti et al. (2014)
	M. galloprovincialis	Fluoxetine	75 ng/L	Oxidative stress, impairment of endocrine system	Gonzalez-Rey and Bebianno (2013)
	Venerupis philippinarum, Venerupis decussatus, M. galloprovincialis	Carbamazepine	0.03, 0.3, 3, 6.0–9.0 μg/L	Oxidative stress, impairment of reproductive capacity	Almeida et al. (2015) and Oliveira et al. (2017)
	M. galloprovincialis	Carbamazepine	0.01, 0.1, 1, 10, 100, 200 μg/L	Oxidative stress, activation of immune responses, and DNA damage	Martin-Diaz et al. (2009) and Tsiaka et al. (2013)
Cardiovascular drugs	*M. galloprovincialis*	Propranolol	0.3, 3, 30, 300, 30,000 ng/L	Alteration of immunological parameters and modulation of cAMP pathway	Franzellitti et al. (2011, 2013)
	Mytilus edulis	Propranolol	1, 100, 1000, 5000, 10,000 μg/L	Decrement of byssus strength and energy available for growth and reproduction	Ericson et al. (2010)

With permission.

leading to fewer unfinished prescriptions. A study about the disposal habits of the American population indicated that only 1.4% of the people returned unused drugs to the drug stores, but 54% threw them away and 35.4% disposed of them in the sink or toilet (Bound and Voulvoulis, 2005).

Wastewater resources (hospital, aquaculture, industrial, and domestic) are considered as one of the most important point sources of pharmaceuticals to release in the aquatic environment (Glassmeyer et al., 2005). Many studies conducted in different countries such as British, Australia, USA, and Spain have been confirmed the presence of 16 to 54 kinds of pharmaceuticals in wastewater (Li, 2014).

Septic tank is a small sewage treatment which mainly collects the waste generated by the domestic household. According to the early study conducted by Godfrey et al. (2007), there are 22 different kinds of pharmaceuticals detected in the septic system and 18 out of them are higher than the detection limit with the caffeine being the most common. Recent studies also found that the septic system in USA contains similar kinds of pharmaceuticals including IBU, paracetamol, salicylic acid, and triclosan (Godfrey et al., 2007; Carrara et al., 2008; Conn et al., 2010).

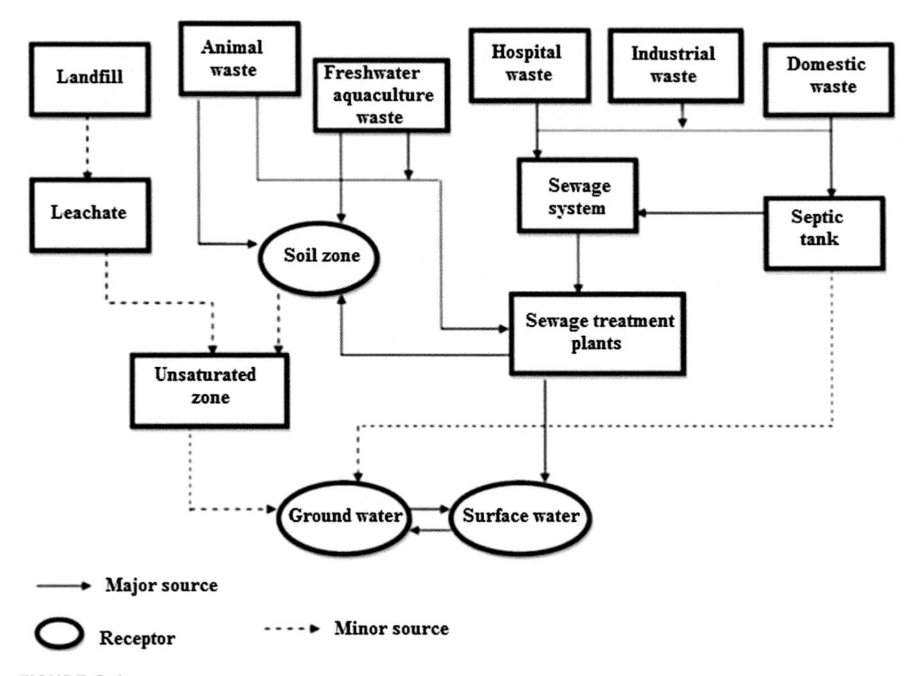

FIGURE 2.1

Potential sources and pathways of pharmaceuticals pollution in soil and water (Li, 2014). *With permission.*

Land application of sewage sludge is a nonpoint source for the pharmaceuticals distribution in soils and the freshwater (Lapworth et al., 2012). It is estimated that 8×10^6 dry tons of sludge produced, and due to having nutrient and organic matter, 50% of them are applied for agricultural application to improve the soil quality (Kinney et al., 2008). Veterinary drugs such as tetracycline and sulfamethazine usually found in farm manure at concentrations of 66 and 40 mg/L, respectively. Veterinary pharmaceuticals could enter the aquatic environments not only through direct application but also indirect application (Kim et al., 2008).

Incomplete remove of drugs during wastewater treatment is a reason for the presence of pharmaceuticals in groundwater and drinking water resources (Sui et al., 2015). For instance, drugs, such as carbamazepine, atenolol (ATE), metoprolol, trimethoprim, or DIC are partiality had removal efficiency of $<10\%$ for most of them and 10%–39% for DIC (Hernando et al., 2006). Also, the occurrence of 93 pharmaceuticals, illicit drugs and their metabolites has been investigated in stabilized sewage sludge from five municipal wastewater treatment plants (WWTPs) in the Slovak Republic. The sludge production from the five tested plants was >8100 tons in 2016, which is approximately 15% of the total Slovak sewage sludge production in 2016. The highest total concentration of all pharmaceuticals was found between 11,800 and 11,300 ng/g dry matter. Among individual pharmaceuticals, the highest concentrations were recorded for fexofenadine (mean 2340 ng/g and maximum of 5600 ng/g dry matter) and telmisartan (mean 1170 ng/g and maximum of 3370 ng/g dry matter) (Ivanova et al., 2018).

2.5 Methods of pharmaceutical removal from water and wastewater

Because of nonbiodegradable feature and pseudopersistence, pharmaceutical removal is ineffective during the conventional wastewater treatments. Biodegradation and persistence of pharmaceuticals is influenced by molecular structures such as chloro-, nitro-, and fluoro- functions attached to aromatic rings (Patel et al., 2019). Remediation efficiencies can be less than 10% in the case of such pharmaceuticals as carbamazepine, ATE, acetylsalicylicacid, DIC, mefenamic acid (MA), propranolol (PRP), ATE, clofibric acid, and lincomycin (Patel et al., 2019).

When these incompletely treated effluents are discharged into the aquatic environments and used by end users for agricultural application, the reused water may impose many hazards to humans, soil, crops, and biota. Hence, pharmaceutical compounds have been known as important environmental pollutants (Kanakaraju et al., 2018). Therefore it is necessary to conduct more research for design the effective and advanced methods to remove pharmaceuticals from wastewater effluents and consequently the environment.

2.5.1 Conventional treatment methods

Treatment processes in conventional drinking water plants are including coagulation/flocculation, clarification, and filtration. These processes may be followed by advanced treatments such as activated carbon, membrane filtrations, ozone oxidation, or advanced oxidation processes (AOPs). Conventional drinking water treatment cannot remove pharmaceuticals from surface water. However, low elimination of pharmaceuticals is possible during disinfection of drinking water by chlorine, chlorine dioxide, and ozone processes.

In clarification unit destabilization of colloid particles occurs to facilitate and promote their flocculation and settling along with sediments and suspended solids. By ferric chloride leading to acid or basic hydrolysis of compounds, hydrophilic pharmaceuticals such as AMP, sulfamethoxazole, and dehydronifedipine decrease to 25%. Clarification only removed minor amounts of carbamazepine, erythromycin·in H_2O, whereas the average removal of caffeine, ethynylestradiol, estrone, estradiol, progesterone, and androstenedione was calculated to be ~6% with alum and ferric salt coagulants. Low percent removal of pharmaceuticals after coagulation process in water treatment is due the reactivity of these compounds with chlorine added to water producing HOCl. Disinfection as a final barrier in water treatment plants can lead to a significant removal of a large number of microbes and contaminants. In the presence of chlorine in disinfected drinking water resources, degradation of some compounds increases, but pharmaceuticals are resistant even after treatment.

Due to its cost effective, chlorination process is mostly used in conventional treatment plants. Hypochlorous acid (HOCl) is the active form of chlorine in water generated by the reaction between water and chlorine [Eqs. (2.1) and (2.2)].

$$Cl_2 + H_2O \leftrightarrow HOCl + H^+ + Cl^- \quad (2.1)$$

Hypochlorous acid, a weak acid, ionizes as follows:

$$HOCl \leftrightarrow H^+ + OCl^- \quad (2.2)$$

The degree of ionization of HOCl depends on the pH and water temperature. Hypochlorous acid also reacts with organic matter to form many by-products including total trihalomethanes (TTHM).

Conventional wastewater treatment is usually conducted in primary and secondary units. In secondary process, activated sludge and microbes cannot completely adsorb and digest many pharmaceuticals, respectively. Also pharmaceuticals prevent from microbial activities in digestion of organic compounds.

Hydrophilic and hydrophobic structures of pharmaceuticals play a key role in their removal efficiency. Polar and semi polar of pharmaceuticals display lower and higher removal efficiency, respectively. When high polarity increases water solubility, adsorption process decreases. Pharmaceutical biodegradation mainly occurs when they are dissolved while pharmaceuticals adsorption requires hydrophobic properties, electrostatic interactions, and suspended solids.

2.5.2 Advanced oxidation processes

Among the water treatment methods employed up to now, advanced oxidation processes (AOPs) shows more potential for removing a wide range of priority and emerging micropollutants (Chavoshani et al., 2018). AOPs is defined based on in situ generation of powerful reactive oxygen species (ROS) with quit low selectivity such as hydroxyl radicals (·OH), H_2O_2, O_3, and superoxide anion radicals ($O_2^{-\cdot}$), leading to complete mineralization of organic compounds to CO_2, H_2O, and inorganic ions or acids. Advanced oxidation processes are including Fenton, photoFenton, photocatalysis, ozonation, microwave radiation, ultrasonic radiation, and electrochemical oxidation. Fig. 2.2 displayed a noticeable increase in the number of studies on AOP application for pharmaceutical removal between 2000 and 2018 (Kanakaraju et al., 2018).

AOPs can initiate by reactive oxygen or free radical species to degrade pharmaceuticals to simple and nontoxic compounds. Such as other free radical species, ·OH radicals are nonselective and powerful oxidizing agents ($E^{\cdot} = 2.80$ V). After fluorine ($E = 3.03$ V), they are ordered second.

Reactions of ·OH radicals with organic compounds can achieve either by hydrogen abstraction [Eq. (2.3)] from C-H, N-H, or OH groups, or radical–radical interactions, e.g., the addition of molecular O_2 leading to the formation of the peroxyl radical [Eq. (2.4)], or through direct electron transfer

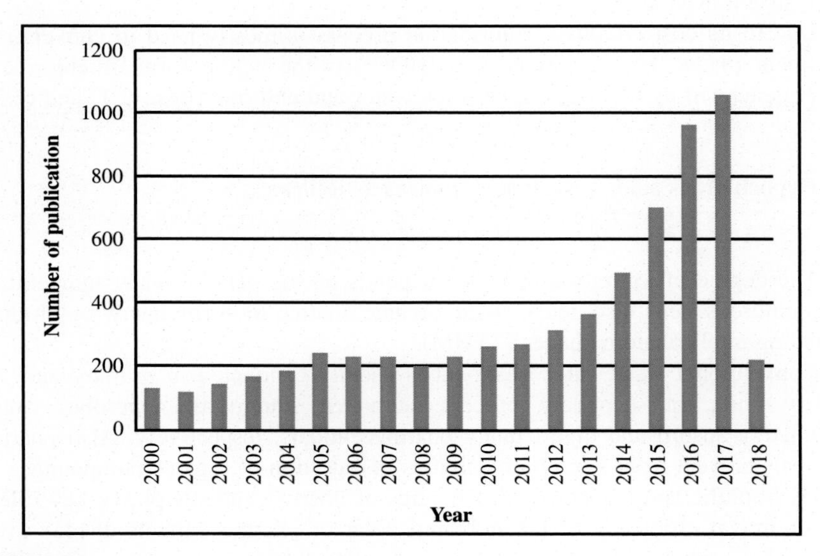

FIGURE 2.2

Statistics of publications (2000–18) on applications of AOPs for pharmaceutical (Kanakaraju et al., 2018). *AOPs*, Advanced oxidation processes.

With permission.

[Eq. (2.5)] yielding oxidized intermediates or, in the case of complete mineralization, the production of CO_2, H_2O and inorganic acids. Despite the high oxidation potential, kinetic rates of interactions between ·OH radicals and organic compounds depend on the affinity of these compounds for the oxidant.

$$\cdot OH + RH \rightarrow R\cdot + H_2O \quad (2.3)$$

$$R\cdot + O_2 \rightarrow RO_2^{\cdot} \quad (2.4)$$

$$\cdot OH + RX \rightarrow RX\cdot + + HO^{-} \quad (2.5)$$

In this section, the most AOPs such as ozonation, Fenton, photoFenton, sonolysis, UV and UV/peroxide oxidation, electrochemical oxidation, radiation, and TiO_2 photocatalysis are discussed. According to Table 2.3, These reactions are classified into three groups including photochemical processes, nonphotochemical processes, and hybrid or combined processes.

2.5.2.1 Ozenation

Ozonation and combinations with other oxidants such as O_3 with H_2O_2 (O_3/H_2O_2), and O_3 with UV (O_3/UV) have been used to remove pharmaceuticals in water. Different variables such as pH, ozone dose, and temperature influence on the removal and mineralization of pharmaceuticals in these treatments. Due to the short half-life of ozone, this method is expensive and need to high energy for real application. With respect to persistent by-products generated during ozonation process, low mineralization of pharmaceuticals via this method is reported. Therefore evaluating the toxicity before and after treatment by ozonation is necessary. The addition of H_2O_2 and irradiation have been suggested to increase the degree of mineralization (Kanakaraju et al., 2018).

Effects of operation variables like pH, ozone dose, water matrix, and presence of organic matter on the removal efficiency of pharmaceuticals have been studied. The effect of different pH ranges (3, 7, and 10) in amoxicillin removal revealed that the highest reaction rate (1.970/min) occurs during the alkaline pH (pH 10) in compared with acidic and neutral conditions. The higher formation of ·OH radicals at alkaline pH is reason of high removal efficiency. In contrast the highest degradation of salicylic acid (40%–50%) has been reported at acidic pH (pH 4) compared with basic pH (pH 8 and 10) with removal efficiency of 10%–20%. Although it is known that the concentration of ·OH radicals increases with increasing pH, the degradation of salicylic acid is happened by direct ozone oxidation instead (Kanakaraju et al., 2018). Due to the presence of natural organic matter (NOM), higher ozone dose is consumed during pharmaceutical degradation from WWTP effluent than synthetic wastewater. Also the presence of functional groups such as carboxyl, aliphatic hydroxyl, and aryl groups leads to difference in the elimination rate constant under ozonation process.

Table 2.3 Overview on selected recent publications on the application of different types of AOPs to the degradation of pharmaceuticals (Kanakaraju et al., 2018).

AOPs applied	Pharmaceuticals	Water matrix	Significant findings	References
Single AOP ozonation	Antibiotics, steroid hormone, lipid regulator, antineoplastic, nonsteroidal antiinflammatory drug, and psychostimulant	Synthetic wastewater, surface water, and effluents of municipal wastewater treatment plant	Specific ozone doses ranging from 0.82 to 2.55 mg O_3/mg DOC resulted in >99.9% removal for most of the studied pharmaceuticals. The increased toxicity for aqueous solutions of acidic pharmaceuticals at a specific ozone dose of 2.24 mg O_3/mg DOC was due to formation of more toxic by-products.	Almomani et al. (2016)
	Indomethacin	Ultrapure water	Ozone doses of 2, 10, 20, and 35 mg/L resulted in complete indomethacin (25 mM) degradation within 7 min in contrast to poor mineralization (TOC), despite extending the reaction time to 30 min.	Zhao et al. (2017)
	Propranolol	Milli-Q water	Complete removal of propranolol was achieved in 8 min. Total organic carbon (TOC) removal did not increase above 5%, despite increased contact time of 60 min. Low dose ozone was inefficient to improve biodegradability of ozonated samples	Dantas et al. (2011)
	Tetracycline	Deionized water	Direct ozonation showed complete degradation of tetracycline with H_2O_2 concentrations with tert butyl alcohol (OH_ radical scavenger) showing no effect on the degradation rate. Only 35% of COD removal was attained after 90 min ozonation	Wang et al. (2011)
	Carbamazepine, diclofenac, sulfamethoxazole, and trimethoprim	Milli-Q water	Carbamazepine, diclofenac, and trimethoprim degraded completely when a lower dose of ozone was applied, 1.6 mg/L, 2.3 mg/L, and 2.8 mg/L, respectively. However, sulfamethoxazole consumed a higher dose, 4.5 mg/L and longer time to achieve complete degradation due to the formation of highly reactive by-products	Alharbi et al. (2016)
	Amoxicillin	Distilled water and ultrapure water	The pseudofirst order reaction rates for amoxicillin by ozonation at pH 3, pH 7, and pH 10 were 0.064/min, 0.321/min, and 1.970/min, respectively, with pH 10 being the optimum one.	Kıdak and Doğan (2018)

(Continued)

Table 2.3 Overview on selected recent publications on the application of different types of AOPs to the degradation of pharmaceuticals (Kanakaraju et al., 2018). *Continued*

AOPs applied	Pharmaceuticals	Water matrix	Significant findings	References
	Salicylic acid	Deionized water	Salicylic acid removal was observed to be more significant and rapid at pH 4 compared to pH 8 and pH 10. At pH 4 and in the presence of 1 mg/L of ozone, about 95% of salicylic acid was removed.	Hu et al. (2016)
	Ibuprofen, acetyl sulfamethoxazole, and metoprolol	Secondary effluent from wastewater treatment plant	Effect of pHs (6.5, 7.0 and 7.5) at a constant temperature, 20°C in the presence of organic matter on the ozonation treatment with an initial concentration of 1.5 mg/L showed that metoprolol degraded at the fastest rate followed by acetyl sulfamethoxazole and ibuprofen at all pHs.	Cai and Lin (2016)
Fenton and photoFenton PhotoFenton	Amoxicillin	Distilled water	Complete and rapid oxidation was attained for amoxicillin in the presence potassium ferrioxalate complex within 5 min, while for $FeSO_4$ 15 min was required in experiments using a solar simulator	Trovó et al. (2011)
Solar photoFenton	Ofloxacin and trimethoprim	Ultrapure water	Comparison of solar photoFenton between acidic pH (pH 2.8–2.9) and neutral (unadjusted pH 7) showed that complete degradation of ofloxacin and trimethoprim was attained likewise at the acidic pH but at a slower rate. Poor DOC removal was observed for both conditions.	Michael et al. (2012)
Solar photoFenton	Nalidixic acid	Demineralized water, saline water, synthetic industrial effluent, real industrial effluent	Although complete degradation was obtained for nalidixic acid, degradation and mineralization was slower in saline water and synthetic industrial effluent with a compound parabolic collector.	Sirtori et al. (2011)
Solar photoFenton	5-Fluorouracil	Ultrapure water	Solar simulated Fenton-like treatment ($Fe^{3+}/S_2O_8{}^{2-}$) resulted in a higher degradation rate and dissolved organic carbon (DOC) removal than Fe^{3+}/H_2O_2 for the degradation of 5-fluorouracil. The degradation rate and DOC removal under $Fe^{3+}/S_2O_8{}^{2-}$ was 0.04/minand 40%, respectively while for Fe^{3+}/H_2O_2 system the values were 0.024/min and 25%.	Koltsakidou et al. (2017)
PhotoFenton	Antipyrine	Aqueous solution	Ferrioxalate induced photoFenton reaction with UVA-LED was effective to degrade antipyrine as a result of the production of more ·OH radicals in the system. The complete degradation of antipyrine was obtained after 2.5 min, while 93% of TOC removal was recorded after 60 min ($[H_2O_2]_o = 100$ mg/L, $[Fe]_o = 2$ mg/L, and $[H_2C_2O_4]_o = 100$ mg/L, pH = 2.8).	Davididou et al. (2017)

(*Continued*)

Table 2.3 Overview on selected recent publications on the application of different types of AOPs to the degradation of pharmaceuticals (Kanakaraju et al., 2018). *Continued*

AOPs applied	Pharmaceuticals	Water matrix	Significant findings	References
Solar and artificial UV photoFenton	Oxacillin	Deionized water	Based on the applied factorial design, removal of oxacillin (203 mmol/L) was found to be optimum when the concentration of Fe^{2+}, H_2O_2, and applied light power were 90 mmol/L, 10 mmol/L, and 30 W, respectively.	Giraldo-Aguirre et al. (2018)
PhotoFenton	15 pharmaceuticals (in combination with other micro pollutants)	Municipal wastewater treatment plant	The highest percentage of micro pollutant degradation at 83% was achieved in the presence of UV (254 nm) using 30 mg/L H_2O_2 and 2 mg/L Fe(III) at natural pH.	de La Cruz et al. (2013)
Solar photoFenton	Mixtures of 15 emerging contaminants (ECs)	Synthetic water, simulated effluent wastewater, real effluent wastewater	Mild solar photoFenton (Fe = 5 mg/L, H_2O_2 = 50 mg/L) was efficient to degrade mixtures of 15 ECs (pharmaceuticals, personal care products, pesticides) without any pH adjustments. But toxicity level increased, with the degradation products formed in real effluent wastewater.	Klamerth et al. (2010)
Solar photoFenton	Carbamazepine, ibuprofen, ofloxacin, flumequine, sulfamethoxazole	Municipal wastewater treatment plant	Solar photoFenton using Fe: ethylenediamine-disuccinic acid (1:2) resulted in >96% removal of pharmaceuticals within 45 min while Fe: citrate (1:5) produced 94% removal after 96 min at neutral pH using nanofiltration concentrated sample.	Miralles-Cuevas et al. (2014)
PhotoFenton	Ciprofloxacin	Milli-Q water	PhotoFenton degradation of low and high concentrations of ciprofloxacin in the presence of different iron sources (iron citrate, iron oxalate, and iron nitrate) and pH (2.5, 4.5, and 6.5) gave different results. For a high concentration of ciprofloxacin (25 mg/L) at pH 4.5, both the iron complexes iron citrate and iron oxalate produced total conversion within 10 min. Higher ciprofloxacin conversion was obtained using iron citrate for low concentration (1 mg/L) of this compound at pH 2.5 after 10 min.	de Lima Perini et al. (2013)
UV and UV/peroxide processes UV and UV/H_2O_2	41 APIs (10 analgesics, 4 antiarrhythmic agents, and 12 antibiotics and 15 others)	Municipal treatment plant	The removal efficiencies by UV and UV/H_2O_2 were highly dependent on the type of pharmaceutical, while H_2O_2 addition during the treatment enhanced the API removal up to 90% as well as DOC removal	Kim et al. (2009)
UV/H_2O_2 and UV	Sulfamethoxazole, sulfamethazine, sulfadiazine, trimethoprim, bisphenol A, and diclofenac	Milli-Q water, lake water, and wastewater treatment plant effluent	Photolysis rate of all the bioactive compounds using low pressure UV photolysis (254 nm) differed at pHs tested, while efficiency of UV/H_2O_2 on the tested bioactive compounds was as follows: diclofenac > sulfamethoxazole > sulfamethazine > sulfadiazine > bisphenol Aztrimethoprim.	Baeza and Knappe (2011)

(*Continued*)

Table 2.3 Overview on selected recent publications on the application of different types of AOPs to the degradation of pharmaceuticals (Kanakaraju et al., 2018). *Continued*

AOPs applied	Pharmaceuticals	Water matrix	Significant findings	References
UV	Sulfasalazine, sulfapyridine and 5-aminosalicyclic acid	Milli-Q water	Sulfasalazine was resistant to direct UV (254 nm) photolysis while sulfapyridine demonstrated the fastest degradation due to its high molar absorption coefficient, 15,241/M/cm	Ji et al. (2018)
UV/H_2O_2 and UVC	Amoxicillin	Distilled deionized water	Degradation of amoxicillin by direct UV and UV/H_2O_2 with a low pressure Hg lamp (254 nm) showed that the degradation of 100 mM of amoxicillin (pH 7, 20°C) followed first-order kinetics and the degradation rate increased with the H_2O_2 concentration. An addition of 10 mM H_2O_2 improved the degradation rate up to sixfold when compared to direct UV.	Jung et al. (2012)
UV	Ketoprofen, carprofen, and diclofenac acid	Ultrapure water and methanol	The photolysis kinetics of ketoprofen, carprofen and diclofenac acid followed pseudofirst order kinetics. Degradation of diclofenac acid was much slower compared to carprofen and ketoprofen. The predicted toxicity revealed that the transformation products of ketoprofen were more toxic than the parent API.	Li et al. (2017)
UV/H_2O_2 and UVC	Amoxicillin	Distilled deionized water	Degradation of amoxicillin by direct UV and UV/H_2O_2 with a low pressure Hg lamp (254 nm) showed that the degradation of 100 mM of amoxicillin (pH 7, 20°C) followed first-order kinetics and the degradation rate increased with the H_2O_2 concentration. An addition of 10 mM H_2O_2 improved the degradation rate up to sixfold when compared to direct UV	Jung et al. (2012)
UV	Ketoprofen, carprofen, and diclofenac acid	Ultrapure water and methanol	The photolysis kinetics of ketoprofen, carprofen and diclofenac acid followed pseudofirst order kinetics. Degradation of diclofenac acid was much slower compared to carprofen and ketoprofen. The predicted toxicity revealed that the transformation products of ketoprofen were more toxic than the parent API	Li et al. (2017)
UVC/H_2O_2 and UVC/S_2O_8	17a-ethinyl estradiol, 17b-estradiol, azithromycin, carbamazepine, dexamethasone, erythromycin and oxytetracycline	Ultrapure water	In the presence of natural organic matter in the UVC/H_2O_2 system, the degradation rates (kapp) of azithromycin, carbamazepine, dexamethasone, and 17a-ethinyl estradiol were enhanced between 3% and 11%, while an inhibitory effect resulted in the case of 17b-estradiol, erythromycin, and oxytetracycline.	Markic et al. (2018)

(*Continued*)

Table 2.3 Overview on selected recent publications on the application of different types of AOPs to the degradation of pharmaceuticals (Kanakaraju et al., 2018). *Continued*

AOPs applied	Pharmaceuticals	Water matrix	Significant findings	References
UVC/H_2O_2	Diclofenac	Ultrapure water	Diclofenac completely degraded in solution within 2 min under UVC/H_2O_2 compared to UVA/TiO_2, which took 156 min to achieve similar degradation. A much higher mineralization (TOC) rate constant, 3.92×10^{-4}/s was obtained from the UVC/H_2O_2 treatment	Perisic et al. (2016)
UV	Sulfamethoxazole and ibuprofen	Deionized water	The direct photolysis (UV 254 nm) of sulfamethoxazole and ibuprofen at pH 3 and pH 7.55 followed pseudofirst order kinetics. The initial reaction rate of the neutral sulfamethoxazole at pH 3 was 0.9149/min higher than anionic sulfamethoxazole at pH 7.55, 0.3558/min in contrast to ibuprofen, where the initial reaction rate was higher for its anionic form at pH 7.55 (0.0263/min) than its neutral form at pH 3 (0.0043/min).	Luo et al. (2018)
Sonolysis	Ciprofloxacin	Deionized water	Degradation of ciprofloxacin at frequency 544 kHz (pH 7, 25°C) fitted pseudofirst-order degradation with a half-life of 102 min. Addition of t-butanol (0.45, 4.5, and 45 mM) slowed down the degradation of ciprofloxacin confirming that t-butanol acts as a radical scavenger and the degradation of ciprofloxacin occurred due to the ·OH radical.	de Bel et al. (2011)
	Diclofenac and carbamazepine	Milli-Q water and urban wastewater treatment plant	Degradation of diclofenac and carbamazepine followed first-order kinetics. The reaction rates were observed to increase with increasing power density from 100 to 400 W/L.	Naddeo et al. (2013)
	Piroxicam	Ultrapure water, bottled water, and surface water	The reaction rates of piroxicam (640 mg/L) at power density of 20, 36, and 60 W/L were 0.1157/min, 0.1695/min, and 0.1967/min, respectively	Lianou et al. (2018)
	Ibuprofen	Ultrapure water	Application of single ultrasonic frequencies, 20 kHz, 40 kHz, 200 kHz, 572 kHz, and 1130 kHz to ibuprofen (50 mM) resulted in 0.033/min, 0.035/min, 0.038/min, 0.234/min, and 0.090/min, respectively, indicating increasing degradation with frequency. Addition of zero-valent iron markedly increased ibuprofen degradation even at low single frequencies, 20, 40, and 200 kHz.	Ziylan-Yavas and Ince, (2018)
	Oxacillin	Distilled water	Sonochemical process (275 KHz) efficiently degraded oxacillin (47.23 mmol/L) and eliminated antimicrobial activity in the presence and absence of additives (calcium carbonate and mannitol).	Serna-Galvis et al. (2016)

(*Continued*)

Table 2.3 Overview on selected recent publications on the application of different types of AOPs to the degradation of pharmaceuticals (Kanakaraju et al., 2018). *Continued*

AOPs applied	Pharmaceuticals	Water matrix	Significant findings	References
	Diclofenac	Milli-Q water	The optimum conditions, initial concentration, pH and frequency ultrasound for DCF degradation was found to be 30 mM, 3.0 and 861 kHz, respectively. Addition of Fe-containing additives improved diclofenac elimination in particular with paramagnetic iron oxide nanoparticles. Mineralization occurred after 60 min of sonolysis in all cases.	Guyer and Ince (2011)
Electrochemical oxidation	Diclofenac, sulfamethoxazole, iopromide, and 17-alphaethinyl estradiol	Deionized water and hospital wastewater treatment plant	When the degradation rates of the four APIs in synthetic wastewater and real wastewater was compared, higher rates were obtained for the latter when current conditions $I = 0.9$ A, initial concentration, $C_o = 0.5$ mg/L, and flow rate = 500 L/h were used. This was attributed to the consumption of less oxidative species by organic matter present in real wastewater compared to those present in synthetic wastewater	
	Carbamazepine	Demineralized water, tap water, and treated wastewater	Comparison of carbamazepine degradation in tap water, demineralized water, and treated municipal wastewater using Nb/BDD anode and 14 mM of NaCl showed that electrolysis resulted better performance in demineralized water (for pH 2 > pH7 > pH 10) followed by tap water and treated municipal wastewater	García-Espinoza et al. (2018)
Radiation	Carbamazepine	Deionized water	The increase of peroxymonosulfate concentration (mole ratio of peroxymonosulfate to carbamazepine from 10:1 to 30:1) increased the degradation of carbamazepine from 80% to 100% within 10 min of treatment time.	Wang and Wang (2018)
	Carbamazepine	Ultrapure water	TOC reduction in carbamazepine solution decreased with increasing H_2O_2 concentrations (0–200 mM) at varying irradiation doses. Carbamazepine solution containing 50mM H_2O_2 produced the highest TOC removal at 41% when the irradiation dose was 20 kGy.	Liu et al. (2016)
	Carbamazepine	River water and ultrapure water	Addition of sulfite ion (SO_3^{2-}) prior to the electron beam radiation led to 85.4% of carbamazepine (75 mg/L) degradation in pure water. Sulfite radical (SO_3^{-}), $e^{-}aq$ and $O^{\cdot-}$ were concluded to be a contributing active species for carbamazepine degradation in the presence of Na_2SO_3.	Zheng et al. (2014)
	Nineteen pharmaceutical compounds	Wastewater sample from WWTP	pharmaceutical compounds (<50 ng/L). The extent of the degradation of the pharmaceuticals was found to be dependent on the type and concentration of the compound.	Reinholds et al. (2017)

(*Continued*)

Table 2.3 Overview on selected recent publications on the application of different types of AOPs to the degradation of pharmaceuticals (Kanakaraju et al., 2018). *Continued*

AOPs applied	Pharmaceuticals	Water matrix	Significant findings	References
	Fluoxetine	Ultrapure water	Electron beam irradiation yielded 90% degradation of fluoxetine at radiation dose of 0.5 kGy whereas doses above 2.5 kGy led to a below detection limit	Silva et al. (2016)
	Piperacillin	Distilled water, synthetic wastewater	The initial value of the calculated radiation chemical yield for the degradation of piperacillin was 0.26 mmol/J. Comparison of electron-beam mediated antimicrobial inactivation in aqueous solution and synthetic wastewater revealed that the adsorbed dose and degradation products affected the findings.	Szabó et al. (2018)
Combined AOPs Ozone/TiO_2 solar photocatalysis	Mixtures of four pharmaceuticals (atenolol, hydrochlorothiazide, ofloxacin, and trimethoprim)	Distilled water and simulated synthetic secondary effluent solution	Four APIs, atenolol, hydrochlorothiazide, ofloxacin, and trimethoprim sequentially treated, by ozonation and solar photocatalytic oxidation revealed that initial ozonation step led to poor removal of TOC (10%), while subsequent solar TiO_2 photocatalysis improved the TOC removal to 80% and 60% in distilled water and secondary effluent, respectively.	Márquez et al. (2014)
Ultrasound/Fenton oxidation (sonoFenton)	Ibuprofen	Distilled water and effluent from municipal wastewater treatment plant	Coupling of Fenton with ultrasound (20 kHz) enhanced the degradation of ibuprofen in the presence of 6.4 mM whereby 95% removal was achieved within 60 min, and mineralization was also improved under the same conditions.	Adityosulindro et al. (2017)
Ultrasound and ozonation	Diclofenac, sulfamethoxazole, and carbamazepine	Distilled water	The combined ultrasound/ozonation process positively enhanced the degradation of three APIs in single and mixed solutions at an ozone flow of 3.3 g/h after 20 min of treatment time when compared to ozonation alone at the same flow.	Naddeo et al. (2015)
Sonolysis and photolysis (UV/H_2O_2)	Diclofenac, paracetamol, salicylic acid, chloramphenicol, etc.	Synthetic pharmaceutical wastewater	Sonophotolysis resulted in the highest TOC removal of 91% in the presence of 900 mg/L H_2O_2, 80 W ultrasonic power and UV (253.7 nm). Two factors, ultrasound power and initial concentration of H_2O_2 were concluded as having the most effect based on the three-level BoxeBehnken experimental design performed.	Ghafoori et al. (2015)

(*Continued*)

Table 2.3 Overview on selected recent publications on the application of different types of AOPs to the degradation of pharmaceuticals (Kanakaraju et al., 2018). *Continued*

AOPs applied	Pharmaceuticals	Water matrix	Significant findings	References
Sonophotocatalysis with TiO_2, sonophotoFenton and sonobiphotocatalysis with TiO_2 and Fe^{2+}	Ibuprofen	Milli-Q water	Sonobiphotocatalysis produced the highest mineralization rate (DOC removal of 98%) with more efficient consumption of H_2O_2. Initial degradation rate was 3.50×10^{-3} mM/min	Mendez-Arriaga et al. (2009)
Photocatalytic ozonation	Diclofenac and amoxicillin	Aqueous solution (not specified) and urban wastewater	Complete mineralization (TOC abatement) was achieved with TiO_2 photocatalytic ozonation for amoxicillin and diclofenac after 30 min and 120 min, respectively.	Moreira et al. (2015)
Ozone/TiO_2/UVB, UVB/TiO_2, O_3/UVB, and single systems (UV, O_3)	Mixtures of nine pharmaceuticals	Water (not specified)	Ozone/TiO_2/UVB (313 nm) yielded the highest TOC removal of 95% within 120 min for the pharmaceutical mixtures (each 10 ppm).	Rivas et al. (2012)
Electroperoxone	Venlafaxine	Milli-Q water, secondary effluent from wastewater treatment plant	Compared to single ozonation and electrolysis treatment, electroperoxone efficiently degraded 20 mg/L of venlafaxine within 3 min of reaction time, when the applied current was increased from 50 mA to 300 mA and the O_3 concentration was fixed at 40 mg/L.	Li et al. (2015)
Ozone and ultrasound	Amoxicillin	Distilled water and ultrapure water	Coupling of ozonation and ultrasound resulted in a higher pseudofirst order degradation rate of 2.5/min at pH and higher TOC removal (45%) than single ozonation treatment with 1.97/min at similar pH.	Kıdak and Doğan (2018)

AOP, *Advanced oxidation processes;* DOC, *dissolved organic carbon;* ECs, *emerging contaminants;* TOC, *total organic carbon.*
With permission.

2.5.2.2 Fenton and photofenton

The Fenton process is according to the use of a combination of iron salts (Fe^{2+}) and H_2O_2, producing $\cdot OH$ radicals under slight acidic conditions [Eq. (2.6)]. The Fe^{2+} ion can be regenerated as shown in Eq. (2.7) or through reactions of Fe^{3+} with other mediators (Bautista et al., 2008)

$$Fe^{2+} + H_2O_2 \rightarrow Fe^{3+} + OH^- + \cdot OH \quad (2.6)$$

$$Fe^{3+} + H_2O_2 \rightarrow Fe^{2+} + HO_2^{\cdot} + H^+ \quad (2.7)$$

The effective degradation of pharmaceuticals has been reported under both Fenton and photoFenton reactions. Combination of photoFenton/UV-Vis processes can initiate more $\cdot OH$ radicals. Solar photoFenton has been demonstrated for the treatment of various pharmaceuticals such as antibiotics, antiinflammatory, analgesic, and antineoplastic drugs. The application of solar photoFenton has been reported for the treatment of different drugs like antibiotics, antiinflammatory, analgesic, and antineoplastic drugs. For instance, it has been reported the photoFenton efficiency using a solar cell simulator for the amoxicillin removal in the presence of two different iron species, ferrioxalate complex, and $FeSO_4$ (Kanakaraju et al., 2018).

Complete removal of amoxicillin was achieved in the presence of the ferrioxalate complex within 5 min, while 15 min was needed when $FeSO_4$ was applied in the presence of 120 mg/L H_2O_2. However, the ferrioxalate complex was found to demonstrate higher toxicity particularly for Daphia magna bioassays. Under solar photoFenton, water composition has great effects on removal of organic compounds especially COD. For example, DOC removal from the simulated industrial effluent, saline water, and demineralized water was 20% (after 240 min), 73% (after 107 min), and 86% (after 92 min), respectively (Kanakaraju et al., 2018).

Although energy consumption of the Fenton process is lower than other AOPs such as O_3 and UV, this process involves low acidic pH. At a pH higher than 3, Fe^{3+} precipitates as $Fe(OH)_3$, while at even higher pH values, the formation of Fe(II) complexes leads to a decline in Fe^{2+} concentration. Much progress has been made in operating Fenton and photoFenton processes at a neutral or near-neutral pH to deal with large-scale operations or real wastewater samples such as hospital wastewater. However, more studies should be conducted and optimized at a different range of pH using natural and wastewater matrices (Kanakaraju et al., 2018).

2.5.2.3 UV and UV/peroxide processes

The UV processes efficiency for the pharmaceuticals removal strongly depends on the UV absorption by pharmaceuticals. The important variables influencing the degradation kinetics of direct UV photolysis include the rate constants, k_{UV}, quantum yield (φ), and molar extinction coefficient (ε) (Luo et al., 2018). UV integrated with H_2O_2 (UV/H_2O_2) usually more improve removal efficiencies of pharmaceuticals with low UV absorption. UV/H_2O_2 processes are run by the H_2O_2 dose, amount of $\cdot OH$ radical, power of UV light, water constituents, chemical structure of the pharmaceutical, and also the solution pH (Kanakaraju et al., 2018).

Removal efficiency of UV process is increased by addition of peroxides, H_2O_2, and peroxydisulfate to the UV system. This phenomenon is due to the presence of the highly reactive free radicals of $\cdot OH$ and $SO_4^{\cdot}$. Comparison of effectiveness of UVC/H_2O_2 and UVC/$S_2O_8^{2-}$ processes on the degradation of 17a-ethinyl estradiol, 17b-estradiol, azithromycin, carbamazepine, dexamethasone, erythromycin, and oxytetracycline in a mixture (in the absence of NOM) has been shown that first-order reaction rates for UVC/$S_2O_8^{2-}$ were higher than that of UVC/H_2O_2. The observed results were attributed to the different reaction mechanisms of the $\cdot OH$ and $SO_4^{\cdot}$ radicals with the pharmaceuticals and also due to the differences in their optical characteristics (Kanakaraju et al., 2018).

Sonolysis is another AOP method that it based on generation of $\cdot OH$ radicals from water because of strong intensity of cavitation bubbles

$$\text{H2O }))) \rightarrow \cdot OH + H\cdot \tag{2.8}$$

$$2\ \cdot OH \rightarrow H_2O_2 \tag{2.9}$$

$$2\ \cdot OH \rightarrow H_2O + O \tag{2.10}$$

where))) refers to the ultrasound irradiation. The removal efficiency of pharmaceuticals in this method is significantly depend on the intensity and frequency of the used ultrasound irradiation (Ince, 2018; Kıdak and Doğan, 2018). The influence of three frequencies (544, 801, and 1081 kHz) on the degradation of ciprofloxacin (15 mg/L) showed that the lowest frequency (544 kHz) generated a high degradation rate constant of 0.0067/min at reaction time of 25°C. It has been reported that increasing the intensity of ultrasound (100–400 W/L) successfully enhanced the DIC and carbamazepine degradation. Also in another studies it is reported that increasing intensity (20–60 W/L) improved the piroxicam degradation due to the production of $\cdot OH$ radicals and higher mixing intensity (Kanakaraju et al., 2018). Although sonolysis method mostly does not require additional chemical compounds, because of energy demand and low mineralization, application of sonolysis method is limited to lab scale. To overcome these problems and reduce the operation costs, it is better that sonolysis is be integrated with other AOPs methods to reduce the operation costs (Kanakaraju et al., 2018).

2.5.2.4 Electrochemical oxidation

Today electrochemical oxidations based on AOPs appear as an attractive selection for pharmaceutical removal via generation of reactive species by electricity and without required for chemicals (Garcia-Segura et al., 2018).

Two oxidation mechanisms are involved in the electrochemical oxidation process. First there is the direct oxidation at the anode whereby direct charge transfer occurs between the pharmaceutical compound and the anode surface. Indirect oxidation occurs via in situ generation of reactive oxygen species by oxidants at the surface of the electrode (Feng et al., 2013).

In comparison with the conventional anodes like IrO_2, Pt, or PbO_2, boron-doped diamond (BDD) anodes are general in the electrochemical oxidations of pharmaceuticals due to their stability to corrosion, high oxygen, and inert surfaces. Many studies have confirmed the BDD anodes efficiency in pharmaceuticals degradation. Although advantages of electrochemical oxidation are remarkable, the electrochemical processes efficiency is greatly limited by the electrode surface due to reduce the pharmaceuticals treatment as a result of the products formation (Kanakaraju et al., 2018).

2.5.2.5 TiO_2 photocatalytic degradation

TiO_2 photocatalysis has resulted in the effective degradation of different groups of pharmaceuticals, thus leading to an increasing interest in its application for the removal of pharmaceuticals in water and wastewater. TiO_2 photocatalytic studies frequently investigate the effects of operational parameters such as catalyst loading, initial concentration of substrate, type of TiO_2 photocatalyst, pH of the solution, wavelength/light intensity, and water matrix on the degradation kinetics of the pharmaceutics (Carbajo et al., 2016). Besides these common parameters, effects of stirring speed, temperature, and gas sparging rates were also examined.

2.5.3 Membrane filtration

Membrane processes are popular in reuse of wastewater and drinking water treatments because of their efficiency in removal of both macro- and micropollutants. Only nanofiltration/reverse osmosis (NF/RO) has been found to be efficient for the removal of pharmaceuticals and other dissolved pollutants from wastewater while microfiltaration (MF) and ultafiltration (UF) processes can only be used as pretreatment process to remove NOM and colloidal materials. According to the treatment method used in WWTPs (activated sludge, membrane bioreactor, coagulation, or adsorption), the WWTPs effluent involves significant NOM amount which has great effect on EMPs rejection during NF/RO process.

Protection of NF/RO membranes is vital to retain the filtration process efficiency during water treatment. NOM in the feedwater for NF/RO processes is a great risk to membranes efficiency because NOM is a main factor for membrane fouling. Although some authors have known increased the pharmaceuticals rejection by NOM deposited in NF, reduction in water flux is not preventable. Three mechanisms are importance in pharmaceuticals rejection by NF/RO membrane including size exclusion or steric hindrance effect, hydrophobic interaction, and electrostatic interaction.

2.6 Advantage and disadvantage of different removal methods

Different water treatment processes can lead to different pharmaceutical removal efficiencies. For example, biological degradations and sorption occurred in activated sludge process result in pharmaceuticals removal from wastewater. It is reported that order removal of pharmaceuticals by activated sludge process follows: stimulant drugs > metabolites > analgesics > antibiotics > antiinflammatory > lipid regulators > nonsteroidal antiinflammatory drugs (NSAIDs) > other pharmaceuticals [such as iopromide, fluoxetine (FLU), ranitidine, omeprazole, and tamoxifen]. With respect to considerable efficiency and low cost, constructed wetland is an excellent choice for wastewater treatment. But need to long retention times and large space are the most disadvantages coupled with constructed wetlands. In the ion-exchange process, the most common ion-exchange methods are softening and deionization. Deionization cannot effectively remove the most organic compounds or microorganisms from water. Microorganisms can attach to the resins bed, thus forming a culture medium for rapid growth of bacterial. Due to reaction with highly reactive functional groups of pharmaceuticals, chlorination as an important stage in the drinking water treatment leads to formation of chlorination by-products that is a major problem. These by-products are very high toxic and carcinogenic.

Although advance membrane filtration such as NF, MF, and UF are able to remove pharmaceuticals with high efficiency, their application is limited by membrane fouling, high operational cost, and high energy demands. Ozonation process has excellent removal efficiency for a wide range of pharmaceuticals and its efficiency increases by ozone combined with other oxidants such as H_2O_2. However, high energy consumption and formation of oxidative by-products limit its applications. Similarly other AOPs processes (H_2O_2/UV, Fe^{2+}/H_2O_2, Fe^{2+}/H_2O_2/UV, O_3/H_2O_2, O_3/UV) run high removal efficiency, but high energy, cost inputs, and formation of oxidative by-products limit their applications.

2.7 Factors influencing the concentration of pharmaceuticals in the aquatic environment

Bulky consumption of pharmaceuticals for human and veterinary use and the improper disposal or nondisposal of medicines at domestic sites are the major reasons for the huge discharge of these compounds in aquatic environments (Klatte et al., 2017; Valenti et al., 2017). It is found that a large amount of drugs are not controlled and are disposed directly into wastewaters. It has been an estimated that $1 billion of prescription drugs are discarded year in the USA, from hospitals, care facilities, and pharmacies (Ruhoy and Daughton, 2007).

These compounds entering WWTPs may be affected by biodegradation, residual suspended or dissolved particles in the water, or attached to biosolids or sewage sludge (Trudeau et al., 2005). Degradation rate of pharmaceuticals in WWTPs is very depending on the properties of pharmaceuticals and aqueous environments. "For example, IBU has a high elimination rate (>90%) and so is rapidly degraded." In contrast, carbamazepine has low elimination rate (4%–8%). Although carbamazepine is not a high consumption pharmaceutical, the slow degradation is a reason for frequently its detection in both WWTPs effluent and surface water (Corcoran et al., 2010). It is considered that pharmaceutical elimination in anaerobic conditions is slower than aerobic conditions. Adsorption of medicines to biosolids in the activated sludge process and sediments is associated with hydrophobic and electrostatic properties of these compounds. For example, at neutral pH, acidic pharmaceuticals such as NSAIDs, clofibric acid, and gemfibrozil are in ion forms and have low adsorption via biosolids and sediments (Corcoran et al., 2010).

In return, these pharmaceuticals are more probably to stay in the water column and thus may more readily discharge into effluent and release into surface waters (Corcoran et al., 2010). Because some pharmaceuticals, such as 17α-ethinyloestradiol (EE2) and fluoroquinolone antibiotics, attend to be adsorbed by sludge and biosolids, they can be detected in sediment and sludge at ng/g concentration. Therefore this facilitates their removal between 70% and 80% from the aqueous phase. Pharmaceuticals adsorbed to sediments and sludge in WWTPs may resulted in greater problem for environment because the sewage sludge usually is applied as fertilizer for land amendment (Corcoran et al., 2010).

It was cleared that a wide range of pharmaceuticals are sun light sensitive. Photo degradation can also be essential in the environmental degradation of some pharmaceuticals such as DIC, sulfamethoxazole, PRP, and ofloxacin (Ahmad et al., 2016; Mahlambi et al., 2015; Corcoran et al., 2010). Photo degradation of pharmaceuticals are decreased by increasing turbidity and reducing sunlight penetration (Mezzelani et al., 2018b).

The correlation analysis revealed the significant association between EMPs removal efficiency and physicochemical parameters such as conductivity, DO, chlorophyll α, TSS, turbidity, TN, and TP (Table 2.4) (You et al., 2015).

Negative correlations have been found between dissolved oxygen and several compounds such as caffeine, salicylic acid, AMP, naproxen, triclosan, BPA, and DEET. In agreement with previous studies (Carr et al., 2011; Ferguson et al., 2013), this phenomenon may be due to the improved biodegradation under aerobic conditions. It was found that caffeine, salicylic acid, AMP, BPA, and triclosan have a negative correlation with Chlorophyll a, an indicator of algal biomass. The high removal efficiencies (>75%) of BPA, estrone, salicylic acid, and triclosan from wastewater have been obtained by freshwater green algae (Zhou et al., 2014). Similarly algal content would possibly increase the biodegradation, photodegradation and uptake of these chemicals. In return, the positively correlation was found between DO and chlorophyll (Veach and Bernot, 2011). Negative correlation ($P = -.554$) between concentrations of salicylic acid and light intensity reveals that photodegradation limits persistence of salicylic acid against photodegradation process (Liu et al., 2012).

Table 2.4 Spearman rank correlations between pharmaceuticals concentrations and environmental factors. Only significant correlations at the <0.05 and <0.01 levels are shown (You et al., 2015).

	Caffeine	Salicylic acid	Acetaminophen	Crotamiton	Sulpiride	Naproxen	BPA	DEET	Triclosan	BP-3	Fipronil	Total
Physicochemical characteristics												
Temperature °C												
Conductivity (μS/cm)	−0.338[a]		−0.413[a]				−0.389[a]		−0.277[b]		−0.412[a]	−0.396[a]
Salinity (ppt)			−0.533[a]								−0.583[a]	
pH	−0.197[b]								−0.216[b]		−0.336[a]	
DO (mg/L)	−0.642[a]	−0.312[a]	−0.613[a]			−0.210[b]	−0.554[a]	−0.319[a]	−0.545[a]			−0.715[a]
Chlorophyll a (μg/L)	−0.575[a]	−0.441[a]	−0.532[a]				−0.252[b]		−0.473[a]	−0.199[b]		−0.555[a]
TSS (mg/L)	0.217[b]							0.205[b]				0.232[b]
Turbidity (NTU)	0.389[a]	0.328[a]	0.385[a]				0.302[a]	0.209[b]	0.395[a]			0.470[a]
TN (mg/L)	0.446[a]		0.371[a]	0.527[a]		0.340[a]	0.289[b]	0.269[b]	0.469[a]	0.227[b]		0.508[a]
TP (mg/L)	0.199[b]				0.413[a]	0.347[a]	0.227[b]		0.249[b]			0.260[b]
TOC (mg/L)					0.313[a]	0.304[a]	0.201[b]					
Light intensity (μmol/m^2 s)		−0.554[a]										
Precipitation												
24 h Cumulative rainfall[c]	0.556[a]							0.519[a]			0.579[a]	0.723[a]
48 h Cumulative rainfall[c]	0.703[a]						0.565[a]	0.474[a]	0.439[a]		0.525[a]	0.747[a]
96 h Cumulative rainfall[c]	0.683[a]						0.525[a]	0.494[a]	0.525[a]		0.636[a]	0.762[a]
Microbial indicator												
E. coli (MPN/100 mL)	0.503[a]	0.293[b]	0.544[a]				0.560[a]		0.356[a]			0.576[a]
Enterococcus (MPN/100 mL)	0.564[a]	0.389[a]	0.591[a]				0.402[a]	0.300[b]	0.460[a]	0.245[b]		0.613[a]

[a]*Correlation is significant at the 0.01 level (2-tailed).*
[b]*Correlation is significant at the 0.05 level (2-tailed).*
[c]*EOC concentration in reservoir was used.*
With permission.

A negative correlation known between numerous compounds and salinity/conductivity of the water may be due to increasing sorption of substances in the presence of dissolved ions. Also amount of deprotonated species at different water pH will also affect sorption kinetics. For instance, in pH ranged from7 to 9.35, the deprotonated EMPs are found to be dominant in water led to decreasing their sorption into biosolids and particulates (You et al., 2015).

Positive correlations between the environmental parameters (i.e., TSS, turbidity, TN, and TP) and total EMP abundance were observed although these parameters were not correlated to concentrations of all the individual compounds. In common, contaminants which have a long half-life ($T_{1/2}$) in surface water can persist and distribute more widely in urban watersheds (Bayen et al., 2013). However, it has been cleared that readily degradable compounds with a relatively short half-life are frequently detected at relatively high concentration in the aquatic environments. This occurrence may be due to high loads of pollutants and/or continuous discharge from resource. For example, because of continuous release from resource, caffeine with low persistent is one of the most contaminant frequently detected in the worldwide urban waterbodies (You et al., 2015).

Also it has been estimated that microbial indicators (*Escherichia coli* and *Enterococcus*) have strongly positive correlations with caffeine, salicylic acid, AMP, BPA, and triclosan referring to this fact that EMPs could potentially integrated with microbial sources (You et al., 2015).

2.8 Most investigated pharmaceuticals in aquatic environment

Since these compounds belong to highly heterogeneous groups with different physical–chemical characteristics, mechanism of action (MOA) and behavior in aquatic ecosystems, scientific outcomes cannot be generalized for all the pharmaceutical classes (Välitalo et al., 2017; Stadlmair et al., 2018). In the next paragraphs, the state of NSAIDs, psychiatric, and cardiovascular drugs and information on other pharmaceutical classes including steroid hormones, antibiotics, and hypocholesterolemic drugs have been reviewed (Välitalo et al., 2017).

2.8.1 Nonsteroidal antiinflammatory drugs

NSAIDs are commonly used to reduce pain, fever, blood clots, and inflammation. NSAIDs may be grouped as salicylates acide (aspirin), arylalkanoic acids (DIC, indomethacin, nabumetone, sulindac), 2-arylpropionic acids or profens [IBU, flurbiprofen, ketoprofen (KET), naproxen], *N*-arylanthranilic acids or fenamic acids (MA, meclofenamic acid), pyrazolidine derivatives (phenylbutazone), oxicams (piroxicam, meloxicam), sulfonanilides [nimesulide (NIM)]. Despite all these

NSAIDs are structurally diverse and differ in pharmacokinetic and pharmacodynamic properties (Mezzelani et al., 2018b).

Among these compounds, the presence of DIC, IBU, KET, MA, naprossen (NAP), NIM, and AMP. Detection of NSAIDs has been reported in effluents of several WWTPs and surface waters (Mezzelani et al., 2018b; Carmona et al., 2014).

Overall in more than 50 studies conducted in 20 countries, the occurrence of DIC and IBU has been reported in more than 90% of examined samples. The high heterogeneity of results in different site and resources do not allow to be established a specific range for different water bodies. For instance, concentration of DIC in Canadian WWTPs effluent has been reported between 70 and 250 ng/L, concentration of 1500 ng/L has been measured in the Mediterranean Sea, and concentration of 4–38 ng/L has been detected in Indian Ocean. The highest concentration of DIC (2400 ng/L) has been measured in a Switzerland WWTPs effluent, while IBU with concentration of 28,000 ng/L has been detected in the effluents of a Spanish plant. More limited data are available for AMP, KET, MA, NAP, and NIM: in the Mediterranean Sea levels ranged between 23 and 200,000 ng/L for AMP and from <8 to 6000 ng/L for KET (Gros et al., 2012; Mezzelani et al., 2018b).

2.8.2 Psychiatric drugs

Psychiatric drugs are widely prescribed worldwide, including antidepressants, antiepileptics, and anxiolytics. Among antidepressants, selective serotonin reuptake inhibitors (SSRIs) are the most commonly used to treat clinical depression, obsessive-compulsive disorders, panic disorders, social phobia, and attention-deficit disorders. Drugs like FLU, CIT, paroxetine (PAR), sertraline (SER), and venlafaxine (VEN) block the serotonin [5-hydroxytryptamine (5-HT)] reuptake sites on cell membranes, enhancing levels of 5-HT and transmission by serotonergic neurons. Environmental presence of antidepressant residues and metabolites [norfluoxetine (norFLU); *N*-desmethylcitalopram (*N*-des-CIT); norsertraline (norSER)] has been documented in a wide range of water samples, with concentrations ranging from 0.15 to 32,228 ng/L in WWTPs, from 0.5 to 8000 ng/L in surface and groundwaters, from 0.5 to 1400 ng/L in drinking waters, and values typically lower than 1 μg/L in seawater (Mezzelani et al., 2018b).

2.8.3 Cardiovascular drugs

Several groups of cardiovascular drugs are commonly used for the therapeutic treatment of heart disorders, such as beta-blockers [ATE, metoprolol (MET), PRP], angiotensin-converting-enzyme (ACE) inhibitors, angiotensin II receptor antagonists [sartans i.e., Valsartan (VAL)] and calcium channel blockers [diltiazem (DLM)]. Beta-blockers and sartans represent the most frequently detected compounds in aquatic environments with concentration up to 10 μg/L. In freshwater organisms ATE was detected in homogenates of two fish species, *Jenynsia multidentata* and *G. affinis* collected in the polluted area of the Suquía River (Córdoba, Argentina) with concentrations of 10 ± 11 and 57 ± 23 ng/g (d. w.), respectively; the latter species exhibited also detectable levels of PRP (17.78 ± 0.04 ng/g d.w.) and corazolol (2.00 ± 1.00 ng/g d.w). The wild sawbelly *Hemiculter leucisculus* from a Chinese river downstream WWTPs contained PRP in the range of 0.2–0.4 and 0.1–0.2 ng/g w.w. in liver and brain, respectively. The same drug was measured in the common snail from an effluent-dependent stream in central Texas, at values of 1.0 ± 0.28 ng/g w.w. Tissues of the freshwater mussels Lasmig on a costata collected from the Grand River (Ontario) contained PRP and DLM at concentration ranges of 2.36–9.39 and 0.26–3.15 ng/g w.w., respectively. VAL was measured in the benthic worms *Erpobdella octoculata* sampled in a small stream close to the effluent of WWTPs in Czech Republic with mean levels of 1.8 ng/g w.w., and roughly 90% of invertebrates and vertebrate's species collected from Taihu Lake (China) showed measurable levels of PRP, with concentrations up to 111 ng/g d.w. (Mezzelani et al., 2018b).

2.8.4 Steroid hormones, antibiotics, and hypocholesterolemic drugs

Steroids were the first pharmaceuticals to be reported in sewage effluents, thus driving the general concern for the presence of such compounds in natural ecosystems. Estrogenic medicines are largely used as oral contraceptives in livestock farming activities to improve productivity and to enhance athletic performances. Estrogens have become a widespread problem due to their high resistance to degradation processes, capability accumulate in sediments and organisms, remarkable toxicological effects as endocrine disruptor compounds (EDCs) in aquatic fauna. The occurrence of 17α-ethynylestradiol (EE2) and other estrogens has been documented by several studies, sediments act as a sink for such compounds with concentrations up to 1000 folds higher compared to overlying water column, where levels typically do not exceed 100 ng/L. Toxic potential of EE2 to aquatic organisms has been extensively documented in both laboratory and field. Effects of EE2 include increment of plasma vitellogenin in male and female fish, enhancement of intersex processes, decrement in eggs and sperm production, reduction of gamete quality, feminization of male fish, reduction of fertility, fecundity, fate of survival, growth of early life stage, and reproductive success were also documented with possible deleterious consequences at population level.

Antibiotics represented one of the greatest medical discoveries of the 20th century, leading to a marked reduction in mortality and morbidity of several infectious diseases. On the other hand, a typical hazardous side effect of this therapeutical class is the development of antibiotic resistance, an adaptive genetic trait possessed or acquired by bacterial subpopulations, enabling them to survive at concentrations of antibiotic agents that would normally inhibit or destroy these microorganisms.

Beside intrinsic characteristics of bacteria, the resistance phenomenon is further exacerbated by different variables including inappropriate and overuse of antibiotics in human, veterinary and aquaculture medicines, along with the related input in natural ecosystems. For many years the environmental concern for antibiotics has been mainly focused on the development of antibiotic resistance, rather than the potential ecotoxicity of such molecules toward non target species. Compared to other therapeutical classes, the ubiquitous presence of these pharmaceuticals in aquatic environment is strongly influenced by the direct input from aquaculture and livestock activities. Despite the use of antibiotics as growth promoters in livestock has been forbidden in EU countries since 2006, some categories are still allowed in specific circumstances to prevent diseases. Water concentrations typically range from a few ng to hundreds of μg/L, up to mg/L in countries such as China and India where no legislation regulates the use of antibiotics. Recently reviewed the ecotoxicology of antibiotic compounds in European aquatic environments, describing a very complex scenario with roughly 11 different subclasses of antibiotics occurring in ecosystems, each constituted by numbers of molecules: among these, sulfonamides represent the most frequently detected class in water column, mainly related to their massive usage in veterinary medicine (Carvalho and Santos, 2016; Song et al., 2017). Some studies reported that sulfonamides can inhibit denitrification in biogeochemical cycles, further stimulating the release of nitrous oxide (N_2O) with consequent enhancement of eutrophication processes and greenhouse effects (Song et al., 2017; Hou et al., 2015). As for many other therapeutical classes, acute toxicity tests were preferentially performed compared to chronic studies on sublethal adverse effects in nontarget organisms underlining the urgent need to fill this gap of knowledge. The clam, *R. philippinarum* exposed to realistic environmental concentrations of trimethoprim, highlighted the alteration of haemocytes parameters, with a significant decrement in lysosomal membrane stability, while oxidative stress responses were only slightly affected. Conversely low levels of amoxicillin, barely influenced haemocytes biomarkers in both *Ruditapes philippinarum* and *Mytilus galloprovincialis* (Matozzo et al., 2016).

Hypocholesterolemic compounds, which mainly include statins, have been only recently investigated for their ecotoxicological effects. Simvastatin (SIM) and atorvastatin (ATO) are among the most prescribed pharmaceuticals especially in Western countries. Despite data on the environmental occurrence and persistence of these compounds are still limited, these molecules were measured in freshwater ecosystems. Physicochemical properties suggest a high bioaccumulation potential and the few ecotoxicological investigations revealed that environmentally realistic levels (hundreds of ng/L) affect aquatic species. In vertebrates, statins decrease cholesterol synthesis trough the competitive inhibition of the enzyme 3-hydroxy-3-methyl-glutaryl-CoA reductase (HMGR), responsible for the rate-limiting step in the mevalonate pathway. The competitive inhibition of HMGR has also been documented in arthropods. A delay in nauplii development of the harpacticoid copepod *Nitocra spinipes* was caused by 160 ng/L of SIM, and a similar effect was reported in the embryo development of the sea urchin

Paracentrotus lividus at a concentration of 2000 ng/L. In the amphipod *Gammarus locusta* exposed for 35 days to SIM (0.32, 1.6, 8 μg/L), severe effects on growth, reproduction, and gonad maturation were observed: reproductive potential was most sensitive endpoint with significant impairment at the lowest exposure dose. A comparative genomic study, modeling homologues and simulating molecular docking on a large number of metazoans suggested that all these taxa might be hampered by statins (Table 2.5 and 2.6).

2.9 Standards and guidelines for pharmaceutical concentrations in the environment

Sufficient knowledge about pharmaceutical toxic effects on flora and fauna is very scare. Existing steps that have been taken by various environmental agencies around the world are discussed here. Environmental assessment of veterinary drugs in USA has been required since 1980. The European Union, in 1990, provided one of the first legal limits for antibiotics in milk (4 – 1500 μg/kg) and other food products of animal origin (25 – 6000 μg/kg). The United States Food and Drug Administration (USFDA) guidelines for human drug assessments, requires US applicants to provide an Environmental Impact Assessment (EIA) report when the concentrations expected for an active pharmaceutical ingredient into an aquatic environment are >1 μg/L. The total soil concentration of veterinary pharmaceuticals has been limited to ≤ 100 μg/kg by the "International Cooperation on Harmonization of Technical Requirements for the Registration of Veterinary Medical Products." This limit has been set based on ecotoxic effects on earthworms, plants, and microbes (Patel et al., 2019).

The EMA issued guidelines in 2006 on ERA for pharmaceuticals based on serious threats these can pose to the environment. The goal was to improve pharmaceutical risk assessment in the environment. The European Water Framework Directives on Priority Substances includes DIC, ibuprofen, α-estradiol, and β-ethinylestradiol. The USEPA also added erythromycin, estrogenic hormones, and several other pharmaceutical ingredients in its third Contaminant Candidate List (CCL_3), and again in CCL_4. Unfortunately no statutory maximum limits are present worldwide for active pharmaceutical compounds in drinking water (Patel et al., 2019).

No drinking water or reclaimed water pharmaceutical guidelines exist. However, according to USFDA rules, ecological evaluation and testing is required in the USA if the PEC in water and soil exceeds 1 μg/L or 100 μg/kg, respectively. This limit was set for possible acute effects based on limited available toxicity knowledge. The European Union has set a threshold value of 0.01 μg/L based on the extrapolation of PECs or on the measure of pharmaceutical concentrations in the environment. The Australian government has taken a first step by introducing pharmaceutical guidelines in drinking water in 2008. These guidelines used the lowest recommended therapeutic doses (LRTD) to set acceptable levels. Pharmaceutical LRTD values are then divided by safety factors to calculate tolerable daily intakes. A safety factor of 1000 has been

Table 2.5 Concentrations of pharmaceuticals measured worldwide in seawater (Mezzelani et al., 2018b).

Compound	Country	Environment	Concentration (ng/L)	References
Acetaminophen	Antarctica	Southern Ocean	nd-48744	González-Alonso et al. (2017)
	Brasil	Atlantic Ocean	<LOQ-34.6	Pereira et al. (2015)
	France	Mediterranean Sea	200,000	Togola and Budzinski (2008)
	Greece	Aegean See	2983	Nödler et al. (2014)
	Portugal	Arade Estuary	88	Gonzalez-Rey et al. (2015)
	Spain	Atlantic Ocean	nd-41.5	Biel-Maeso et al. (2018)
Diclofenac	Antarctica	Southern Ocean	nd-15087	González-Alonso et al. (2017)
	Brasil	Atlantic Ocean	<LOQ-19.4	Pereira et al. (2015)
	France	Mediterranean Sea	1500	Togola and Budzinski (2008)
	Greece	Mediterranean Sea	<1.4–16.3	Alygizakis et al. (2016)
	Ireland	Atlantic Ocean	220–550	McEneff et al. (2014)
	Portugal	Arade Estuary	31	Gonzalez-Rey et al. (2015)
	Spain	Mediterranean Sea	4	Gros et al. (2012)
	Spain	Mediterranean Sea	nd- <LOQ	Rodríguez-Navas et al. (2013)
	Spain	Atlantic Ocean	nd-31.9	Biel-Maeso et al. (2018)
	Singapore	Indian Ocean	<1.5–11.6	Biel-Maeso et al. (2018)
	Singapore	Indian Ocean	4–38	Wu et al. (2010)
	Taiwan	Indian Ocean	<2.5–53.6	Fang et al. (2012)
	United Kingdom	Estuaries	<8–195	Thomas and Hilton (2004)
	United Kingdom	North Sea	<0.12	Nebot et al. (2007)
Ibuprof	Antarctica	Southern Ocean	nd-10053	González-Alonso et al. (2017)
	Brasil	Atlantic Ocean	326.1–2094.4	Pereira et al. (2016)
	France	Mediterranean Sea	1500	Togola and Budzinski (2008)
	Italy	Adriatic Sea	<0.049–1.146	Loos et al. (2013)
	Norway	Norwegian Sea	<0.07–0.7	Weigel et al. (2004)

(Continued)

Table 2.5 Concentrations of pharmaceuticals measured worldwide in seawater (Mezzelani et al., 2018b). *Continued*

Compound	Country	Environment	Concentration (ng/L)	References
	Portugal	Aveiro Lagoon	242	Paíga et al. (2013)
	Portugal	Arade Estuary	28	Gonzalez-Rey et al. (2015)
	Singapore	Indian Ocean	41–121	Wu et al. (2010)
	Singapore	Seawater	<2.2–9.1	Bayen et al. (2013)
	Spain	Mediterranean Sea	16	Gros et al. (2012)
	Spain	Atlantic Ocean	nd-1219.7	Biel-Maeso et al. (2018)
	Taiwan	Indian Ocean	<2.5–57.1	Fang et al. (2012)
	United Kingdom	North Sea	<0.52	Nebot et al. (2007)
	United Kingdom	Estuaries	<8–928	Thomas and Hilton (2004)
Ketoprofen	France	Mediterranean Sea	6000	Togola and Budzinski (2008)
	Spain	Mediterranean Sea	<8	Gros et al. (2012)
	Spain	Atlantic Ocean	nd-2.6	Biel-Maeso et al. (2018)
	Sweden	STP Effluent	330	Bendz et al. (2005)
	Taiwan	Indian Ocean	<1.7–6.59	Fang et al. (2012)
Nimesulide	Italy	Po river basin	nd-0.15	Ferrari et al. (2011)
Salicylic acid	Belgium	North sea	855	Wille et al. (2010)
	Greece	Mediterranean Sea	0.4–53.3	Alygizakis et al. (2016)
	Spain	Mediterranean Sea	nd-16	Moreno-González et al. (2015)
	Spain	Atlantic Ocean	<LOQ-977.2	Biel-Maeso et al. (2018)
Citalopram	Greece	Mediterranean Sea	<0.06–8.0	Alygizakis et al. (2016)
Fluoxetine	Greece	Mediterranean Sea	nd	Alygizakis et al. (2016)
	Norway	Norwegian Sea	1.4–4.8	Vasskog et al. (2008)
	Portugal	Arade estuary	1.0–2.0	Gonzalez-Rey et al. (2015)
				Gros et al. (2012)
	Spain	Atlantic Ocean	nd-1.6	Biel-Maeso et al. (2018)

(*Continued*)

Table 2.5 Concentrations of pharmaceuticals measured worldwide in seawater (Mezzelani et al., 2018b). *Continued*

Compound	Country	Environment	Concentration (ng/L)	References
	Spain	Mediterranean Sea	nd	Moreno-González et al. (2015)
	United States	Chesapeake bay	3	Pait et al. (2006)
Paroxetine	Greece	Mediterranean Sea	nd	Alygizakis et al. (2016)
	Norway	Norwegian Sea	0.6–1.4	Vasskog et al. (2008)
	Spain	Mediterranean Sea	nd	Gros et al. (2012)
Sertraline	Greece	Mediterranean Sea	nd	Alygizakis et al. (2016)
	Norway	Norwegian Sea	<0.16	Vasskog et al. (2008)
Alprazolam	Greece	Mediterranean Sea	nd	Alygizakis et al. (2016)
	Antarctica	Southern Ocean	nd-2.55	González-Alonso et al. (2017)
	Spain	Mediterranean Sea	nd	Gros et al. (2012)
	Spain	Mediterranean Sea	0–1.0	Moreno-González et al. (2015)
Diazepam	Greece	Mediterranean Sea	nd	Alygizakis et al. (2016)
	Spain	Mediterranean Sea	<LOD	Gros et al. (2012)
	Spain	Mediterranean Sea	0–1.0	Moreno-González et al. (2015)
Lorazepam	Spain	Mediterranean Sea	3.2–41.8	Moreno-González et al. (2015)
Carbamazepine	Belgium	North sea	nd-321	Wille et al. (2010)
	France	Mediterranean Sea	10.0–40.0	Togola and Budzinski (2008)
	Greece	Mediterranean Sea	nd-1.4	Alygizakis et al. (2016)
	Ireland	Irish coast	0.21–1.56	McEneff et al. (2014)
	Portugal	Arade estuary	31	Gonzalez-Rey et al. (2015)
	Singapore	Indian Ocean	0.28–10.9	Bayen et al. (2013)
	Spain	Mediterranean Sea	8	Gros et al. (2012)
	Spain	Mediterranean Sea	15–31	Moreno-González et al. (2015)
	Spain	Atlantic Ocean	nd-31.1	Biel-Maeso et al. (2018)
	United States	Pacific Ocean	5.2–44.2	Klosterhaus et al. (2013)

(*Continued*)

Table 2.5 Concentrations of pharmaceuticals measured worldwide in seawater (Mezzelani et al., 2018b). *Continued*

Compound	Country	Environment	Concentration (ng/L)	References
Atenolol	Greece/Turkey	Aegean Sea	190	Nödler et al. (2014)
	Spain	Atlantic Ocean	0.4–138.9	Biel-Maeso et al. (2018)
	Spain	Mediterranean Sea	6	Gros et al. (2012)
	USA	Pacific Ocean	37	Klosterhaus et al. (2013)
Diltiazem	Belgium	North sea	290	Wille et al. (2010)
	Singapore	Indian Ocean	1.7	Bayen et al. (2013)
Irbesartan	Spain	Mediterranean Sea	4	Gros et al. (2012)
	Spain	Mediterranean Sea	21	Gros et al. (2012)
	Spain	Mediterranean Sea	0.1–16.8	Moreno-González et al. (2015)
	USA	Pacific Ocean	3.5	Klosterhaus et al. (2013)
Losartan	Brasil	Atlantic Ocean	<LOQ-16.4	Pereira et al. (2015)
	Spain	Mediterranean Sea	4	Gros et al. (2012)
Metoprolol	China	Jiulong River estuary	0.4–11	Lv et al. (2014)
	Germany	Baltic Sea	160	Nödler et al. (2014)
	Spain	Mediterranean Sea	0.1–0.73	Moreno-González et al. (2015)
	Spain	Atlantic Ocean	nd-5.1	Biel-Maeso et al. (2018)
	USA	Pacific Ocean	26	Klosterhaus et al. (2013)
Propranolol	Belgium	North sea	24	Wille et al. (2010)
	China	Jiulong River estuary	0.8	Lv et al. (2014)
	Spain	Atlantic Ocean	nd-5.9	Biel-Maeso et al. (2018)
	Portugal	Douro River estuary	3.2	Madureira et al. (2010)
	UK	Tyne estuary	110	Roberts and Thomas (2006)
	UK	Estuaries	56	Thomas and Hilton (2004)
Nadolol	Spain	Mediterranean Sea	0.3	Gros et al. (2012)
	Spain	Atlantic Ocean	nd-1.6	Biel-Maeso et al. (2018)

(*Continued*)

Table 2.5 Concentrations of pharmaceuticals measured worldwide in seawater (Mezzelani et al., 2018b). *Continued*

Compound	Country	Environment	Concentration (ng/L)	References
Sotalol	Spain	Mediterranean Sea	2	Gros et al. (2012)
	China	Jiulong River estuary	0.8	Lv et al. (2014)
	Greece/Turkey	Aegean Sea	67	Nödler et al. (2014)
Valsartan	Brasil	Atlantic Ocean	<LOQ-75.0	Pereira et al. (2015)
	Spain	Mediterranean Sea	29	Gros et al. (2012)
	USA	Pacific Ocean	92	Klosterhaus et al. (2013)
	Greece	Mediterranean Sea	<0.8–3.7	Alygizakis et al. (2016)
	Spain	Mediterranean Sea	0.6–38	Moreno-González et al. (2015)

With permission.

Table 2.6 Concentrations of pharmaceuticals measured worldwide in different aquatic organisms (Mezzelani et al., 2018b).

Compound	Species	Tissues	Environment	Concentration (ng/L)	References
Acetaminophen	*Mytilus edulis*	Whole body	Belgian coast	≤115	Wille et al. (2011)
	Mytilus galloprovincialis	Whole body	Adriatic coast	<LOD	Mezzelani et al. (2016a,b)
Diclofenac	*M. edulis*	Whole body	Irish coast	nd	McEneff et al. (2014)
	M. galloprovincialis	Whole body	Adriatic coast	<LOD-16.11	Mezzelani et al. (2016a,b)
	Mytilus spp	Whole body	Portuguese coast	0.5–4.5	Cunha et al. (2017)
	Liza aurata	Muscle	Mar Menor lagoon	nd-1.3	Mezzelani et al. (2016a,b)
	L. aurata	Liver	Mar Menor lagoon	nd- ≤2.2	Mezzelani et al. (2016a,b)
	Hemiculter leucisculus	Muscle	China rivers	0.4–1.2	Liu et al. (2015)
		Gill	China rivers	1.3–1.5	Liu et al. (2015)
		Brain	China rivers	1.4–3.6	Liu et al. (2015)
		Liver	China rivers	1.4–8.6	Liu et al. (2015)
	Carassius auratus	Muscle	China rivers	0.1–0.2	Liu et al. (2015)
		Gill	China rivers	0.1–0.7	Liu et al. (2015)
		Brain	China rivers	1.0–1.6	Liu et al. (2015)
		Liver	China rivers	1.9–4.6	Liu et al. (2015)
	Barbus graellsii	Whole body	Mediterranean rivers	8.8	Huerta et al. (2013)
	Micropterus salmoides	Whole body	Mediterranean rivers	4.1	Huerta et al. (2013)
	Phytoplankton	Whole body	Taihu Lake (China)	nd-25	Xie et al. (2017)
	Zooplankton	Whole body	Taihu Lake (China)	nd-31	Xie et al. (2017)
	Anodonta	Whole body	Taihu Lake (China)	nd-19	Xie et al. (2017)
	Bellamya sp.	Whole body	Taihu Lake (China)	2.9–16	Xie et al. (2017)
	Corbiculidae	Whole body	Taihu Lake (China)	1.2–31	Xie et al. (2017)

(*Continued*)

Table 2.6 Concentrations of pharmaceuticals measured worldwide in different aquatic organisms (Mezzelani et al., 2018b). *Continued*

Compound	Species	Tissues	Environment	Concentration (ng/L)	References
	Macrobranchium nipponense	Muscle	Taihu Lake (China)	nd-27	Xie et al. (2017)
	Exopalaemon modestus	Muscle	Taihu Lake (China)	1.7–22	Xie et al. (2017)
	Reganisalanx brachyrostralis	Muscle	Taihu Lake (China)	nd-16	Xie et al. (2017)
	Coilia ectenes	Muscle	Taihu Lake (China)	nd-14	Xie et al. (2017)
	Hypophthalmichthys molitrix	Muscle	Taihu Lake (China)	1.8–5.9	Xie et al. (2017)
		Gill	Taihu Lake (China)	5.4–12	Xie et al. (2017)
		Brain	Taihu Lake (China)	5.6–16	Xie et al. (2017)
		Liver	Taihu Lake (China)	10–27	Xie et al. (2017)
	Cyprinus carpio	Muscle	Taihu Lake (China)	nd-13	Xie et al. (2017)
		Gill	Taihu Lake (China)	nd-19	Xie et al. (2017)
		Brain	Taihu Lake (China)	2.8–29	Xie et al. (2017)
		Liver	Taihu Lake (China)	4.0–62	Xie et al. (2017)
	C. auratus	Muscle	Taihu Lake (China)	0.97–8.2	Xie et al. (2017)
		Gill	Taihu Lake (China)	nd-17	Xie et al. (2017)
		Brain	Taihu Lake (China)	5.2–29	Xie et al. (2017)
		Liver	Taihu Lake (China)	8.9–46	Xie et al. (2017)
	Cultrichthys erythropterus	Muscle	Taihu Lake (China)	nd-4.1	Xie et al. (2017)
		Gill	Taihu Lake (China)	nd-7.4	Xie et al. (2017)
		Brain	Taihu Lake (China)	nd-68	Xie et al. (2017)
	Pelteobagrus fulvidraco	Liver	Taihu Lake (China)	0.8–7.3	Xie et al. (2017)
		Gill	Taihu Lake (China)	nd-15	Xie et al. (2017)

(Continued)

Table 2.6 Concentrations of pharmaceuticals measured worldwide in different aquatic organisms (Mezzelani et al., 2018b). *Continued*

Compound	Species	Tissues	Environment	Concentration (ng/L)	References
		Brain	Taihu Lake (China)	nd-28	Xie et al. (2017)
		Liver	Taihu Lake (China)	2.5–42	Xie et al. (2017)
Ibuprofen	*Geukensia demissa*	Whole body	San Francisco Bay	<LOQ	Klosterhaus et al. (2013)
	M. galloprovincialis	Whole body	Adriatic coast	<LOD-9.39	Mezzelani et al. (2016a,b)
	Gambusia holbrooki	Whole body	Florida WTPs effluent	10	Wang and Gardinali (2013)
	H. leucisculus	Muscle, gill, brain, liver	China rivers	nd	Liu et al. (2015)
	C. auratus	Muscle, gill, brain, liver	China rivers	nd	Liu et al. (2015)
	Phytoplankton	Whole body	Taihu Lake (China)	nd-28	Xie et al. (2017)
	Zooplankton	Whole body	Taihu Lake (China)	1.3–24	Xie et al. (2017)
	Anodonta	Whole body	Taihu Lake (China)	nd-82	Xie et al. (2017)
	Bellamya sp.	Whole body	Taihu Lake (China)	nd-47	Xie et al. (2017)
	Corbiculidae	Whole body	Taihu Lake (China)	5.0–44	Xie et al. (2017)
	Macrobranchium nipponense	Muscle	Taihu Lake (China)	nd-22	Xie et al. (2017)
	E. modestus	Muscle	Taihu Lake (China)	4.0–26	Xie et al. (2017)
	R. brachyrostralis	Muscle	Taihu Lake (China)	nd-40	Xie et al. (2017)
	C. ectenes	Muscle	Taihu Lake (China)	3.8–28	Xie et al. (2017)
	H. molitrix	Muscle	Taihu Lake (China)	nd-43	Xie et al. (2017)
		Gill	Taihu Lake (China)	6.7–57	Xie et al. (2017)
		Brain	Taihu Lake (China)	nd-39	Xie et al. (2017)

(Continued)

Table 2.6 Concentrations of pharmaceuticals measured worldwide in different aquatic organisms (Mezzelani et al., 2018b). *Continued*

Compound	Species	Tissues	Environment	Concentration (ng/L)	References
		Liver	Taihu Lake (China)	2.2–39	Xie et al. (2017)
	C. carpio	Muscle	Taihu Lake (China)	2.8–40	Xie et al. (2017)
		Gill	Taihu Lake (China)	15–62	Xie et al. (2017)
		Brain	Taihu Lake (China)	12–51	Xie et al. (2017)
		Liver	Taihu Lake (China)	10–48	Xie et al. (2017)
	C. auratus	Muscle	Taihu Lake (China)	nd-24	Xie et al. (2017)
		Gill	Taihu Lake (China)	2.5–42	Xie et al. (2017)
		Brain	Taihu Lake (China)	nd-34	Xie et al. (2017)
		Liver	Taihu Lake (China)	nd-31	Xie et al. (2017)
	C. erythropterus	Muscle	Taihu Lake (China)	nd-40	Xie et al. (2017)
		Gill	Taihu Lake (China)	1.7–60	Xie et al. (2017)
		Brain	Taihu Lake (China)	3.0–62	Xie et al. (2017)
		Liver	Taihu Lake (China)	2.0–49	Xie et al. (2017)
	P. fulvidraco	Muscle	Taihu Lake (China)	nd-24	Xie et al. (2017)
		Gill	Taihu Lake (China)	nd-45	Xie et al. (2017)
		Brain	Taihu Lake (China)	nd-58	Xie et al. (2017)
		Liver	Taihu Lake (China)	nd-46	Xie et al. (2017)
Ketoprofen	*M. galloprovincialis*	Whole body	Adriatic coast	<LOD	Mezzelani et al. (2016a,b)
Mefenamic acid	*M. edulis*	Whole body	Irish coast	nd- ≤ 23	McEneff et al. (2014)
Naprossen	*G. demissa*	Whole body	San Francisco Bay	<LOQ	Klosterhaus et al. (2013)
Nimesulide	*M. galloprovincialis*	Whole body	Adriatic coast	2.99–6.04	Mezzelani et al. (2016a,b)

(Continued)

Table 2.6 Concentrations of pharmaceuticals measured worldwide in different aquatic organisms (Mezzelani et al., 2018b). *Continued*

Compound	Species	Tissues	Environment	Concentration (ng/L)	References
Salicylic acid	*M. edulis*	Whole body	Belgian coast	≤490	Wille et al. (2011)
Citalopram	*M. galloprovincialis*	Whole body	Po delta	20.6	Álvarez-Muñoz et al. (2015)
	Mytilus spp.	Whole body	Tagus estuary	<0.06	Álvarez-Muñoz et al. (2015)
	M. galloprovincialis	Whole body	Ebro delta	<0.06	Álvarez-Muñoz et al. (2015)
	Chamalea gallina	Whole body	Ebro delta	<0.21	Álvarez-Muñoz et al. (2015)
	Crassostrea gigas	Whole body	Ebro delta	1.9	Álvarez-Muñoz et al. (2015)
	L. aurata	Muscle	Tagus estuary	<0.41	Álvarez-Muñoz et al. (2015)
	Platichthys flesus	Muscle	Scheldt estuary	<0.41	Álvarez-Muñoz et al. (2015)
	M. galloprovincialis	Whole body	Portouguese Atlantic coast	3.79–4.97	Silva et al. (2017)
	Pimephales promelas	Whole body	Grand River (Ontario)	<LOQ-2.90	Metcalfe et al. (2010)
	Catostomus commersoni	Brain	WTPs downstream (Texas)	nd-0.195	Schultz et al. (2010)
	Elliptio complanata	Whole body	WTPs downstream (North Carolina)	2.0	Bringolf et al. (2010)
N-des-Citalopram	*M. galloprovincialis*	Whole body	Portouguese Atlantic coast	3.38–10.17	Silva et al. (2017)
	P. promelas	Whole body	Grand River (Ontario)	nd-4.12	Metcalfe et al. (2010)
Fluoxetine	*M. galloprovincialis*	Whole body	Portouguese Atlantic coast	2.25–9.93	Silva et al. (2017)
	P. promelas	Whole body	Grand River (Ontario)	nd-4.12	Metcalfe et al. (2010)

(*Continued*)

Table 2.6 Concentrations of pharmaceuticals measured worldwide in different aquatic organisms (Mezzelani et al., 2018b). *Continued*

Compound	Species	Tissues	Environment	Concentration (ng/L)	References
Fluoxetine	*M. galloprovincialis*	Whole body	Portouguese Atlantic coast	2.25–9.93	Silva et al. (2017)
	P. promelas	Whole body	Grand River (Ontario)	nd- <LOQ	Metcalfe et al. (2010)
	C. commersoni	Brain	WTPs downstream (Texas)	nd-1.648	Schultz et al. (2010)
	E. complanata	Whole body	WTPs downstream (North Carolina)	9.8	Bringolf et al. (2010)
Norfluoxetine	*M. galloprovincialis*	Whole body	Portouguese Atlantic coast	9.03–21.66	Silva et al. (2017)
	P. promelas	Whole body	Grand River (Ontario)	nd-3.22	Metcalfe et al. (2010)
	C. commersoni	Brain	WTPs downstream (Texas)	nd-3.567	Schultz et al. (2010)
Paroxetine	*C. commersoni*	Brain	WTPs downstream (Texas)	nd-0.113	Schultz et al. (2010)

applied for all pharmaceuticals. This factor takes account of the following contributors: (i) 10—for the differences in the responses between humans(intrahuman variations taking into account sensitive humans); (ii) 10—for protecting sensitive subgroups, which includes infants and children; (iii) 10—for the lowest recommended therapeutic doses per day above a no-effect level (NOEL). Additionally, a factor of 10 is introduced for cytotoxic drugs and synthetic and endogenous hormones, due to their higher level of associated toxicities and unwanted hormonal effects, respectively. Thus, the applied safety factor lies between 1000 and 10,000 (Patel et al., 2019).

2.10 Conclusion and recommendation for future studies

The presence of pharmaceuticals has been found from long time ago but their detection and dangerous impacts only have discovered during the past three decades. Despite many articles published on pharmaceuticals, their acute and chronic effects on human and other living organisms are not well understood. There are not any global legal maximum environmental concentrations for pharmaceuticals compounds. Pharmaceutical limits based on human health considerations have not been conducted except for the Australian groundwater guidelines. Conventional water and wastewater treatment plants generally are unable to completely remove pharmaceuticals, leading to their release into aquatic environments. Thus advanced and effective treatment methods are vital to increase the pharmaceuticals removal efficiency. Therefore it is necessary:

- Strict regulations must be employed for effluent discharge from industrial and hospital point sources.
- Environmental friendly technologies should be used for production, development, and use of pharmaceuticals.
- Continuous longitude research should be conducted on toxic effects of chronic or acute exposure to pharmaceuticals in aquatic environments
- Standards for environmental protection should be strike to limit pharmaceuticals presence in aquatic environments
- More effective technologies for pharmaceutical removal should be developed for fast application on a large scale and low cost.

References

Adityosulindro, S., Barthe, L., González-Labrada, K., Haza, U.J.J., Delmas, H., Julcour, C., 2017. Sonolysis and sono-Fenton oxidation for removal of ibuprofen in (waste) water. Ultrason. Sonochem. 39, 889–896.

Aguirre-Martínez, G.V., Buratti, S., Fabbri, E., Delvalls, A.T., Martín-Díaz, M.L., 2013. Using lysosomal membrane stability of haemocytes in *Ruditapes philippinarum* as a biomarker of cellular stress to assess contamination by caffeine, ibuprofen, carbamazepine and novobiocin. J. Environ. Sci. 25, 1408–1418.

Aguirre-Martínez, G.V., Delvalls, T.A., Martín-Díaz, M.L., 2016. General stress, detoxification pathways, neurotoxicity and genotoxicity evaluated in *Ruditapes philippinarum* exposed to human pharmaceuticals. Ecotoxicol. Environ. Saf. 124, 18–31.

Aherne, G., Briggs, R., 1989. The relevance of the presence of certain synthetic steroids in the aquatic environment. J. Pharm. Pharmacol. 41, 735–736.

Ahmad, I., Ahmed, S., Anwar, Z., Sheraz, M.A., Sikorski, M., 2016. Photostability and photostabilization of drugs and drug products. Int. J. Photoenergy 2016.

Aitken, M., Kleinrock, M., 2015. Global Medicines Use in 2020: Outlook and Implications. IMS Institute for Healthcare Informatics, Parsippany, NJ.

Alharbi, S.K., Price, W.E., Kang, J., Fujioka, T., Nghiem, L.D., 2016. Ozonation of carbamazepine, diclofenac, sulfamethoxazole and trimethoprim and formation of major oxidation products. Desali. Water Treat. 57, 29340–29351.

Almeida, Â., Freitas, R., Calisto, V., Esteves, V.I., Schneider, R.J., Soares, A.M., et al., 2015. Chronic toxicity of the antiepileptic carbamazepine on the clam *Ruditapes philippinarum*. Comp. Biochem. Physiol. Part. C: Toxicol. Pharmacol. 172, 26–35.

Almomani, F.A., Shawaqfah, M., Bhosale, R.R., Kumar, A., 2016. Removal of emerging pharmaceuticals from wastewater by ozone-based advanced oxidation processes. Environ. Prog. Sustain. Energy 35, 982–995.

Álvarez-Muñoz, D., Rodríguez-Mozaz, S., Maulvault, A., Tediosi, A., Fernández-Tejedor, M., Van Den Heuvel, F., et al., 2015. Occurrence of pharmaceuticals and endocrine disrupting compounds in macroalgaes, bivalves, and fish from coastal areas in Europe. Environ. Res. 143, 56–64.

Alygizakis, N.A., Gago-Ferrero, P., Borova, V.L., Pavlidou, A., Hatzianestis, I., Thomaidis, N.S., 2016. Occurrence and spatial distribution of 158 pharmaceuticals, drugs of abuse and related metabolites in offshore seawater. Sci. Total Environ. 541, 1097–1105.

Ashton, D., Hilton, M., Thomas, K., 2004. Investigating the environmental transport of human pharmaceuticals to streams in the United Kingdom. Sci. Total Environ. 333, 167–184.

Baeza, C., Knappe, D.R., 2011. Transformation kinetics of biochemically active compounds in low-pressure UV photolysis and UV/H_2O_2 advanced oxidation processes. Water Res. 45, 4531–4543.

Balakrishna, K., Rath, A., Praveenkumarreddy, Y., Guruge, K.S., Subedi, B., 2017. A review of the occurrence of pharmaceuticals and personal care products in Indian water bodies. Ecotoxicol. Environ. Saf. 137, 113–120.

Balaram, V., 2016. Recent advances in the determination of elemental impurities in pharmaceuticals–status, challenges and moving frontiers. Trac. Trends Anal. Chem. 80, 83–95.

Bautista, P., Mohedano, A., Casas, J., Zazo, J., Rodriguez, J., 2008. An overview of the application of Fenton oxidation to industrial wastewaters treatment. J. Chem. Technol. Biotechnol.: Int. Res. Process, Environ. Clean. Technol. 83, 1323–1338.

Bayen, S., Zhang, H., Desai, M.M., Ooi, S.K., Kelly, B.C., 2013. Occurrence and distribution of pharmaceutically active and endocrine disrupting compounds in Singapore's marine environment: influence of hydrodynamics and physical–chemical properties. Environ. Pollut. 182, 1–8.

Bebianno, M., Mello, A., Serrano, M., Flores-Nunes, F., Mattos, J., Zacchi, F., et al., 2017. Transcriptional and cellular effects of paracetamol in the oyster *Crassostrea gigas*. Ecotoxicol. Environ. Saf. 144, 258–267.

Bendz, D., Paxeus, N.A., Ginn, T.R., Loge, F.J., 2005. Occurrence and fate of pharmaceutically active compounds in the environment, a case study: Höje River in Sweden. J. Hazard. Mater. 122, 195–204.

Biel-Maeso, M., Baena-Nogueras, R.M., Corada-Fernández, C., Lara-Martín, P.A., 2018. Occurrence, distribution and environmental risk of pharmaceutically active compounds (PhACs) in coastal and ocean waters from the Gulf of Cadiz (SW Spain). Sci. Total Environ. 612, 649–659.

Bound, J.P., Voulvoulis, N., 2005. Household disposal of pharmaceuticals as a pathway for aquatic contamination in the United Kingdom. Environ. Health Perspect. 113, 1705–1711.

Bound, J.P., Voulvoulis, N., 2006. Predicted and measured concentrations for selected pharmaceuticals in UK rivers: implications for risk assessment. Water Res. 40, 2885–2892.

Bringolf, R.B., Heltsley, R.M., Newton, T.J., Eads, C.B., Fraley, S.J., Shea, D., et al., 2010. Environmental occurrence and reproductive effects of the pharmaceutical fluoxetine in native freshwater mussels. Environ. Toxicol. Chem. 29, 1311–1318.

Bu, Q., Shi, X., Yu, G., Huang, J., Wang, B., 2016. Assessing the persistence of pharmaceuticals in the aquatic environment: challenges and needs. Emerg. Contam. 2, 145–147.

Cai, M.-J., Lin, Y.-P., 2016. Effects of effluent organic matter (EfOM) on the removal of emerging contaminants by ozonation. Chemosphere 151, 332–338.

Carbajo, J., Jimenez, M., Miralles, S., Malato, S., Faraldos, M., Bahamonde, A., 2016. Study of application of titania catalysts on solar photocatalysis: influence of type of pollutants and water matrices. Chem. Eng. J. 291, 64–73.

Carmona, E., Andreu, V., Picó, Y., 2014. Occurrence of acidic pharmaceuticals and personal care products in Turia River Basin: from waste to drinking water. Sci. Total Environ. 484, 53–63.

Carr, D.L., Morse, A.N., Zak, J.C., Anderson, T.A., 2011. Microbially mediated degradation of common pharmaceuticals and personal care products in soil under aerobic and reduced oxygen conditions. Water Air Soil Pollut. 216, 633–642.

Carrara, C., Ptacek, C.J., Robertson, W.D., Blowes, D.W., Moncur, M.C., Sverko, E., et al., 2008. Fate of pharmaceutical and trace organic compounds in three septic system plumes, Ontario, Canada. Environ. Sci. Technol. 42, 2805–2811.

Carvalho, I.T., Santos, L., 2016. Antibiotics in the aquatic environments: a review of the European scenario. Environ. Int. 94, 736–757.

Celiz, M.D., Tso, J., Aga, D.S., 2009. Pharmaceutical metabolites in the environment: analytical challenges and ecological risks. Environ. Toxicol. Chem. 28, 2473–2484.

Chavoshani, A., Amin, M.M., Asgari, G., Seidmohammadi, A., Hashemi, M., 2018. Microwave/hydrogen peroxide processes. Adv. Oxi. Pro. Waste Water Treatment. Elsevier.

Chen, H., Li, X., Zhu, S., 2012. Occurrence and distribution of selected pharmaceuticals and personal care products in aquatic environments: a comparative study of regions in China with different urbanization levels. Environ. Sci. Pollut. Res. 19, 2381–2389.

Conn, K.E., Lowe, K.S., Drewes, J.E., Hoppe-Jones, C., Tucholke, M.B., 2010. Occurrence of pharmaceuticals and consumer product chemicals in raw wastewater and septic tank effluent from single-family homes. Environ. Eng. Sci. 27, 347–356.

Corcoran, J., Winter, M.J., Tyler, C.R., 2010. Pharmaceuticals in the aquatic environment: a critical review of the evidence for health effects in fish. Crit. Rev. Toxicol. 40, 287–304.

Cunha, S., Pena, A., Fernandes, J., 2017. Mussels as bioindicators of diclofenac contamination in coastal environments. Environ. Pollut. 225, 354–360.

Dantas, R.F., Sans, C., Esplugas, S., 2011. Ozonation of propranolol: transformation, biodegradability, and toxicity assessment. J. Environ. Eng. 137, 754–759.

Davididou, K., Monteagudo, J.M., Chatzisymeon, E., Durán, A., Expósito, A.J., 2017. Degradation and mineralization of antipyrine by UV-A LED photo-Fenton reaction intensified by ferrioxalate with addition of persulfate. Sep. Purif. Technol. 172, 227–235.

de Bel, E., Janssen, C., De Smet, S., Van Langenhove, H., Dewulf, J., 2011. Sonolysis of ciprofloxacin in aqueous solution: influence of operational parameters. Ultrason. Sonochem. 18, 184–189.

de La Cruz, N., Esquius, L., Grandjean, D., Magnet, A., Tungler, A., De Alencastro, L., et al., 2013. Degradation of emergent contaminants by UV, UV/H_2O_2 and neutral photo-Fenton at pilot scale in a domestic wastewater treatment plant. Water Res. 47, 5836–5845.

de Lima Perini, J.A., Perez-Moya, M., Nogueira, R.F.P., 2013. Photo-Fenton degradation kinetics of low ciprofloxacin concentration using different iron sources and pH. J. Photochem. Photobiol. A: Chem. 259, 53–58.

Debaveye, S., De soete, W., De Meester, S., Vandijck, D., Heirman, B., Kavanagh, S., et al., 2016. Human health benefits and burdens of a pharmaceutical treatment: discussion of a conceptual integrated approach. Environ. Res. 144, 19–31.

Deblonde, T., Hartemann, P., 2013. Environmental impact of medical prescriptions: assessing the risks and hazards of persistence, bioaccumulation and toxicity of pharmaceuticals. Public Health 127, 312–317.

Di Poi, C., Darmaillacq, A.-S., Dickel, L., Boulouard, M., Bellanger, C., 2013. Effects of perinatal exposure to waterborne fluoxetine on memory processing in the cuttlefish Sepia officinalis. Aquat. Toxicol. 132, 84–91.

Ebele, A.J., Abdallah, M.A.-E., Harrad, S., 2017. Pharmaceuticals and personal care products (PPCPs) in the freshwater aquatic environment. Emerg. Contam. 3, 1–16.

Ericson, H., Thorsen, G., Kumblad, L., 2010. Physiological effects of diclofenac, ibuprofen and propranolol on Baltic Sea blue mussels. Aquat. Toxicol. 99, 223–231.

Fang, T.-H., Nan, F.-H., Chin, T.-S., Feng, H.-M., 2012. The occurrence and distribution of pharmaceutical compounds in the effluents of a major sewage treatment plant in Northern Taiwan and the receiving coastal waters. Mar. Pollut. Bull. 64, 1435–1444.

Feng, L., Van Hullebusch, E.D., Rodrigo, M.A., Esposito, G., Oturan, M.A., 2013. Removal of residual anti-inflammatory and analgesic pharmaceuticals from aqueous systems by electrochemical advanced oxidation processes. A review. Chem. Eng. J. 228, 944–964.

Ferguson, P.J., Bernot, M.J., Doll, J.C., Lauer, T.E., 2013. Detection of pharmaceuticals and personal care products (PPCPs) in near-shore habitats of southern Lake Michigan. Sci. Total Environ. 458, 187–196.

Ferrari, F., Gallipoli, A., Balderacchi, M., Ulaszewska, M.M., Capri, E., Trevisan, M., 2011. Exposure of the main Italian river basin to pharmaceuticals. J. Toxicol. 2011.

Franzellitti, S., Buratti, S., Valbonesi, P., Capuzzo, A., Fabbri, E., 2011. The β-blocker propranolol affects cAMP-dependent signaling and induces the stress response in Mediterranean mussels, *Mytilus galloprovincialis*. Aquat. Toxicol. 101, 299–308.

Franzellitti, S., Buratti, S., Valbonesi, P., Fabbri, E., 2013. The mode of action (MOA) approach reveals interactive effects of environmental pharmaceuticals on *Mytilus galloprovincialis*. Aquat. Toxicol. 140, 249–256.

Franzellitti, S., Buratti, S., Capolupo, M., Du, B., Haddad, S.P., Chambliss, C.K., et al., 2014. An exploratory investigation of various modes of action and potential adverse outcomes of fluoxetine in marine mussels. Aquat. Toxicol. 151, 14–26.

Garcia, S.N., Foster, M., Constantine, L.A., Huggett, D.B., 2012. Field and laboratory fish tissue accumulation of the anti-convulsant drug carbamazepine. Ecotoxicol. Environ. Saf. 84, 207–211.

García-Espinoza, J.D., Mijaylova-Nacheva, P., Aviles-Flores, M., 2018. Electrochemical carbamazepine degradation: effect of the generated active chlorine, transformation pathways and toxicity. Chemosphere 192, 142–151.

Garcia-Segura, S., Ocon, J.D., Chong, M.N., 2018. Electrochemical oxidation remediation of real wastewater effluents—a review. Process. Saf. Environ. Prot. 113, 48–67.

Garrison, A., 1976. GC/MS analysis of organic compounds in domestic wastewaters. Identification and analysis of organic pollutants in water.

Ghafoori, S., Mowla, A., Jahani, R., Mehrvar, M., Chan, P.K., 2015. Sonophotolytic degradation of synthetic pharmaceutical wastewater: statistical experimental design and modeling. J. Environ. Manag. 150, 128–137.

Giraldo-Aguirre, A.L., Serna-Galvis, E.A., Erazo-Erazo, E.D., Silva-Agredo, J., Giraldo-Ospina, H., Flórez-Acosta, O.A., et al., 2018. Removal of β-lactam antibiotics from pharmaceutical wastewaters using photo-Fenton process at near-neutral pH. Environ. Sci. Pollut. Res. 25, 20293–20303.

Glassmeyer, S.T., Furlong, E.T., Kolpin, D.W., Cahill, J.D., Zaugg, S.D., Werner, S.L., et al., 2005. Transport of chemical and microbial compounds from known wastewater discharges: potential for use as indicators of human fecal contamination. Environ. Sci. Technol. 39, 5157–5169.

Godfrey, E., Woessner, W.W., Benotti, M.J., 2007. Pharmaceuticals in on-site sewage effluent and ground water, western Montana. Groundwater 45, 263–271.

Gonzalez-Rey, M., Bebianno, M.J., 2011. Non-steroidal anti-inflammatory drug (NSAID) ibuprofen distresses antioxidant defense system in mussel *Mytilus galloprovincialis* gills. Aquat. Toxicol. 105, 264–269.

Gonzalez-Rey, M., Bebianno, M.J., 2012. Does non-steroidal anti-inflammatory (NSAID) ibuprofen induce antioxidant stress and endocrine disruption in mussel *Mytilus galloprovincialis*? Environ. Toxicol. Pharmacol. 33, 361–371.

Gonzalez-Rey, M., Bebianno, M.J., 2013. Does selective serotonin reuptake inhibitor (SSRI) fluoxetine affects mussel *Mytilus galloprovincialis*? Environ. Pollut. 173, 200–209.

Gonzalez-Rey, M., Bebianno, M.J., 2014. Effects of non-steroidal anti-inflammatory drug (NSAID) diclofenac exposure in mussel *Mytilus galloprovincialis*. Aquat. Toxicol. 148, 221–230.

González-Alonso, S., Merino, L.M., Esteban, S., De Alda, M.L., Barceló, D., Durán, J.J., et al., 2017. Occurrence of pharmaceutical, recreational and psychotropic drug residues in surface water on the northern Antarctic Peninsula region. Environ. Pollut. 229, 241–254.

Gonzalez-Rey, M., Tapie, N., Le Menach, K., Devier, M.-H., Budzinski, H., Bebianno, M. J., 2015. Occurrence of pharmaceutical compounds and pesticides in aquatic systems. Mar. Pollut. Bull. 96, 384–400.

Gros, M., Petrović, M., Barceló, D., 2006. Development of a multi-residue analytical methodology based on liquid chromatography–tandem mass spectrometry (LC–MS/MS) for screening and trace level determination of pharmaceuticals in surface and wastewaters. Talanta 70, 678–690.

Gros, M., Rodríguez-Mozaz, S., Barceló, D., 2012. Fast and comprehensive multi-residue analysis of a broad range of human and veterinary pharmaceuticals and some of their metabolites in surface and treated waters by ultra-high-performance liquid chromatography coupled to quadrupole-linear ion trap tandem mass spectrometry. J. Chromatogr. A 1248, 104–121.

Guyer, G.T., Ince, N.H., 2011. Degradation of diclofenac in water by homogeneous and heterogeneous sonolysis. Ultrason. Sonochem. 18, 114–119.

Hernando, M.D., Mezcua, M., Fernández-Alba, A.R., Barceló, D., 2006. Environmental risk assessment of pharmaceutical residues in wastewater effluents, surface waters and sediments. Talanta 69, 334–342.

Hou, X., Shen, J., Zhang, S., Jiang, H., Coats, J.R., 2003. Bioconcentration and elimination of sulfamethazine and its main metabolite in sturgeon (Acipenser schrenkii). J. Agric. Food Chem. 51, 7725–7729.

Hou, L., Yin, G., Liu, M., Zhou, J., Zheng, Y., Gao, J., Zong, H., Yang, Y., Gao, L., Tong, C., 2015. Effects of sulfamethazine on denitrification and the associated N2O release in estuarine and coastal sediments. Environ. Sci. Technol. 49, 326–333.

Howard, P.H., Muir, D.C., 2011. Identifying new persistent and bioaccumulative organics among chemicals in commerce II: pharmaceuticals. Environ. Sci. Technol. 45, 6938–6946.

Hu, R., Zhang, L., Hu, J., 2016. Study on the kinetics and transformation products of salicylic acid in water via ozonation. Chemosphere 153, 394–404.

Huerta, B., Jakimska, A., Gros, M., Rodriguez-Mozaz, S., Barcelo, D., 2013. Analysis of multi-class pharmaceuticals in fish tissues by ultra-high-performance liquid chromatography tandem mass spectrometry. J. Chromatogr. A 1288, 63–72.

Ince, N.H., 2018. Ultrasound-assisted advanced oxidation processes for water decontamination. Ultrason. Sonochem. 40, 97–103.

Ivanová, L., Mackulak, T., Grabic, R., Golovko, O., Koba, O., Stanová, A.V., Szabová, P., Grenciková, A., BoDíK, I., 2018. Pharmaceuticals and illicit drugs–a new threat to the application of sewage sludge in agriculture. Sci. Total Environ. 634, 606–615.

Ji, Y., Yang, Y., Zhou, L., Wang, L., Lu, J., Ferronato, C., et al., 2018. Photodegradation of sulfasalazine and its human metabolites in water by UV and UV/peroxydisulfate processes. Water Res. 133, 299–309.

Jones, O., Voulvoulis, N., Lester, J., 2002. Aquatic environmental assessment of the top 25 English prescription pharmaceuticals. Water Res. 36, 5013–5022.

Jung, Y.J., Kim, W.G., Yoon, Y., Kang, J.-W., Hong, Y.M., Kim, H.W., 2012. Removal of amoxicillin by UV and UV/H_2O_2 processes. Sci. Total Environ. 420, 160–167.

Kalyva, M., 2017. Fate of pharmaceuticals in the environment-A review.

Kanakaraju, D., Glass, B.D., Oelgemöller, M., 2018. Advanced oxidation process-mediated removal of pharmaceuticals from water: a review. J. Environ. Manag. 219, 189–207.

Kıdak, R., Doğan, Ş., 2018. Medium-high frequency ultrasound and ozone based advanced oxidation for amoxicillin removal in water. Ultrason. Sonochem. 40, 131–139.

Kim, S.-C., Carlson, K., 2006. Occurrence of ionophore antibiotics in water and sediments of a mixed-landscape watershed. Water Res. 40, 2549–2560.

Kim, Y., Jung, J., Kim, M., Park, J., Boxall, A.B., Choi, K., 2008. Prioritizing veterinary pharmaceuticals for aquatic environment in Korea. Environ. Toxicol. Pharmacol. 26, 167–176.

Kim, I., Yamashita, N., Tanaka, H., 2009. Performance of UV and UV/H_2O_2 processes for the removal of pharmaceuticals detected in secondary effluent of a sewage treatment plant in Japan. J. Hazard. Mater. 166, 1134–1140.

Kinney, C.A., Furlong, E.T., Kolpin, D.W., Burkhardt, M.R., Zaugg, S.D., Werner, S.L., et al., 2008. Bioaccumulation of pharmaceuticals and other anthropogenic waste indicators in earthworms from agricultural soil amended with biosolid or swine manure. Environ. Sci. Technol. 42, 1863–1870.

Klamerth, N., Rizzo, L., Malato, S., Maldonado, M.I., Aguera, A., FERNáNdez-Alba, A. R., 2010. Degradation of fifteen emerging contaminants at μg L − 1 initial concentrations by mild solar photo-Fenton in MWTP effluents. Water Res. 44, 545–554.

Klatte, S., Schaefer, H.-C., Hempel, M., 2017. Pharmaceuticals in the environment—a short review on options to minimize the exposure of humans, animals and ecosystems. Sustain. Chem. Pharm. 5, 61–66.

Klosterhaus, S.L., Grace, R., Hamilton, M.C., Yee, D., 2013. Method validation and reconnaissance of pharmaceuticals, personal care products, and alkylphenols in surface waters, sediments, and mussels in an urban estuary. Environ. Int. 54, 92–99.

Koltsakidou, A, Antonopoulou, M., Sykiotou, M., Evgenidou, E, Konstantinou, I., Lambropoulou, D., 2017. Photo-Fenton and Fenton-like processes for the treatment of the antineoplastic drug 5-fluorouracil under simulated solar radiation. Environ. Sci. Pollut. Res. 24, 4791–4800.

Kuster, A., Adler, N., 2014. Pharmaceuticals in the environment: scientific evidence of risks and its regulation. Philos. Trans. R. Soc. B: Biol. Sci. 369, 20130587.

Lahti, M., Brozinski, J.M., Jylhä, A., Kronberg, L., Oikari, A., 2011. Uptake from water, biotransformation, and biliary excretion of pharmaceuticals by rainbow trout. Environ. Toxicol. Chem. 30, 1403–1411.

Lapworth, D., Baran, N., Stuart, M., Ward, R., 2012. Emerging organic contaminants in groundwater: a review of sources, fate and occurrence. Environ. Pollut. 163, 287–303.

Li, W.C., 2014. Occurrence, sources, and fate of pharmaceuticals in aquatic environment and soil. Environ. Pollut. 187, 193–201.

Li, X., Wang, Y., Zhao, J., Wang, H., Wang, B., Huang, J., et al., 2015. Electro-peroxone treatment of the antidepressant venlafaxine: operational parameters and mechanism. J. Hazard. Mater. 300, 298–306.

Li, J., Ma, L.Y., Li, L.S., Xu, L., 2017. Photodegradation kinetics, transformation, and toxicity prediction of ketoprofen, carprofen, and diclofenac acid in aqueous solutions. Environ. Toxicol. Chem. 36, 3232–3239.

Lianou, A., Frontistis, Z., Chatzisymeon, E., Antonopoulou, M., Konstantinou, I., Mantzavinos, D., 2018. Sonochemical oxidation of piroxicam drug: effect of key operating parameters and degradation pathways. J. Chem. Technol. Biotechnol. 93, 28–34.

Liu, G., Liu, H., Zhang, N., Wang, Y., 2012. Photodegradation of salicylic acid in aquatic environment: effect of different forms of nitrogen. Sci. Total Environ. 435, 573–577.

Liu, J., Lu, G., Xie, Z., Zhang, Z., Li, S., Yan, Z., 2015. Occurrence, bioaccumulation and risk assessment of lipophilic pharmaceutically active compounds in the downstream rivers of sewage treatment plants. Sci. Total Environ. 511, 54–62.

Liu, N., Lei, Z.-D., Wang, T., Wang, J.-J., Zhang, X.-D., Xu, G., et al., 2016. Radiolysis of carbamazepine aqueous solution using electron beam irradiation combining with hydrogen peroxide: efficiency and mechanism. Chem. Eng. J. 295, 484–493.

Loos, R., Tavazzi, S., Paracchini, B., Canuti, E., Weissteiner, C., 2013. Analysis of polar organic contaminants in surface water of the northern Adriatic Sea by solid-phase extraction followed by ultrahigh-pressure liquid chromatography–QTRAP® MS using a hybrid triple-quadrupole linear ion trap instrument. Anal. Bioanal. Chem. 405, 5875–5885.

Luo, S., Wei, Z., Spinney, R., Zhang, Z., Dionysiou, D.D., Gao, L., et al., 2018. UV direct photolysis of sulfamethoxazole and ibuprofen: an experimental and modelling study. J. Hazard. Mater. 343, 132–139.

Lv, M., Sun, Q., Hu, A., Hou, L., Li, J., Cai, X., et al., 2014. Pharmaceuticals and personal care products in a mesoscale subtropical watershed and their application as sewage markers. J. Hazard. Mater. 280, 696–705.

Madureira, T.V., Barreiro, J.C., Rocha, M.J., Rocha, E., Cass, Q.B., Tiritan, M.E., 2010. Spatiotemporal distribution of pharmaceuticals in the Douro River estuary (Portugal). Sci. Total Environ. 408, 5513–5520.

Mahlambi, M.M., Ngila, C.J., Mamba, B.B., 2015. Recent developments in environmental photocatalytic degradation of organic pollutants: the case of titanium dioxide nanoparticles—a review. J. Nanomater. 2015, 5.

Mandaric, L., Kalogianni, E., Skoulikidis, N., Petrovic, M., Sabater, S., 2019. Contamination patterns and attenuation of pharmaceuticals in a temporary Mediterranean river. Sci. Total Environ. 647, 561–569.

Maria, V.L., Amorim, M.J., Bebianno, M.J., Dondero, F., 2016. Transcriptomic effects of the non-steroidal anti-inflammatory drug Ibuprofen in the marine bivalve *Mytilus galloprovincialis* Lam. Mar. Environ. Res. 119, 31–39.

Markic, M., Cvetnic, M., Ukic, S., Kusic, H., Bolanca, T., Bozic, A.L., 2018. Influence of process parameters on the effectiveness of photooxidative treatment of pharmaceuticals. J. Environ. Sci. Health Part. A 53, 338–351.

Márquez, G., Rodríguez, E.M., Maldonado, M.I., Álvarez, P.M., 2014. Integration of ozone and solar TiO2-photocatalytic oxidation for the degradation of selected pharmaceutical compounds in water and wastewater. Sep. Purif. Technol. 136, 18–26.

Martin-Diaz, L., Franzellitti, S., Buratti, S., Valbonesi, P., Capuzzo, A., Fabbri, E., 2009. Effects of environmental concentrations of the antiepilectic drug carbamazepine on biomarkers and cAMP-mediated cell signaling in the mussel *Mytilus galloprovincialis*. Aquat. Toxicol. 94, 177–185.

Matozzo, V., Rova, S., Marin, M.G., 2012. The nonsteroidal anti-inflammatory drug, ibuprofen, affects the immune parameters in the clam *Ruditapes philippinarum*. Mar. Environ. Res. 79, 116–121.

Matozzo, V., BattIstara, M., Marisa, I., Bertin, V., Orsetti, A., 2016. Assessing the effects of amoxicillin on antioxidant enzyme activities, lipid peroxidation and protein carbonyl content in the clam ruditapes philippinarum and the mussel mytilus galloprovincialis. Bulletin of Environ. Contam.Toxicol 97, 521–527.

McEneff, G., Barron, L., Kelleher, B., Paull, B., Quinn, B., 2014. A year-long study of the spatial occurrence and relative distribution of pharmaceutical residues in sewage effluent, receiving marine waters and marine bivalves. Sci. Total Environ. 476, 317–326.

Mehinto, A.C., Hill, E.M., Tyler, C.R., 2010. Uptake and biological effects of environmentally relevant concentrations of the nonsteroidal anti-inflammatory pharmaceutical diclofenac in rainbow trout (Oncorhynchus mykiss). Environ. Sci. Technol. 44, 2176–2182.

Mendez-Arriaga, F., Torres-Palma, R., Petrier, C., Esplugas, S., Gimenez, J., Pulgarin, C., 2009. Mineralization enhancement of a recalcitrant pharmaceutical pollutant in water by advanced oxidation hybrid processes. Water Res. 43, 3984–3991.

Meredith-Williams, M., Carter, L.J., Fussell, R., Raffaelli, D., Ashauer, R., Boxall, A.B., 2012. Uptake and depuration of pharmaceuticals in aquatic invertebrates. Environ. Pollut. 165, 250–258.

Metcalfe, C.D., Chu, S., Judt, C., Li, H., Oakes, K.D., Servos, M.R., et al., 2010. Antidepressants and their metabolites in municipal wastewater, and downstream exposure in an urban watershed. Environ. Toxicol. Chem. 29, 79–89.

Mezzelani, M., Gorbi, S., Da Ros, Z., Fattorini, D., D'errico, G., Milan, M., et al., 2016a. Ecotoxicological potential of non-steroidal anti-inflammatory drugs (NSAIDs) in marine organisms: bioavailability, biomarkers and natural occurrence in *Mytilus galloprovincialis*. Mar. Environ. Res. 121, 31–39.

Mezzelani, M., Gorbi, S., Fattorini, D., D'errico, G., Benedetti, M., Milan, M., et al., 2016b. Transcriptional and cellular effects of Non-Steroidal Anti-Inflammatory Drugs (NSAIDs) in experimentally exposed mussels, *Mytilus galloprovincialis*. Aquat. Toxicol. 180, 306–319.

Mezzelani, M., Gorbi, S., Fattorini, D., D'errico, G., Consolandi, G., Milan, M., et al., 2018a. Long-term exposure of *Mytilus galloprovincialis* to diclofenac, Ibuprofen and Ketoprofen: insights into bioavailability, biomarkers and transcriptomic changes. Chemosphere 198, 238–248.

Mezzelani, M., Gorbi, S., Regoli, F., 2018b. Pharmaceuticals in the aquatic environments: evidence of emerged threat and future challenges for marine organisms. Mar. Environ. Res. 140, 41–60.

Michael, I., Hapeshi, E., Michael, C., Varela, A., Kyriakou, S., Manaia, C., et al., 2012. Solar photo-Fenton process on the abatement of antibiotics at a pilot scale: degradation kinetics, ecotoxicity and phytotoxicity assessment and removal of antibiotic resistant enterococci. Water Res. 46, 5621–5634.

Milan, M., Pauletto, M., Patarnello, T., Bargelloni, L., Marin, M.G., Matozzo, V., 2013. Gene transcription and biomarker responses in the clam *Ruditapes philippinarum* after exposure to ibuprofen. Aquat. Toxicol. 126, 17–29.

Mimeault, C., Woodhouse, A., Miao, X.-S., Metcalfe, C., Moon, T., Trudeau, V., 2005. The human lipid regulator, gemfibrozil bioconcentrates and reduces testosterone in the goldfish, *Carassius auratus*. Aquat. Toxicol. 73, 44–54.

Miralles-Cuevas, S., Oller, I., Perez, J.S., Malato, S., 2014. Removal of pharmaceuticals from MWTP effluent by nanofiltration and solar photo-Fenton using two different iron complexes at neutral pH. Water Res. 64, 23–31.

Moreira, N.F., Orge, C.A., Ribeiro, A.R., Faria, J.L., Nunes, O.C., Pereira, M.F.R., et al., 2015. Fast mineralization and detoxification of amoxicillin and diclofenac by photocatalytic ozonation and application to an urban wastewater. Water Res. 87, 87–96.

Moreno-González, R., Rodriguez-Mozaz, S., Gros, M., Barceló, D., León, V., 2015. Seasonal distribution of pharmaceuticals in marine water and sediment from a mediterranean coastal lagoon (SE Spain). Environ. Res. 138, 326–344.

Naddeo, V., Landi, M., Scannapieco, D., Belgiorno, V., 2013. Sonochemical degradation of twenty-three emerging contaminants in urban wastewater. Desalin. Water Treat. 51, 6601–6608.

Naddeo, V., Uyguner-Demirel, C.S., Prado, M., Cesaro, A., Belgiorno, V., Ballesteros, F., 2015. Enhanced ozonation of selected pharmaceutical compounds by sonolysis. Environ. Technol. 36, 1876–1883.

Nakamura, Y., Yamamoto, H., Sekizawa, J., Kondo, T., Hirai, N., Tatarazako, N., 2008. The effects of pH on fluoxetine in Japanese medaka (Oryzias latipes): acute toxicity in fish larvae and bioaccumulation in juvenile fish. Chemosphere 70, 865–873.

Nallani, G.C., Paulos, P.M., Constantine, L.A., Venables, B.J., Huggett, D.B., 2011. Bioconcentration of ibuprofen in fathead minnow (*Pimephales promelas*) and channel catfish (*Ictalurus punctatus*). Chemosphere 84, 1371–1377.

Nebot, C., Gibb, S.W., Boyd, K.G., 2007. Quantification of human pharmaceuticals in water samples by high performance liquid chromatography–tandem mass spectrometry. Anal. Chim. Acta 598, 87–94.

Nödler, K., Voutsa, D., Licha, T., 2014. Polar organic micropollutants in the coastal environment of different marine systems. Mar. Pollut. Bull. 85, 50–59.

Oliveira, P., Almeida, Â., Calisto, V., Esteves, V.I., Schneider, R.J., Wrona, F.J., et al., 2017. Physiological and biochemical alterations induced in the mussel *Mytilus galloprovincialis* after short and long-term exposure to carbamazepine. Water Res. 117, 102–114.

Öllers, S., Singer, H.P., FäSsler, P., Muller, S.R., 2001. Simultaneous quantification of neutral and acidic pharmaceuticals and pesticides at the low-ng/l level in surface and waste water. J. Chromatogr. A 911, 225–234.

Paíga, P., Santos, L.H., Amorim, C.G., Araújo, A.N., Montenegro, M.C.B., Pena, A., et al., 2013. Pilot monitoring study of ibuprofen in surface waters of north of Portugal. Environ. Sci. Pollut. Res. 20, 2410–2420.

Pait, A.S., Warner, R.A., Hartwell, S.I., Pacheco, P.A., Mason, A.L., 2006. Human Use Pharmaceuticals in the Estuarine Environment: A Survey of the Chesapeake Bay. Biscayne Bay and Gulf of the Farallones.

Patel, M., Kumar, R., Kishor, K., Mlsna, T., Pittman Jr, C.U., Mohan, D., 2019. Pharmaceuticals of emerging concern in aquatic systems: chemistry, occurrence, effects, and removal methods. Chem. Rev. 119, 3510–3673.

Peng, X., Yu, Y., Tang, C., Tan, J., Huang, Q., Wang, Z., 2008. Occurrence of steroid estrogens, endocrine-disrupting phenols, and acid pharmaceutical residues in urban riverine water of the Pearl River Delta, South China. Sci. Total. Environ. 397, 158–166.

Pereira, A.M., Silva, L.J., Meisel, L.M., Lino, C.M., Pena, A., 2015. Environmental impact of pharmaceuticals from Portuguese wastewaters: geographical and seasonal occurrence, removal and risk assessment. Environ. Res. 136, 108–119.

Pereira, C.D.S., Maranho, L.A., Cortez, F.S., Pusceddu, F.H., Santos, A.R., Ribeiro, D.A., et al., 2016. Occurrence of pharmaceuticals and cocaine in a Brazilian coastal zone. Sci. Total Environ. 548, 148–154.

Perisic, D.J., Kovacic, M., Kusic, H., Stangar, U.L., Marin, V., Bozic, A.L., 2016. Comparative analysis of UV-C/H 2 O 2 and UV-A/TiO 2 processes for the degradation of diclofenac in water. React. Kinet. Mech. Catal. 118, 451–462.

Rahman, M., Yanful, E., Jasim, S., 2009. Occurrences of endocrine disrupting compounds and pharmaceuticals in the aquatic environment and their removal from drinking water: challenges in the context of the developing world. Desalination 248, 578–585.

Rahman, M.A., Hasegawa, H., Lim, R.P., 2012. Bioaccumulation, biotransformation and trophic transfer of arsenic in the aquatic food chain. Environ. Res. 116, 118–135.

Reinholds, I., Pugajeva, I., Perkons, I., Lundanes, E., Rusko, J., Kizane, G., et al., 2017. Decomposition of multi-class pharmaceutical residues in wastewater by exposure to ionising radiation. Int. J. Environ. Sci. Technol. 14, 1969–1980.

Rivas, F.J., BELTRáN, F.J., Encinas, A., 2012. Removal of emergent contaminants: integration of ozone and photocatalysis. J. Environ. Manag. 100, 10–15.

Roberts, P.H., Thomas, K.V., 2006. The occurrence of selected pharmaceuticals in wastewater effluent and surface waters of the lower Tyne catchment. Sci. Total Environ. 356, 143–153.

Rodríguez-Navas, C., Björklund, E., Bak, S.A., Hansen, M., Krogh, K.A., Maya, F., et al., 2013. Pollution pathways of pharmaceutical residues in the aquatic environment on the island of Mallorca, Spain. Arch. Environ. Contam. Toxicol. 65, 56–66.

Ruhoy, I.S., Daughton, C.G., 2007. Types and quantities of leftover drugs entering the environment via disposal to sewage—revealed by coroner records. Sci. Total Environ. 388, 137–148.

Sangion, A., Gramatica, P., 2016. PBT assessment and prioritization of contaminants of emerging concern: pharmaceuticals. Environ. Res. 147, 297–306.

Schultz, M.M., Furlong, E.T., Kolpin, D.W., Werner, S.L., Schoenfuss, H.L., Barber, L.B., et al., 2010. Antidepressant pharmaceuticals in two US effluent-impacted streams: occurrence and fate in water and sediment, and selective uptake in fish neural tissue. Environ. Sci. Technol. 44, 1918–1925.

Schwaiger, J., Ferling, H., Mallow, U., Wintermayr, H., Negele, R., 2004. Toxic effects of the non-steroidal anti-inflammatory drug diclofenac: part I: histopathological alterations and bioaccumulation in rainbow trout. Aquat. Toxicol. 68, 141–150.

Serna-Galvis, E.A., Silva-Agredo, J., Giraldo-Aguirre, A.L., Flórez-Acosta, O.A., Torres-Palma, R.A., 2016. High frequency ultrasound as a selective advanced oxidation process to remove penicillinic antibiotics and eliminate its antimicrobial activity from water. Ultrason. Sonochem. 31, 276–283.

Serrano, M.A., Gonzalez-Rey, M., Mattos, J.J., Flores-Nunes, F., Mello, Á.C., Zacchi, F.L., et al., 2015. Differential gene transcription, biochemical responses, and cytotoxicity assessment in Pacific oyster *Crassostrea gigas* exposed to ibuprofen. Environ. Sci. Pollut. Res. 22, 17375–17385.

Silva, V.H.O., Dos Santos Batista, A.P., Teixeira, A.C.S.C., Borrely, S.I., 2016. Degradation and acute toxicity removal of the antidepressant Fluoxetine (Prozac®) in aqueous systems by electron beam irradiation. Environ. Sci. Pollut. Res. 23, 11927–11936.

Silva, L.J., Pereira, A.M., Rodrigues, H., Meisel, L.M., Lino, C.M., Pena, A., 2017. SSRIs antidepressants in marine mussels from Atlantic coastal areas and human risk assessment. Sci. Total Environ. 603, 118–125.

Sirtori, C., Zapata, A., Gernjak, W., Malato, S., Lopez, A., Aguera, A., 2011. Solar photo-Fenton degradation of nalidixic acid in waters and wastewaters of different composition. Analytical assessment by LC–TOF-MS. Water Res. 45, 1736–1744.

Snyder, S.A., Westerhoff, P., Yoon, Y., Sedlak, D.L., 2003. Pharmaceuticals, personal care products, and endocrine disruptors in water: implications for the water industry. Environ. Eng. Sci. 20, 449–469.

Song, C., Li, L., Zhang, C., Qiu, L., Fan, L., Wu, W., Meng, S., Hu, G., Chen, J., Liu, Y., 2017. Dietary risk ranking for residual antibiotics in cultured aquatic products around tai lake, China. Ecotoxicol. Environ. Safety 144, 252–257.

Stadlmair, L.F., Letzel, T., Drewes, J.E., Grassmann, J., 2018. Enzymes in removal of pharmaceuticals from wastewater: a critical review of challenges, applications and screening methods for their selection. Chemosphere 205, 649–661.

Sui, Q., Cao, X., Lu, S., Zhao, W., Qiu, Z., Yu, G., 2015. Occurrence, sources and fate of pharmaceuticals and personal care products in the groundwater: a review. Emerg. Contam. 1, 14–24.

Szabó, L., Gyenes, O., Szabó, J., Kovács, K., Kovács, A., Kiskó, G., et al., 2018. Electron beam treatment for eliminating the antimicrobial activity of piperacillin in wastewater matrix. J. Ind. Eng. Chem. 58, 24–32.

Thomas, K.V., Hilton, M.J., 2004. The occurrence of selected human pharmaceutical compounds in UK estuaries. Mar. Pollut. Bull. 49, 436–444.

Tischler, L., Buzby, M., Finan, D.S., Cunningham, V.L., 2013. Landfill disposal of unused medicines reduces surface water releases. Integr. Environ. Assess. Manag. 9, 142–154.

Togola, A., Budzinski, H., 2008. Multi-residue analysis of pharmaceutical compounds in aqueous samples. J. Chromatogr. A 1177, 150–158.

Trenholm, R.A., Vanderford, B.J., Holady, J.C., Rexing, D.J., Snyder, S.A., 2006. Broad range analysis of endocrine disruptors and pharmaceuticals using gas chromatography and liquid chromatography tandem mass spectrometry. Chemosphere 65, 1990–1998.

Trovó, A.G., Nogueira, R.F.P., Aguera, A., Fernandez-Alba, A.R., Malato, S., 2011. Degradation of the antibiotic amoxicillin by photo-Fenton process–chemical and toxicological assessment. Water Res. 45, 1394–1402.

Trudeau, V.L., Metcalfe, C.D., Mimeault, C., Moon, T.W., 2005. Pharmaceuticals in the Environment: Drugged Fish? Biochemistry and Molecular Biology of Fishes. Elsevier.

Tsiaka, P., Tsarpali, V., Ntaikou, I., Kostopoulou, M.N., Lyberatos, G., Dailianis, S., 2013. Carbamazepine-mediated pro-oxidant effects on the unicellular marine algal species Dunaliella tertiolecta and the hemocytes of mussel *Mytilus galloprovincialis*. Ecotoxicology 22, 1208–1220.

Valenti, S., David, J., Yang, S., Cappellaro, E., Tartaglia, L., Corsi, A., et al., 2017. The discovery of the electromagnetic counterpart of GW170817: kilonova AT 2017gfo/DLT17ck. Astrophys. J. Lett. 848, L24.

Välitalo, P., Kruglova, A., Mikola, A., Vahala, R., 2017. Toxicological impacts of antibiotics on aquatic micro-organisms: a mini-review. Int. J. Hyg. Environ. Health 220, 558–569.

Vasskog, T., Anderssen, T., Pedersen-Bjergaard, S., Kallenborn, R., Jensen, E., 2008. Occurrence of selective serotonin reuptake inhibitors in sewage and receiving waters at Spitsbergen and in Norway. J. Chromatogr. A 1185, 194–205.

Veach, A.M., Bernot, M.J., 2011. Temporal variation of pharmaceuticals in an urban and agriculturally influenced stream. Sci. Total Environ. 409, 4553–4563.

Wang, J., Gardinali, P.R., 2013. Uptake and depuration of pharmaceuticals in reclaimed water by mosquito fish (*Gambusia holbrooki*): a worst-case, multiple-exposure scenario. Environ. Toxicol. Chem. 32, 1752–1758.

Wang, S., Wang, J., 2018. Degradation of carbamazepine by radiation-induced activation of peroxymonosulfate. Chem. Eng. J. 336, 595–601.

Wang, Y., Zhang, H., Zhang, J., Lu, C., Huang, Q., Wu, J., et al., 2011. Degradation of tetracycline in aqueous media by ozonation in an internal loop-lift reactor. J. Hazard. Mater. 192, 35–43.

Weigel, S., Berger, U., Jensen, E., Kallenborn, R., Thoresen, H., Huhnerfuss, H., 2004. Determination of selected pharmaceuticals and caffeine in sewage and seawater from Tromsø/Norway with emphasis on ibuprofen and its metabolites. Chemosphere 56, 583–592.

Wille, K., Noppe, H., Verheyden, K., Bussche, J.V., De Wulf, E., Van Caeter, P., et al., 2010. Validation and application of an LC-MS/MS method for the simultaneous quantification of 13 pharmaceuticals in seawater. Anal. Bioanal. Chem. 397, 1797–1808.

Wille, K., Kiebooms, J.A., Claessens, M., Rappe, K., Bussche, J.V., Noppe, H., et al., 2011. Development of analytical strategies using U-HPLC-MS/MS and LC-ToF-MS for the quantification of micropollutants in marine organisms. Anal. Bioanal. Chem. 400, 1459–1472.

Wu, J., Qian, X., Yang, Z., Zhang, L., 2010. Study on the matrix effect in the determination of selected pharmaceutical residues in seawater by solid-phase extraction and ultra-high-performance liquid chromatography–electrospray ionization low-energy collision-induced dissociation tandem mass spectrometry. J. Chromatogr. A 1217, 1471–1475.

Wu, M., Xiang, J., Que, C., Chen, F., Xu, G., 2015. Occurrence and fate of psychiatric pharmaceuticals in the urban water system of Shanghai, China. Chemosphere 138, 486–493.

Xie, Z., Lu, G., Yan, Z., Liu, J., Wang, P., Wang, Y., 2017. Bioaccumulation and trophic transfer of pharmaceuticals in food webs from a large freshwater lake. Environ. Pollut. 222, 356–366.

You, L., Nguyen, V.T., Pal, A., Chen, H., He, Y., Reinhard, M., et al., 2015. Investigation of pharmaceuticals, personal care products and endocrine disrupting chemicals in a tropical urban catchment and the influence of environmental factors. Sci. Total Environ. 536, 955–963.

Yu, Y., Wu, L., 2012. Analysis of endocrine disrupting compounds, pharmaceuticals and personal care products in sewage sludge by gas chromatography–mass spectrometry. Talanta 89, 258–263.

Zenker, A., Cicero, M.R., Prestinaci, F., Bottoni, P., Carere, M., 2014. Bioaccumulation and biomagnification potential of pharmaceuticals with a focus to the aquatic environment. J. Environ. Manag. 133, 378–387.

Zhao, Y., Kuang, J., Zhang, S., Li, X., Wang, B., Huang, J., et al., 2017. Ozonation of indomethacin: kinetics, mechanisms and toxicity. J. Hazard. Mater. 323, 460–470.

Zheng, M., Xu, G., Pei, J., He, X., Xu, P., Liu, N., et al., 2014. EB-radiolysis of carbamazepine: in pure-water with different ions and in surface water. J. Radioanal. Nucl. Chem. 302, 139–147.

Zhou, G.-J., Ying, G.-G., Liu, S., Zhou, L.-J., Chen, Z.-F., Peng, F.-Q., 2014. Simultaneous removal of inorganic and organic compounds in wastewater by freshwater green microalgae. Environ. Sci.: Process. Impacts 16, 2018–2027.

Ziylan-Yavas, A., Ince, N.H., 2018. Single, simultaneous and sequential applications of ultrasonic frequencies for the elimination of ibuprofen in water. Ultrason. Sonochem. 40, 17–23.

CHAPTER

3

Personal care products as an endocrine disrupting compound in the aquatic environment

Afsane Chavoshani[1], Majid Hashemi[2,3], Mohammad Mehdi Amin[1,4] and Suresh C. Ameta[5]

[1]*Department of Environmental Health Engineering, School of Health, Isfahan University of Medical Sciences, Isfahan, Iran*

[2]*Environmental Health Engineering, School of Public Health, Kerman University of Medical Sciences, Kerman, Iran*

[3]*Environmental Health Engineering Research Center, Kerman University of Medical Sciences, Kerman, Iran*

[4]*Environment Research Center, Research Institute for Primordial Prevention of Non-Communicable Disease, Isfahan University of Medical Sciences, Isfahan, Iran*

[5]*Department of Chemistry, PAHER University, Udaipur, India*

3.1 Introduction

Water pollution by emerging micropollutants (EMPs) has taken a considerable interest since 1990; however, due to the lack of suitable controlling systems and high analytical cost, they are out of regulatory monitoring scope (Cabeza et al., 2012). The Environmental Protection Agency of US (US-EPA) defined these chemicals as new compounds without regulatory status which their impacts on the environment and human health are still poorly understood. These are mostly discharged to the environment from anthropogenic sources. EMPs include a wide range of compounds such as personal care products (PCPs), pharmaceuticals, nanoparticles, pesticides, antibiotic resistant genes, and industrial compounds and other trace elements (Montes-Grajales et al., 2017).

PCPs refers to products other than pharmaceuticals that is consumed or used by an individual for personal health, hygiene, or cosmetic reasons. Home and personal care products (HPCPs) are a general term that describes a group of anthropogenic chemicals included in different products widely used in daily human life (such as toothpaste, shampoo, cosmetics, and even in food), being used in considerable quantities (Tolls et al., 2009). PCPs are everyday consumer products used to cleanse, enhance, or alter the appearance of the body (Paulsen, 2015). Biocides

Micropollutants and Challenges. DOI: https://doi.org/10.1016/B978-0-12-818612-1.00003-9

such as triclocarban (TCC), and triclosan (TCS), and parabens are usually used in personal hygiene products and cosmetics as preservatives or antimicrobials (Zhang et al., 2015; Amin et al., 2019).

As emerging contaminants HPCPs have taken especial attention in recent years due to their potential adverse effects (environmental persistent, bioactive, bioaccumulative, and endocrine disrupting compounds) on human health and aquatic organisms (Zhang et al., 2015; Montes-Grajales et al., 2017).

These compounds also are known as endocrine disruption compounds (EDCs) (Kasprzyk-Hordern et al., 2008). Due to different side effects, the presence of PCPs in gray wastewater limits the application of recycling gray wastewater for urinal and toilet flushing, vehicle washing, fire protection, concrete production, and garden or crop irrigation (Eriksson et al., 2003).

Many researches show that PCPs occurs at trace concentrations (<0.1 μg/L) in the aquatic media (Yoon et al., 2010). It is generally accepted that EDCs acts as an estrogenic (compounds which mimic or block natural estrogen), androgenic (compounds which mimic or block natural testosterone), and thyroidal compounds with direct or indirect impacts on the thyroid gland (Yoon et al., 2010). But based on other definition, an EDCs is an exogenous substance or a mixture that changes function(s) of the endocrine system and consequently leads to adverse health effects in an intact organism, or its progeny, or (sub) populations. 1000 Chemical compounds have been found as EDCs. Biomonitoring studies reveal that EDCs can be detected in breast milk, urine, human serum, and placenta samples (Jeong et al., 2017).

HPCPs and their metabolites end up in wastewater treatment plant (WWTPs) (Rodil et al., 2008; Negreira et al., 2012). There they are partially eliminated and either retained in the sludge or released to the aquatic environment such as rivers, lakes, groundwater, surface water, and drinking water resources (Negreira et al., 2012; Peng et al., 2017). According to their lipophilicity futures indicated by octanol–water partition coefficients (log Kow), sediment can be an important media for accumulation of some PCPs such as TCC, TCS, BPA, parabens, and several UVAs (Peng et al., 2017).

In the last 20 years, the concern about the potential hazardous and risk associated to PCPs and their by-products, which can be more persistent and toxic (Bester, 2009), has been increased. The International Nomenclature of Cosmetic Ingredients (INCI) is the official dictionary for cosmetic ingredients adopted by many countries in the world since it was first established in 1970s by the PCPs Council in the USA. Many countries require manufacturers of PCPs to use the INCI nomenclature and to submit all new ingredients for registration in the INCI. In the following, the origin, use and fate of different hazardous compounds involved in the PCPs formulation are described. Many of the considered compounds have been used for decades worldwide, and there is not, in many cases, reliable data about their production rates. Emission inventories are mostly collected for scientific and administrative purposes, with great differences in their spatial and temporal coverage. Scientific studies often require data on other features, and many efforts have been undertaken to estimate source emission levels, environmental occurrence, and fate (Murray et al., 2010; Díaz-Cruz et al., 2009; Gago-Ferrero et al., 2012).

While pharmaceutical compounds are designed for internal use, PCPs are for external use; therefore they are not influenced by metabolic reactions and the regular usage of wide range of PCPs allowed them to enter into the environment without any change (Ternes et al., 2004). Therefore their global occurrence is prevalence.

The Global Beauty Market (GBM) is commonly classified into five main business groups: hair care, skin care, fragrances, color (makeup), and toiletries. The world's largest companies of perfumery and cosmetics are located in European countries. Among European countries Germany, France, UK, Italy, and Spain are the hub of the cosmetic market. These countries are frontrunners in field of production, import, and export of PCPs (Tanwar et al., 2014). In the early 1990s in Germany alone, the yearly production of PCPs has been estimated higher than 550,000 metric tons (Daughton and Ternes, 1999). Between 1998 and 2010, total sales of beauty and PCPs increased from 166.1 billion USD to 382.3 billion USD. During the last decade, the growth of skin care products has been influenced by aging and sun-protecting agents. Data obtained on bulky production of PCPs reveal the need for the management of these compounds in the aquatic environments because water is highly sensitive to pollutants and has a key role in health problems. Groundwater could directly or indirectly be contaminated by storage tanks, septic systems, uncontrolled hazardous waste, landfills, chemicals, road salts, and atmospheric contaminants (Tanwar et al., 2014). According to review study by Montes-Grajales et al. (2017), 72 PCPs have been documented as EMPs at concentrations ranging from 0.029 ng/L to 7.811×10^6 ng/L. Fragrances, antiseptics, and sunscreens were the most groups detected in aquatic environments (Montes-Grajales et al., 2017). Owing to the worldwide use of PCPs and their potential adverse effects on human and other living things, numerous researches study the occurrence of these compounds in environment media, while there are a few studies regarding the presence of PCPs in the aquatic environments (Brausch and Rand, 2011; Cruz and Barceló, 2015; Ebele et al., 2017; Montes-Grajales et al., 2017). Therefore the purpose of this chapter is to obtain a comprehensive sight on occurrence, fate, analysis methods, and management of PCPs in aquatic environments.

3.2 Classification of home and personal care products

HPCPs can be classified into multiple groups according to their properties and purposes. The typical classification of HPCPs is summarized in the following paragraphs (Cruz and Barceló, 2015).

Table 3.1 Summary of measured concentration of personal care products in surface water (ng/L) (Brausch and Rand, 2011).

Compound	Class	n^a	Range (ng/L)	Median (ng/L)
Triclosan	Disinfectant	710	<0.1–2300	48
Methyl triclosan	Disinfectant	4	0.5–74	–
Triclocarban	Disinfectant	29	19–1425	95
Musk ketone	Fragrance	178	4.8–390	11
Musk xylene	Fragrance	93	1.1–180	9.8
Celestolide	Fragrance	73	3.1–520	3.2
Galaxolide	Fragrance	282	64–12,470	160
Tonalide	Fragrance	245	52–6780	88
DEET	Insect repellant	188	13–660	55
Paraben[b]	Preservative	6	15–400	–
4MBC	UV filter	19	2.3–545	10.2
BP3	UV filter	18	2.5–175	20.5
EHMC	UV filter	21	2.7–224	6.1
OC	UV filter	22	1.1–4450	1.9

DEET, N,N-*diethyl-metatoluamide;* 4MBC, *4-methyl-benzilidine-camphor;* BP3, *benzophenone-3;* EHMC, *2-ethyl-hexyl-4-trimethoxycinnamate;* OC, *octocrylene; PCPs, personal care products.*
[a]n, *Number of samples.*
[b]*Includes all parabens.*
With permission.

3.2.1 Triclosan and triclocarban

TCS and TCC are biphenyl ethers used as antimicrobials in soaps, deodorants, skin creams, toothpaste, and plastics (McAvoy et al., 2002). TCS and TCC are among top 10 most commonly detected organic wastewater compounds for frequency and concentration (Kolpin et al., 2002; Halden and Paull, 2005). TCS has been identified in WWTP effluent at concentrations greater than 10 μg/L. A USGS study monitoring 95 compounds in surface water throughout the United States, found TCS to be one of the most frequently detected compounds with surface water concentrations as high as 2.3 μg/L For all published studies conducted to date TCS has been detected in 56.8% of surface water samples with a median concentration of about 50 ng/L (Table 3.1) (Brausch and Rand, 2011).

Furthermore bioaccumulation of TCS has not been observed in aquatic plants although methyl triclosan (M-TCS) has been observed to bioaccumulate after 28 days in *Sesbania herbacea*. However, other studies have contradicted these findings, demonstrating TCS bioaccumulates to a much greater degree in algae than M-TCS. It is currently unknown whether phylogenetic differences in physiology across trophic groups can influence bioaccumulation, and what factors are most important for uptake and accumulation to occur. One possible explanation for differences in TCS bioaccumulation is due to potential ionization of TCS. In typical environmental conditions TCS ranges from completely protonated (pH = 5.4) to totally deprotonated (pH = 9.2) based on a pK_a values of 7.8 at normal pH ranges in surface waters. These variations in ionization cause differences in Dow values relating to differences in bioaccumulation. Although this has not been investigated for TCS studies with the pharmaceutical fluoxetine (Prozac) indicate vast differences in bioaccumulation at different pH values. Based on these results at higher pH values TCS would be expected to accumulate more due to its pK_a value of 7.8 whereas at lower pH values M-TCS would be expected to accumulate to higher levels (Brausch and Rand, 2011).

TCC has been used in PCPs since 1957 and has been observed in surface water at concentrations up to 6.75 μg/L. It is believed that TCC occurs as frequently in WWTPs effluent and surface water as TCS; however, until 2004 TCC could not be detected at low levels (ng/L). However, TCC has been detected at higher concentrations and more frequently in WWTP effluent and surface water than TCS or M-TCS over the last 5 years. Additionally TCC has demonstrated a propensity to bioaccumulate more than either TCS or M-TCS in aquatic organisms. Other disinfectants (phenol, 4-metyhl phenol, and biphenylol) are also commonly used in households and have the potential to be released into aquatic environments. These compounds have been identified in surface water or WWTP effluent with phenol found more often than 4-methyl phenol and biphenylol, as well as in greater concentrations (as high as 1.3 μg/L) (Brausch and Rand, 2011).

Acute toxicity of TCS and biphenylol has been examined in invertebrates, fish, amphibians, algae, and plants. TCS is more toxic to similarly studied trophic groups in comparison to other disinfectants. For all disinfectants studied invertebrates are only slightly more sensitive than fish for acute time periods (Table 3.2). Additionally for longer-term exposure, fish and vascular plants appear to be less sensitive to TCS exposure whereas algae and invertebrates are more sensitive (Table 3.3). High TCS sensitivity in algae is likely due to TCS antibacterial characteristics, through disruption of lipid synthesis through the FabI (fatty acid synthesis) and FASII (enoyl acyl carrier protein reductase) pathways, membrane destabilization, or uncoupling of oxidative phosphorylation, which are similar between algae and bacteria. Toxicity of TCS to animal species is likely due to nonspecific narcosis as no common receptors are known to exist. Acute toxicity to amphibians has been studied in four different amphibians using a modified FETAX assay. Amphibians were more sensitive than fish; however, they were not as sensitive as algae during short-term exposures. For longer exposure duration algae appears to be the most sensitive trophic group. Algal growth was the most sensitive endpoint and was affected at concentrations less than 1 μg/L. Aquatic plants, invertebrates, and fish were not highly sensitive to chronic exposure of TCS (Table 3.3). Only minimal aquatic toxicity data exist for TCC, but recent studies indicate TCC is slightly more toxic to aquatic invertebrate and fish for both short- and long-term exposures than TCS (Tables 3.2 and 3.3). M-TCS toxicity to aquatic organisms (*Daphnia magna* and *Scenedesmus subspicatus*)

Table 3.2 Acute toxicity data for personal care products (Brausch and Rand, 2011).

Compound	Category	Species	Trophic group	Endpoint/duration	LC_{50} (mg/L)	Additional toxicity values
Biphenylol	Antimicrobial	*Daphnia magna*	Invertebrates	48 h Mobility	3.66	
		D. magna	Invertebrates	48 h Survival	3.66	
		Tetrahymena pyriformis	Invertebrates	48 h Survival	5.7–8.26	
		T. pyriformis	Invertebrates	60 h Survival	5.7–8.26	
		Cyprinus carpio				
Triclosan	Antimicrobial	*D. magna*	Invertebrates	48 h	0.39	
		Ceriodaphnia dubia	Invertebrates	24, 48 h (pH = 7.0)	0.2, _125	
		Pimephales promelas	Fish	24, 48, 72, 96 h	0.36, 0.27, 0.27, 0.26	
		Lepomis macrochirus	Fish	24, 48, 96 h	0.44, 0.41, 0.37	
		Oryzias latipes	Fish	96 h	0.602 (larvae), 0.399 (embryos)	
		Xenopus laevis	Amphibian	96 h	0.259	
		Acris blanchardii	Amphibian	96 h	0.367	
		Bufo woodhousii	Amphibian	96 h	0.152	
		Rana sphenocephala	Amphibian	96 h	0.562	
		Pseudokirch-neriella subcapitata	Algae	72 h Growth	0.53 (μg/L)	
Triclocarban	Antimicrobial	*D. magna*	Invertebrates	48 h	0.01	
		C. dubia	Invertebrates	48 h	0.0031	
		Mysidopsis bahia				
		Salmo gairdneri	Fish	96 h	0.120	
		L. macrochirus	Fish	96 h	0.097	
		Scenedesmus	Algae	72 h Growth	0.02	
		P. subcapitata	Algae	72 h Growth	0.017 (μg/L)	

(*Continued*)

Table 3.2 Acute toxicity data for personal care products (Brausch and Rand, 2011). *Continued*

Compound	Category	Species	Trophic group	Endpoint/duration	LC_{50} (mg/L)	Additional toxicity values
Benzophenone	Fixative	*Caenorahbditis elegans*	Nematode	24 h	56.8	
		P. promelas	Fish	96 h	10.89	
1,4-dichlorobenzenea	Insect repellant	*D. magna*	Invertebrates	24, 48 h mobilization	1.6, 0.7	
		Artemia salina	Invertebrates	24 h	14	
		Palaemonetes pugio	Invertebrates	96 h	60	
		M. bahia	Invertebrates	96 h	1.99	
		Danio rerio	Fish	24 h, 96 h	4.25, 2.1	
		Jordanella floridae	Fish	96 h	2.05	
		P. promelas	Fish	96 h	4.2	
		O. mykiss	Fish	24 h	1.18	
		L. macrochirus	Fish	96 h	4.3	
		Cyprinodon variegatus	Fish	96 h	7.4	
		Selenastrum capricornutum	Algae	96 h Growth	0.57	
		Scenedesmus	Algae	72 h Growth	31	
		Scenedesmus	Algae	48 h Growth, biomass	38, 28	
		Skeletonema costatum	Algae	96 h Growth	59.1	
N,N-diethyl-*m*-toluamide (DEET)	Insect repellant	*D. magna*	Invertebrates	48 h, 96 h	160, 108	
		Gammarus fasciatus	Invertebrates	96 h	100	
		P. promelas	Fish	96 h	110	
		Gambusia affinis	Fish	24–48 h	235	
			Fish	96 h	71.3	

(Continued)

Table 3.2 Acute toxicity data for personal care products (Brausch and Rand, 2011). *Continued*

Compound	Category	Species	Trophic group	Endpoint/duration	LC_{50} (mg/L)	Additional toxicity values
		Oncorhynchus mykiss				
		Chlorella protothecoides	Algae	24 h Photosynthesis	388	
Musk ambrette (MA)	Nitro musk	*Vibrio fischeri*	Bacteria	Microtox	>Sol.	
		Pseudokirch-neriella subcapitata	Algae	72 h	>Sol.	
Musk ketone (MK)	Nitro musk	*V. fischeri*	Bacteria	Microtox	>Sol.	
		Nitocra spinipes	Invertebrates	96 h	>1.0	
		Acartia tonsa	Invertebrates	48 h	1.32	$LC_{10} = 0.40$
		D. magna	Invertebrates	24, 48 h	>Sol., 5.6	
		D. magna	Invertebrates	48 h	>0.46	
		D. rerio	Fish	96 h Survival, hatching	>0.4	
		P. subcapitata	Algae	72 h	>Sol.	
Musk moskene (MM)	Nitro musk	*V. fischeri*	Bacteria	Microtox	>Sol.	
		D. magna	Invertebrates	24 h	>Sol.	
		Danio rerio	Fish	96 h Survival, hatching	>0.4	
		P. subcapitata	Algae	72 h	>Sol.	
Musk Tibetene (MT)	Nitro musk	*V. fischeri*	Bacteria	Microtox	>Sol.	
Musk xylene (MX)	Nitro musk	*V. fischeri*	Bacteria	Microtox	>Sol.	
		D. magna	Invertebrates	24, 48 h Mobility	$EC_{50} >= $ Sol.	
		Oncorhynchus. mykiss	Fish	96 h	>1000	

(Continued)

Table 3.2 Acute toxicity data for personal care products (Brausch and Rand, 2011). *Continued*

Compound	Category	Species	Trophic group	Endpoint/duration	LC_{50} (mg/L)	Additional toxicity values
		L. macrochirus	Fish	96 h	1.2	
		D. rerio	Fish	96 h Survival, hatching	>0.4	
Celestolide (ADBI)	Polycyclic musk	*P. subcapitata*	Algae	72 h	>Sol.	
		N. spinipes	Invertebrates	96 h	>2.0	
		A. tonsa	Invertebrates	48 h	>2.0	LC_{10}>2.0
		D. rerio	Fish	96 h Survival, hatching	>1.0	
Galaxolide (HHCB)	Polycyclic musk	*D. rerio*	Fish	96 h Malformation	LOEC ~ 0.65	
		O. latipes	Fish	96 h Survival	1.97	
		N. spinipes	Invertebrates	96 h	1.90	
		A. tonsa	Invertebrates	48 h	0.47	LC_{10} = 0.12
		Lampsilis cardium	Benthic invertebrates	24, 48 h	1.0, 0.99	
		D. rerio	Fish	96 h Survival, hatching	>0.67	
Tonalide (AHTN)	Polycyclic musk	*D. rerio*	Fish	96 h Malformations	LOEC ~ 0.45	
		O. latipes	Fish	96 h Survival	0.95	
		N. spinipes	Invertebrates	96 h	0.61	
		A. tonsa	Invertebrates	48 h	0.71	LC_{10} = 0.45
		L. cardium	Benthic invertebrates	24, 48 h	0.45, 0.28	
		D. rerio	Fish	96 h Malformation	LOEC ~ 0.1	
		D. rerio	Fish	96 h Survival, Hatching	>0.67	

(*Continued*)

Table 3.2 Acute toxicity data for personal care products (Brausch and Rand, 2011). *Continued*

Compound	Category	Species	Trophic group	Endpoint/duration	LC_{50} (mg/L)	Additional toxicity values
		O. latipes	Fish	96 h Survival	1.00	
Traseolide (ATII)		*O. latipes*	Fish	96 h Survival	0.95	
Phantolide (AHMI)		*O. latipes*	Fish	96 h Survival	1.22	
		O. latipes	Fish	96 h Survival	11.6	
Benzylparaben	Preservative	*T. thermophila*	Protozoa	24 h, 28 h	4.3, 5.7	LOEC = 0.48
		V. fisheri	Bacteria	15 min, 30 min Illuminescence	0.11, 0.11	LOEC = 0.02
		Photobacterium leiognathi	Bacteria	15 min, 30 min	1.3, 1.6	LOEC = 0.25
		D. magna	Invertebrates	48 h	4.0	
		D. magna	Invertebrates	24 h, 48 h Mobility	5.2, 6	LOEC = 1.2
		P. promelas	Fish	48 h	3.3	
Butylparaben	Preservative	*T. thermophila*	Protozoa	24 h, 28 h	5.3, 7.3	LOEC = 2.5
		V. fisheri	Bacteria	15 min, 30 min Illuminescence	2.5, 2.8	LOEC = 0.7
		P. leognathi	Bacteria	15 min, 30 min Illuminescence	3.7, 4.3	LOEC = 1.12
		D. magna	Invertebrates	48 h	5.3	
		D. magna	Invertebrates	24 h, 48 h Mobility	6.2, 6	LOEC = 3.2
		P. promelas	Fish	48 h		4.2
Ethylparaben	Preservative	*T. thermophila*	Protozoa	24 h, 28 h	25, 30	LOEC = 10.7
		V. fisheri	Bacteria	15 min, 30 min Illuminescence	2.5, 2.7	LOEC = 0.55
		P. leognathi	Bacteria	15 min, 30 min Illuminescence	19, 24	LOEC = 5.5

(*Continued*)

Table 3.2 Acute toxicity data for personal care products (Brausch and Rand, 2011). *Continued*

Compound	Category	Species	Trophic group	Endpoint/duration	LC_{50} (mg/L)	Additional toxicity values
		D. magna		48 h	18.7	
		D. magna	Invertebrates	24 h, 48 h Mobility	25, 23	LOEC = 12
		P. promelas	Fish	48 h	34.3	
Isobutylparaben	Preservative	*D. magna*	Invertebrates	48 h	7.6	
		P. promelas	Fish	48 h	6.9	
Isopropylparaben	Preservative	*D. magna*	Invertebrates	48 h	8.5	
		P. promelas	Fish	48 h	17.5	
Methylparaben	Preservative	*T. thermophila*	Protozoa	24 h, 28 h	54, 58	LOEC = 11.5
		V. fisheri	Bacteria	15 min, 30 min Illuminescence	9.6, 10	LOEC = 2.9
		P. leognathi	Bacteria	15 min, 30 min Illuminescence	31, 35	LOEC = 8.5
		D. magna	Invertebrates	48 h	24.6	
		D. magna	Invertebrates	24 h, 48 h Mobility	32, 21	LOEC = 15
		P. promelas	Fish	48 h	>Sol.	
		P. leognathi	Bacteria	15 min, 30 min Illuminescence	21, 25	LOEC = 4.5
		D. magna	Invertebrates	48 h	12.3	
		D. magna	Invertebrates	24 h, 48 h Mobility	13, 7	LOEC = 6
		P. promelas	Fish	48 h	9.7	
Benzophenone-3	UV filter	*D. magna*	Invertebrates	48 h Immobility		1.9
Benzophenone-4	UV filter	*D. magna*	Invertebrates	48 h Immobility	50	
4-Methylbenzy-lidene camphor	UV filter	*D. magna*	Invertebrates	48 h Immobility	0.56	
2-Ethyl-hexyl-4-trimethoxycinnamate	UV filter	*D. magna*	Invertebrates	48 h Immobility	0.29	

With permission.

Table 3.3 Chronic toxicity data for personal care products (Brausch and Rand, 2011).

Compound	Category	Species	Trophic level	Endpoint/duration	LOEC (μg/L)	NOEC (μg/L)
Triclosan	Antimicrobial	*Daphnia magna*	Invertebrates	21 day Survival, reproduction	Repro. = 200 (LOEC)	Survival = 200 (NOEC)
		C. dubia	Invertebrates	7 day Survival, reproduction		50, 6
		C. dubia	Invertebrates	7 day Survival, reproduction	$IC_{25} = 170$	
		Chironomus riparius	Invertebrates	28 day Survival, Emergence		440
4		*Chironomus tentans*	Invertebrates	10 day Survival, growth	$LC_{25} = 100$	
		Hyalella azteca	Invertebrates	10 day Survival, growth	$LC_{25} = 60$	
		O. mykiss	Fish	96 day ELSc hatching, survival	No Effect, 71.3	
		O. latipes	Fish	21 day Growth, fecundity, HSI and GSId, VTGe	200, No Effect, 200, 20	
		O. latipes	Fish	14 day Hatchability	$IC_{25} = 290$	
		G. affinis	Fish	35 day Sperm count, VTG	101.3	
		Danio rerio	Fish	9 day Hatchability	$IC_{25} = 160$	
		X. laevis	Amphibian	21 day Metamorphosis	No effect (200)	
		Rana catesbeiana	Amphibian	18 day Development	300	
		Rana pipiens	Amphibian	24 day Survival, growth	230, 2.3	
		Bufo americanus	Amphibian	14 day Survival, growth	No effect (230)	
		S. capricornutum	Algae	96 h Growth	$EC_{50} = 4.46$	$EC_{25} = 2.44$
		S. subspicatus	Algae	96 h Biomass, growth rate	$EC_{50} = 1.2, 1.4$	$EC_{50} = 0.5, 0.69$

(Continued)

Table 3.3 Chronic toxicity data for personal care products (Brausch and Rand, 2011). *Continued*

Compound	Category	Species	Trophic level	Endpoint/duration	LOEC (μg/L)	NOEC (μg/L)
		S. costatum	Algae	96 h Growth rate	$EC_{50} >= 66$	$EC_{25} > 66$
		A. flos-aquae	Algae	96 h Biomass	$EC_{50} = 0.97$	$EC_{25} = 0.67$
		P. subcapitata	Algae	72 h Growth	$EC_{25} = 3.4$	0.2
		Navicula pelliculosa	Algae	96 h Growth rate	$EC_{50} = 19.1$	$EC_{25} = 10.7$
		Natural algal assemblage	Algae	96 h Biomass	0.12	
		Closterium ehrenbergii	Algae	96 h Growth		250
		Dunaliella tertiolecta	Algae	96 h Growth		1.6
		L. gibba	Plant	7 day Growth	$EC_{25} >= 62.5$	$EC_{25} >= 62.5$
		Sesbania herbacea	Plant	28 day Seed germination, morphology	100 germination, 10	
		E. prostrata	Plant	28 day Seed germination, morphology	No effect, 1000	
		B. frondosa	Plant	28 day Seed germination, morphology		100, 10
Triclocarban	Antimicrobial	*D. magna*	Invertebrates	21 day Growth	4.7	2.9
		M. bahia	Invertebrates	28 day Reproduction	0.13	0.06
		P. subcapitata	Algae	14 day Growth	10 000	$EC_{50} = 36\,000$
Benzophenone	Fixative	*Pimephales promelas*	Fish	7 day Survival, growth	9240, 3100	5860, 2100
		P. promelas	Fish	7 day ELS (Survival, growth)	6400, 1800	3300, 1000
1,4-dichlorobenzene	Insect repellant	*D. magna*	Invertebrates	28 day Growth		0.22
		D. magna	Invertebrates	21 day Reproduction		0.3
		Jordanella floridae	Fish	28 day Growth		>0.35

(*Continued*)

Table 3.3 Chronic toxicity data for personal care products (Brausch and Rand, 2011). *Continued*

Compound	Category	Species	Trophic level	Endpoint/duration	LOEC (μg/L)	NOEC (μg/L)
		O. mykiss	Fish	60 day Growth		>0.122
		P. promelas	Fish	33 day Growth		0.57
		D. rerio	Fish	28 day Growth		1.0
Musk ketone (MK)	Nitro musk	*D. magna*	Invertebrates	21 day Development	340	
		D. magna	Invertebrates	21 day Survival	$LC_{50} = 338–675$	
		A. tonsa	Invertebrates	5 day Developmental rate	$EC_{50} = 66$	$EC_{10} = 10$
		A. tonsa	Invertebrates	5 day Juvenile survival	2000	800
		N. spinipes	Invertebrates	7 day Developmental rate, survival	30	
		N. spinipes	Invertebrates	26 day Population Growth Rate	100	
		D. rerio	Fish	ELS 24–48 h Tail extension, coagulated eggs, edema, circulation	1000	330
		D. rerio	Fish	ELS 24–48 h Movement, tail extension	330	100
		D. rerio	Fish	ELS 48 h Heart rate	10	3.3
		D. rerio	Fish	ELS 48 h Survival	33	10
		O. mykiss	Fish	21 day Reproduction	$EC_{50} = 169–338$	
		L. macrochirus	Fish	21 day Survival	$LC_{50} > = 500$	
		D. rerio	Fish	8 week Reproduction	33	
		P. promelas	Fish	96 h Teratogenesis	$EC_{50} > = 400$	
		X. laevis	Amphibian	96 h Teratogenesis	>4000	

(*Continued*)

Table 3.3 Chronic toxicity data for personal care products (Brausch and Rand, 2011). *Continued*

Compound	Category	Species	Trophic level	Endpoint/duration	LOEC (μg/L)	NOEC (μg/L)
		P. subcapitata	Algae	72 h Growth, biomass	$EC_{50} = 244$, 118	
Musk moskene (MM)	Nitro musk	*D. magna*	Invertebrates	21 day Survival	$LC_{50} >= $ Sol	
		O. mykiss	Fish	21 day Reproduction	$EC_{50} >= $ Sol	
		X. laevis	Amphibian	96 h FETAX	$EC_{50} >= 400$	
Musk xylene (MX)	Nitro musk	*D. magna*	Invertebrates	21 day Survival	$LC_{50} = 680$	
		D. rerio	Fish	ELS 24–48 h Tail extension, coagulated eggs, edema, circulation, movement	1000	330
		D. rerio	Fish	ELS 48 h Heart rate, survival	330	10
		D. rerio	Fish	14 day Survival	$LC_{50} >= 400$	
		X. laevis	Amphibian	96 h FETAXb	>400	
		P. subcapitata	Algae	72 h Growth, biomass	$EC_{50} >= $ Sol	
		M. aeruginosa	Algae	5 day Cell count	>10 000	
Celestolide (ADBI)	Polycyclic musk	*N. spinipes*	Invertebrates	7 day Developmental rate, survival	100	
		A. tonsa	Invertebrates	5 day Developmental rate	$EC_{50} = 160$	$EC_{10} = 36$
		A. tonsa	Invertebrates	5 day Juvenile survival	600	240
		X. laevis	Amphibian	96 h FETAX	$EC_{50} >= 1000$	
Galaxolide (HHCB)	Polycyclic musk	*D. magna*	Invertebrates	21 day Development, Reproduction	282 (EC_{50})	

(*Continued*)

Table 3.3 Chronic toxicity data for personal care products (Brausch and Rand, 2011). *Continued*

Compound	Category	Species	Trophic level	Endpoint/duration	LOEC (μg/L)	NOEC (μg/L)
		D. magna		21 day Growth, survival	205	11
		D. magna	Invertebrates	21 day Survival	$LC_{50} = 293$	
		A. tonsa	Invertebrates	5 day Developmental rate	$EC_{50} = 59$	$EC_{10} = 37$
		A. tonsa	Invertebrates	5 day Juvenile survival		300
		N. spinipes	Invertebrates	7 day Developmental rate, survival	20	
		L. cardium	Benthic	96 h Growth	$EC_{50} = 153–563$	
		Capitella sp.	Benthic Invertebrates	119 day Survival, growth, development	123 mg/kg, No effect, 168 mg kg	
		Potamopyrgus	Benthic	94 day Adult and juvenile	100 Time to 1st reproduction, 10 number of offspring	
		L. macrochirus	Fish	21 day Growth, survival	182 $LC_{50} = 452$	182
		P. promelas	Fish	36 day Hatch, survival, growth, development	>140, 68, 68, 68	140
		O. mykiss	Fish	21 day Reproduction	$EC_{50} = 282$	
		D. rerio	Fish	21 day Survival	$LC_{50} = 452$	
		O. latipes	Fish	72 h VTG, ERaf	500	
		X. laevis	Amphibian	96 h FETAX	$EC_{50} >= 100$	
		X. laevis	Amphibian	32 day Survival	$LC_{50} >= 140$	
		P. subcapitata	Algae	72 h Growth, biomass	466	201
		P. subcapitata	Algae	72 h Growth, biomass	$EC_{50} >= 854$, 723	

(*Continued*)

Table 3.3 Chronic toxicity data for personal care products (Brausch and Rand, 2011). *Continued*

Compound	Category	Species	Trophic level	Endpoint/duration	LOEC (μg/L)	NOEC (μg/L)
Tonalide (AHTN)	Polycyclic musk	*D. magna*	Invertebrates	21 day Growth, survival	184–401	89–196
		D. magna	Invertebrates	21 day Development, reproduction	244 (EC_{50})	
		A. tonsa	Invertebrates	5 day Developmental rate	$EC_{50} = 26$	$EC_{10} = 7.2$
		A. tonsa	Invertebrates	5 day Juvenile Survival	160	60
		N. spinipes	Invertebrates	7 day Developmental rate, survival	>60	
		L. cardium	Benthic Invertebrates	96 h Growth	$EC_{50} = 108–708$	
		D. rerio	Fish	ELS 24–48 h heart rate	33	10
		L. macrochirus	Fish	21 day Growth, survival	184 $LC_{50} = 314$	89
		P. promelas	Fish	36 day Hatch, survival, growth, development	>140, 140, 67, 67	>140, 67, 35, 35
		O. mykiss	Fish	21 day Reproduction	$EC_{50} = 244$	
		D. rerio	Fish	21 day Survival	$LC_{50} = 314$	
		O. latipes	Fish	72 h VTG, ERaf	500	
		X. laevis	Amphibian	96 h FETAX	$EC_{50} > = 1000$	
		X. laevis	Amphibian	32 day Survival	$LC_{50} = 100$	
		P. subcapitata	Algae	72 h Growth, biomass	797–835	204–438
		P. subcapitata	Algae	72 h Growth, biomass	$EC_{50} > = 709$, 468	
Benzylparaben	Preservative	*D. magna*	Invertebrates	7 day Growth, reproduction	200, 2600	
		P. promelas	Fish	7 day growth	1700	
Butylparaben	Preservative	*D. magna*		7 day Growth, reproduction	200, 2600	

(Continued)

Table 3.3 Chronic toxicity data for personal care products (Brausch and Rand, 2011). *Continued*

Compound	Category	Species	Trophic level	Endpoint/duration	LOEC (μg/L)	NOEC (μg/L)
		D. magna		7 day Growth, reproduction	200, 2600	
		S. trutta	Fish	10 day VTG	134	76
Ethylparaben	Preservative	*D. magna*	Invertebrates	7 day Growth, reproduction	9000, 2300	
Isobutylparaben	Preservative	*D. magna*	Invertebrates	7 day Growth, reproduction	300, 2000	
		P. promelas	Fish	7 day Growth	3500	
		P. promelas	Fish	7 day Growth	9000	
Methylparaben	Preservative	*D. magna*	Invertebrates	7 day Growth, reproduction	6000, 1500	
		P. promelas	Fish	7 day Growth	25 000	
Isopropylparaben	Preservative	*D. magna*	Invertebrates	7 day Growth, reproduction	4000, 2000	
		P. promelas	Fish	7 day Growth	2500	
		O. latipes	Fish	7 day VTG	9900	
Benzophenone-1	UV filter	*P. promelas*	Fish	14 day VTG	4919.4	
		O. mykiss	Fish	14 day VTG, growth	4919	
Benzophenone-2	UV filter	*P. promelas*	Fish	14 day VTG	8782.9	
		O. mykiss	Fish	14 day VTG, Growth	8783	
Benzophenone-3	UV filter	*O. mykiss*	Fish	14 day Growth	3900	
Benzophenone-4	UV filter	*O. mykiss*	Fish	14 day Growth	4897	

(*Continued*)

Table 3.3 Chronic toxicity data for personal care products (Brausch and Rand, 2011). *Continued*

Compound	Category	Species	Trophic level	Endpoint/duration	LOEC (μg/L)	NOEC (μg/L)
3-benzylidene camphor	UV filter	*Potamopyrgus*	Benthic	56 day Reproduction	0.28 mg/kg sediment	
		Lumbriculus variegatus	Benthic invertebrates	28 day Reproduction	6.47 mg/kg sediment	
		P. promelas	Fish	14 day VTG, reproduction, gonad histology	434.6, 74, 74	
		P. promelas	Fish	14, 21 day VTG	435, 74	
		O. mykiss	Fish	14 day VTG, Growth	453	
		O. mykiss	Fish	10 day Injection	68 mg/kg	
		X. laevis	Amphibian	35 day Metamorphosis	No effect	
3- (40-methylbenzy-lidene camphor)	UV filter	*Potamopyrgus antipodarum*	Benthic invertebrates	56 day Reproduction	1.71 mg/kg sediment	
		Lumbriculus variegatus	Benthic invertebrates.	28 day Reproduction	22.3 mg/kg sediment	
		O. mykiss	Fish	14 day Growth	415	
Oxybenzone	UV filter	*O. mykiss*	Fish	14 day VTG	749	
		O. latipes	Fish	21 day VTG, Hatching	620	
Ethyl-4-aminobenzoate	UV filter	*P. promelas*	Fish	14 day VTG	4394	

With permission.

exposed for short time periods is considerably less than the parent compound TCS. In addition to ionization state affecting bioaccumulation, ionization state is also important when examining toxicity. It is found that determined unionized TCS was slightly more toxic at pH ranging between 8.17 and 8.21 than ionized TCS at higher pH values. This effect is not observed for TCC as TCC only ionizes at extreme pH values outside environmentally relevant ranges (Brausch and Rand, 2011).

In addition to typical acute and chronic studies a number of studies have investigated effects of TCS exposure on swimming behavior of fish. TCS has induced alterations in swimming performance of *Oncorhynchus mykiss*, *Danio rerio*, and *Oryzias latipes* at concentrations as low as 71 μg/L which is considerably greater than other endpoints, indicating behavior is not a sensitive endpoint for identifying TCS effects (Brausch and Rand, 2011).

Evidence suggests TCS is weakly estrogenic, likely due to its similarities in structure to the nonsteriodal estrogen diethylstilbestrol. TCS exposure has been implicated in changes in fin length and sex ratios of medaka (*O. latipes*). TCS has also been demonstrated to induce vitellogenin (VTG) synthesis in male *O. latipes* after 21 days exposure and decreased sperm counts and VTG synthesis after 35 days exposure in *Gambusia affinis*. TCS has also been investigated for endocrine effects in *Xenopus laevis* and *Rana catesbeiana* due to similarities in structure to thyroid hormone (TH). TCS had no effects on *X. laevis* metamorphosis and produced only slight effects in *R. catesbeiana*, suggesting TCS only minimally affects TH and development in amphibians. To date, no studies have looked at potential endocrine disruption of other disinfectants with similar structures (i.e., TCC and M-TCS) (Table 3.3).

Based on toxicity data, algae appear to be the most sensitive trophic group to environmental concentrations of TCS and other disinfectants. However, it is possible that TCS, M-TCS, and TCC could affect benthic invertebrates at environmentally relevant concentrations due to disinfectants potential sorption to sediment although, heretofore, no studies have investigated acute or chronic effects in benthic invertebrates. Additionally TCS has been observed to cause development of antimicrobial strains of resistant bacteria as well as antibiotic resistant bacteria through development of crossresistance. The potential environmental impacts of antimicrobial resistance in aquatic ecosystems are low although it could have major implications on human health and aquaculture. Results of a probabilistic risk assessment indicate minimal effects of TCS on aquatic ecosystems as the 95th percentile of environmental concentrations is below the fifth percentile of sensitive species. Numerous uncertainties remain including effects on benthic organisms and effects of ionization and dissociation on partitioning, toxicity, and bioaccumulation that need to be determined to conduct a comprehensive risk assessment on TCS (Brausch and Rand, 2011)

3.2.2 Fragrances

Fragrances are perhaps the most widely studied class of PCPs and are believed to be ubiquitous contaminants in the environment. The most commonly used fragrances are synthetic musks. Synthetic musks are fragrances used in a wide range of products including deodorants, soaps, and detergents. Synthetic musks are either nitro musks, which were introduced in the late 1800s, or polycyclic musks, introduced in 1950s. The most commonly used nitro musks are musk xylene (MX) and musk ketone (MK) whereas musk ambrette (MA), musk moskene (MM), and musk tibetene (MT) are used less frequently. Nitro musks, however, are slowly being phased out due to their environmental persistence and potential toxicity to aquatic species Polycyclic musks are currently used in higher quantities than nitro musks with celestolide (ABDI), galaxolide (HHCB), and toxalide (AHTN) being used most commonly and traseloide (ATII), phantolide (AHMI), and cashmeran (DPMI) being used less often; HHCB and AHTN production alone has been estimated at about 1 million pounds per year and has thus been placed on the High Production Volume List by the USEPA (Peck, 2006).

Yamagishi et al. (1983) first identified nitro musks in the environment and conducted the first major monitoring study on MX and MK. MX and MK were found in greater than 80% of samples from river water, WWTP effluent, freshwater fish, and shellfish in Japan. Concentrations were highest in WWTP effluent ranging from 25 to 36 ng/L and 140–410 ng/L for MX and MK, respectively (Yamagishi et al., 1983). For all studies conducted in which fragrance concentrations were reported, MX and MK have been detected in 83%–90% of WWTP effluents and approximately 50% of surface waters. Furthermore Winkler et al. (1998) and Moldovan (2006) identified both nitro musks and polycylic musks in Elbe River (Germany) and Somes River (Romania) water samples ranging between 2 and 10 ng/L for nitro musks and 2–300 ng/L for polycyclic musks, with the polycyclic musk HHCB being detected at the highest concentration (Winkler et al., 1998; Moldovan, 2006). Polycyclic musks have been observed more often in surface water (78.3% and 84.6% of samples for AHTN and HHCB, respectively) as well as in greater concentrations than nitro musk compounds worldwide. Käfferlein et al. (1998) and Geyer et al. (1994) published extensive reviews of musks found in the environment and more specifically in biological compartments (Käfferlein et al., 1998; Geyer et al., 1994). The highest concentrations of polycylclic musks reported to date has occurred in surface waters in Berlin, Germany that receive substantial input from WWTP at concentrations approaching 10 μg/L (Table 3.1) (Heberer et al., 1999).

Nitro and polycyclic musks are water soluble, but high octanol–water coefficients (log Kow = 3.8 for MK and 5.4–5.9 for polycyclic musks) (Schramm et al., 1996; Balk and Ford, 1999) indicate high potential for bioaccumulation in aquatic species (Geyer et al., 1994; Winkler et al., 1998). This potential has been realized by numerous researchers having identified high concentrations of musks in lipids from fresh- and salt-water fish and mollusks (Schramm et al., 1996). Median concentrations of synthetic musks in biota range from approximately 0.1 to 3 mg/kg of lipid for MK and AHTN, respectively. Dietrich and Hitzfeld (2004) compiled bioconcentration (BCF) and bioaccumulation factors (BAF) for synthetic musks and found nitro musks, specifically MX, bioconcentrate and bioaccumulate more than polycyclic musks (Dietrich and Hitzfeld, 2004). MX has been observed to bioconcentrate up to 6700x in common carp (Gatermann et al., 2002), whereas AHTN bioconcentrates much less with BCF values ranging between 597 and 1069 in aquatic species (Fromme et al., 2001; Dietrich and Chou, 2001).

Nitro musks have relatively low or no propensity to cause acute toxicity to aquatic taxa studied to date. Furthermore, only three studies examining MK and MX found acute toxicity to occur at levels below water solubility limits (0.15 mg/L for MX, 0.46 for MK) (Table 3.2). However, nitro musks are potentially toxic to aquatic organisms over longer time periods with *D. rerio* (zebrafish) early life stage (ELS) studies being most sensitive (Table 3.3). Additionally *D. rerio* in general are the most sensitive species studied to date, whereas amphibians do not appear sensitive to nitro musk exposure (Chou and Dietrich, 1999; Carlsson and Norrgren, 2004). It has been suggested that nitro musk transformation products have potential to be highly toxic to aquatic organisms although only minimal data exists (Daughton and Ternes, 1999). Polycyclic musks are more acutely toxic than nitro musks based on published literature. HHCB and AHTN are toxic to aquatic invertebrates from ppb to low ppm levels although they are relatively nontoxic to fish (Table 3.2), and for longer exposure periods, invertebrates also appear more sensitive to polycyclic musks than fish (Table 3.3). Similar to nitro musks, polycyclic musks are nontoxic to amphibians. Developmental rates of invertebrates and growth and development for ELS of fish are the most sensitive endpoints studied to date for polycyclic musks. VTG synthesis, indicating potential endocrine effects, was not a sensitive endpoint, suggesting musks do not induce estrogenic effects (Dietrich and Chou, 2001).

Based on the highest reported concentrations of synthetic musks in aquatic environments, only AHTN would have the potential to cause adverse effects in wildlife. EC_{10} values, based on 5 day developmental rates, of the saltwater copepod *Acartia tonsa* are just slightly below the highest environmental concentrations observed (Heberer et al., 1999; Wollenberger et al., 2003) resulting in a hazard quotient of close to 1. However, limited research exists on effects of musks on algae and benthic invertebrates, and therefore potential risk cannot be accurately determined. Because synthetic musks possess high octanol–water coefficients, benthic invertebrates are likely exposed to high concentrations of synthetic musks in sediment and should be tested to evaluate potential toxicity of musks released in WWTP effluent. Only a handful of studies have investigated synthetic musk toxicity to sediment/soil organisms, indicating there is potential risk of musk exposure to benthic invertebrates (Balk and Ford, 1999). The polychaete species Capitella and snail species *Potamopyrgus antipodarum* were exposed to HHCB for 2 weeks and 96 days, respectively (Pedersen et al., 2009; Ramskov et al., 2009). Both studies found adult organisms to be insensitive to HHCB although juveniles were much more sensitive for both species. Both studies also examined potential effects for population, and both studies concluded there were no population level effects at environmentally relevant concentrations.

Up to eight additional fragrances (acetophenone, camphor, dlimonene, ethyl citrate, indole, isoborneol, isoquinolone, and skatol) have been observed in surface water; however, all fragrances except ethyl citrate have been detected in only a small number of samples (Kolpin et al., 2002; Glassmeyer et al., 2005). Ethyl citrate is a tobacco additive that has frequently been detected in surface water throughout the USA (Kolpin et al., 2002; Glassmeyer et al., 2005). Acute and chronic toxicity is not expected to occur with any of these compounds tested individually; however, additional research needs to be conducted.

3.2.3 Insect repellents

Insect repellents are substances that discourage insects from approaching to an applied surface (Rodil and Moeder, 2008). As some insects act as a vector for some diseases, using insect repellents is critical when other forms of protection are not available. They are widely used in tropical regions, being able to heavily influence the infection rates of some pathogens (Antwi et al., 2008). There is little information about their long-term effects in the aquatic environment; however, they have been detected worldwide in wastewaters, groundwater, surface, and drinking water (Rodil and Moeder, 2008; Quednow and Puttmann, 2009; Costanzo et al., 2007; Tay et al., 2009). *N,N*-diethyl-metatoluamide (DEET) is a commonly used broad-range spectrum insect repellent (Costanzo et al., 2007). It was first formulated in 1946 and was registered for commercial use in 1957 (Antwi et al., 2008). It is estimated that only in the USA one-third of the population has used DEET. Although the actual repellent mechanism involved is not well understood, DEET shows a high repellent potential against mites, tsetse flies, Aedes vigilax, and mosquitoes (Murphy et al., 2000), being used in all kinds of insect repellent formulations worldwide. Residues of DEET have been detected in effluent wastewater (Glassmeyer et al., 2005; Sui et al., 2010) and surface water (Glassmeyer et al., 2005; Rodil and Moeder, 2008; Quednow and Puttmann, 2009; Kolpin et al., 2002), being quite persistent in the aquatic environment (Costanzo et al., 2007).

DEET is the most common active ingredient in insect repellants (Costanzo et al., 2007) and is routinely detected in surface waters throughout the United States (Glassmeyer et al., 2005). DEET was developed in 1940s and functions by interfering with insects ability to detect lactic acid on hosts (Davis, 1985). DEET is currently registered for use in 225 products in the US, and it is estimated that annual usage exceeds 1.8 million kg. DEET has been detected in WWTP effluent (Glassmeyer et al., 2005; Sui et al., 2010) and surface water worldwide (Kolpin et al., 2002; Glassmeyer et al., 2005). DEET is relatively persistent in the aquatic environment, but unlike many other PCPs (i.e., fragrances and UV filters) DEET has a low BCF and is likely not accumulated into aquatic organisms (Costanzo et al., 2007). DEET has been regularly detected in effluent (95% of analyzed samples) and surface water (65% of all analyzed samples) with median concentrations of approximately 0.2 μg/L and 55 ng/L, respectively. The only other insect repellant detected in WWTP effluent or surface water is 1,4-dichlorobenzene. 1,4-dichlorobenzene has been detected in surface water (40% of surface water screened) receiving significant inputs of WWTPs effluent throughout the US at concentrations up to 0.28 μg/L (Table 3.1) (Glassmeyer et al., 2005).

3.2.4 Preservatives

Synthetic preservatives are a wide family of compounds used to prevent bacterial and fungal growth and oxidation and also inhibit natural ripening of fruits and vegetables. Some authors also include bactericide agents in this group. They are widely used in many goods (e.g., pharmaceuticals, soaps, gels, creams, food, etc.) (Brausch and Rand, 2011).

Parabens (alkyl-*p*-hydroxybenzoates) are antimicrobial preservatives used in cosmetics, toiletries, pharmaceuticals, and food (Daughton and Ternes, 1999; Amin et al., 2019). There are currently seven different types of parabens in use (benzyl, butyl, ethyl, isobutyl, isopropyl, methyl, and propyl). In 1987 over 7000 kg of parabens were used in cosmetics and toiletries alone (Soni et al., 2005), and that number has been expected to increase over the last 20 years. Methyl and propylparaben are the most commonly used in cosmetics and are typically coapplied to increase preservative effects (Peck, 2006). To date only a handful of studies have examined paraben concentrations in WWTP and surface water. Greatest concentrations of parabens have been identified in surface water with concentrations ranging from 15 to 400 ng/L, whereas effluent had lower concentrations ranging from 50 to 85 ng/L (Table 3.1) (Benijts et al., 2004; Lee et al., 2005; Loraine and Pettigrove, 2006; Kasprzyk-Hordern et al., 2008; Jonkers et al., 2010). Of the seven different types of parabens currently in use, benzylparaben appears to be most acutely toxic (Madsen et al., 2001; Terasaki et al., 2009; Bazin et al., 2010) (Table 3.2). Methyl- and ethylparaben appear to be least acutely toxic with LC_{50} values approximately 3% greater than benzylparaben for all trophic groups studied (Table 3.3) (Bazin et al., 2010). It has previously been reported that increasing chain length of parabens' substituents can increase paraben acute toxicity to bacteria (Dymicky and Huhtanen, 1979; Eklund, 1980), and this appears to be true for other trophic groups as well. There is currently a lack of information on the chronic effects of parabens to aquatic organisms with only a single known study examining toxicity in *D. magna* and *Pimephales promelas* (Dobbins et al., 2009). These authors found benzyl- and butylparaben were most toxic to invertebrates and fish whereas methyl and ethylparaben appeared least toxic. This corresponds directly with results of acute studies, as well as previous studies indicating increased chain length of parabens increases toxicity. In addition to increasing chain length, chlorination also substantially increases toxicity of parabens to both bacteria and *D. magna* (Terasaki et al., 2009).

Based on limited environmental concentration and toxicity data, it appears benzyl-, butyl-, and propylparaben could potentially cause adverse effects to aquatic organisms. Dobbins et al. (2009) concluded parabens only pose limited hazard to aquatic organisms; however, parabens, specifically benzyl-, butyl-, and propylparaben, can elicit low-level estrogenic responses. In vitro studies conducted with fish MCF-7 cell lines and yeast estrogenic screening assays demonstrated that parabens can elicit estrogenic responses at low levels (Routledge et al., 1998; Darbre et al., 2002). Furthermore, Inui et al. (2003) and Bjerregaard et al. (2008) demonstrated parabens can cause VTG synthesis in fish when exposed to low concentrations. Therefore low level exposure to parabens could potentially cause estrogenic effects at environmentally relevant concentrations (Inui et al., 2003; Bjerregaard et al., 2008). Additional studies have been conducted examining effects of parabens on sexual endpoints including spermatogenesis and serum testosterone in male rats. Both butyl- and propylparaben significantly inhibited spermatogenesis, but did not affect serum testosterone. Golden et al. (2005) reviewed paraben endocrine activity in rats and determined butyl-, isobutyl-, and benzylparaben demonstrate estrogenic activity although their potency is much less than estrogen itself (Golden et al., 2005). These results indicate that there are potential affects in aquatic organisms continually exposed to parabens. Preliminary data on environmental concentrations, however, suggest only minimal risk to aquatic organisms as effect concentrations are generally 1000 × higher than what has been observed in surface water (Brausch and Rand, 2011).

3.2.5 Ultraviolet filters (sunscreens)

Growing concern over effects of ultraviolet (UV) radiation in humans has caused an increased usage of UV filters. UV filters are used in sunscreen products and cosmetics to protect from UV radiation and can be either organic (absorb UV radiation, e.g., methylbenzylidene camphor) or inorganic micropigments (reflect UV radiation, e.g., ZnO, TiO_2); however, in this review only organic compounds will be discussed. Typically three to eight separate UV filters are found in sunscreens and cosmetics and can make up greater than 10% of products by mass (Schreurs et al., 2002). There are 16 compounds that are currently certified for use as sunscreen agents (SSA's) and 27 certified UV filters in cosmetics, plastics, among others in the US (Fent et al., 2008). UV filters enter the environment in two ways, either indirectly via WWTP effluent or directly from sloughing off while swimming and other recreational activities. A study in Switzerland estimated the input of four commonly applied UV filters into WWTPs to be as high as 118 g of 2-ethyl-hexyl-4-trimethoxycinnamate (EHMC), 49 g of 4-methyl-benzilidine-camphor (4MBC), 69 g of benzophenone-3 (BP3), and 28 g of octocrylene (OC) per 10,000 people per day in high use times (Balmer et al., 2005). Additionally Poiger et al. (2004) estimated up to 1263 mg of UV filters are applied per person daily resulting in up to 966 kg of UV filters released directly into a small lake in Switzerland per year. Although UV filters are used at high levels and are likely to enter into aquatic environments, very little is known about their environmental concentrations due to a lack of analytical methods (Poiger et al., 2004).

Balmer et al. (2005) examined presence of four UV filters (4MBC, BP3, EHMC, and OC) in wastewater effluent, surface water, and fish tissue in Switzerland. WWTP effluent had the greatest concentrations of UV filters with 4MBC being detected at the highest concentrations (2.7 μg/L) (Balmer et al., 2005) and was also detected at the highest concentrations in surface water (35 ng/L) and fish tissue (123 ng/g lipid tissue). Poiger et al. (2004) found similar results in Swiss lakes with BP3 being detected at the highest concentrations (5–125 ng/L) (Table 3.1). Overall 4MBC has been detected most frequently in WWTP effluent and surface water worldwide (95% and 86% of samples, respectively) whereas OC has been detected much less frequently in both (77% of WWTP effluent samples and 14% of surface water samples). The majority of additional environmental concentration data that exists pertains to bioaccumulation of UV filters in aquatic organisms. UV filters are known to bioaccumulate in fish at levels similar to PCBs and DDT (Daughton and Ternes, 1999) due to their high lipophilicity (log Kow = 3–7) and stability in the environment (Balmer et al., 2005; Poiger et al., 2004). UV filters have been found in lipid tissue in fish at concentrations up to 2 ppm (Nagtegaal, 1997). Additionally UV filters were identified to have BAF greater than 5000 in fish (21 μg/kg in whole fish vs 0.004 μg/L in water) (Hany and Nagel, 1995). In the only known laboratory study 3-benzylidene camphor (3BC) was found to have a BCF of 313 in *P. promelas* exposed for 21 days (Kunz et al., 2006).

A single study indicates UV filters do not appear to be acutely toxic to aquatic organisms (Fent et al., 2008). *D. magna* were observed to be most sensitive over short-term exposures (48 h) to EHMC whereas they were least sensitive to benzophenone-4 (BP4) (Fent et al., 2008) (Table 3.2). The majority of studies pertaining to UV filters have focused on long-term exposures. Schmitt et al. (2008) investigated effects in benthic invertebrates and observed a significant reduction in *P. antipodarum* reproduction and increased mortality when exposed to 3BC and 4MBC for 56 days and significant decreases in reproduction and increased mortality in *Lumbriculus variegutes* when exposed for 28 days to the same compounds (Schmitt et al., 2008) (Table 3.3).

UV filters are well known to bioaccumulate and recent studies have also indicated the potential for estrogenic activity. In vitro assays using fish MCF-7 cell lines indicate five UV-A and UV-B sunscreens [BP3, homosalate (HMS), 4MBC, octyl-methoxycinnamate, and octyl-dimethyl-PABA] have potential to cause estrogenic effects (Schlumpf and Lichtensteiger, 2001; Kunz and Fent, 2006). Additionally Kunz et al. (2006) identified 10 UV filters had estrogenic effects using a recombinant yeast assay with rainbow trout ERa. Benzophenone-1 (BP1) was the most potent UV filter with 40 hydroxybenzophenone (4HB) being the only other compounds that demonstrated estrogenicity below 1 mg/L exposure (Kunz et al., 2006). Aquatic studies using fish (*P. promelas* and *O. mykiss*) indicate numerous UV filters have the potential to cause estrogenic effects and also adversely affect fecundity and reproduction (Table 3.3). 3BC appears to be the most estrogenic compound inducing VTG in *O. mykiss* and *P. promelas* after 14 and 21 day exposure, respectively (Kunz et al., 2006; Fent et al., 2008). Other UV filters (BP1 and BP2) also induce VTG in male fathead minnows but at concentrations 10-fold higher than 3BC. Oxybenzone induces VTG production in both *O. mykiss* and *O. latipes* at similar concentrations as 3BC and also significantly decreases fertilized eggs hatchability in *O. latipes* (Coronado et al., 2008). Based on a single study amphibians do not appear as sensitive as fish as a result of 3BC exposure (Kunz et al., 2006). Recombinant yeast assays using fish hERa also indicate some UV filters also possess antiestrogenic activity (e.g., 4MBC and 3BC), androgenic activity (e.g., BP2 and HMS), and/or antiandrogenic activity [e.g., 4-hydroxy benzophenone (4HB)] (Kunz and Fent, 2006); however, no studies have investigated these effects in vivo in aquatic organisms. In vivo studies using rats have indicated 4-MBC affected the hypothalamic–pituitary–gonadal system in male rats, thus altering gonadal weight and steroid hormone production. 3-BC also affected development of sex organs in male rats after 12 weeks of exposure (Schlumpf et al., 2004). These data substantiate the in vitro data and indicate potential risk in aquatic species; however, the extent of risk of UV filters in WWTP effluent and surface water is currently unknown based on the scarcity of environmental concentration data. Additionally the number of species used to identify toxic effects is minimal and therefore does not allow for a comprehensive risk profile to be developed.

3.2.6 Biocide compounds

Antiseptic and disinfectant compounds are extensively used in many activities such as health care and hospitals for a variety of topical or hard-surface applications. A wide variety of chemicals with biocide properties are found in all kind of products, many of them known for hundreds of years, such as alcohols, iodine and chlorine, demonstrating a wide range of antimicrobial activity. However, the current knowledge about the processes that provide these active chemicals is really scarce. The exposure through diverse goods to these widespread chemical compounds has raised some speculation on the development of microbial resistance and on the possibility of these compounds of being able to induce antibiotic resistance. In this category benzotriazole, TCS, and TCC are the most commonly used compounds (Brausch and Rand, 2011).

3.2.7 Benzotriazole

Benzotriazole (1-H-benzotriazole) is a very versatile compound widely used by their anticorrosive, antifreeze, coolant, vapor phase inhibitor, photographic developer, drug precursor, and biocide properties (Asimakopoulos et al., 2013; Weiss et al., 2006; Domínguez et al., 2012). Its extensive use raises concerns about its presence in the environment. Benzotriazole is a very polar substance, and conventional wastewater treatment technologies are not efficient for its removal (Reemtsma et al., 2010). As a consequence these compounds if not efficiently eliminated reach the aquatic environment and ultimately may reach the drinking water supply (Weiss and Reemtsma, 2005).

3.2.8 Surfactants

Surfactants are a key group of chemicals in a large number of applications such as in the manufacture of detergents, the formulation of herbicides, in textile industry, and as stabilising agent for fragrances in cosmetics. With a high production value estimated over 18 million tons (da Silva et al., 2014), their wide use generates the disposal of large amounts of these compounds in WWTPs or improperly directly into the aquatic environment without any kind of treatment. Their amphoteric character allows them to be accumulated in sediments, sludge and biota, generating concern about the potential related hazard to the environment (Olmez-Hanci et al., 2011).

Despite increasing effort to monitor other surfactants there is still scarcity of environmental data on their concentrations. For example, secondary alkane sulfonates (SAS), commonly used in dishwashing liquids and personal care products (e.g., cosmetics), were only recently monitored in an estuary from SW Spain, there reaching concentrations up to 990 ng/L (Baena-Nogueras et al., 2013). On the other hand other ingredients in personal care products, such as the two antibacterials, TCS and TCC, have been also widespread detected in surface waters at concentrations up to 2.3 and 5.6 μg/L in streams from the USA and 27.2 and 5.8 ng/L in an estuary from the southeast of China, respectively (Kolpin et al., 2002; Lv et al., 2014).

The alkylphenols 4-nonylphenol (NP) and 4-tert-octylphenol (OP) exist mainly as intermediates in the manufacturing industry; NP and OP are also degradation products of nonionic surfactants alkylphenolsethoxylate used in industrial and institutional formulations. Alkylphenols, OP, NP, and nonylphenol ethoxylates (NPEs) have been shown to be exist in the environment such as river water and sewage sludge, and in fish tissue. In addition, the estrogenic activity of OP and NP has been extensively evaluated in a variety of assays (Bina et al., 2018; Inoue et al., 2003).

The wide range of products that can contain NPEs include fabrics, paper processing, paints, resins, and protective coatings. It is also widely used in loads of domestic uses as a component in cleaning products, degreasers, detergents and cosmetics. Despite being restricted in the EU as a hazard to human and environmental safety, its regulated use it is still allowed in countries worldwide. Nonylphenol and its ethoxylates have been detected on surface water (Jonkers et al., 2010), sediment (Shang et al., 1999), wastewater (Petrovic et al., 2003), and sludge (Pryor et al., 2002).

3.2.9 Phthalates

Phthalates are present in many consumer products because of their property as flexibiliser of rigid polymers such as PVC. They are used in the production of a wide range of products such as food wrappings, medical devices, children's toys, wood finishers, paints, and plastic products. Besides that in cosmetic products, phthalate esters are used as solvents or fragrances (Api, 2001), suspension agents, antifoaming agents, skin emollients, plasticisers in nail polishes, and fingernail elongators (Hubinger and Havery, 2006). In 2002 a study found that 52 out of the 72 cosmetic products investigated contained phthalates at concentrations ranging from 50 μg/g to nearly 3% of the product. Of the 52 cosmetics none had the phthalates listed in their product label (Houlihan et al., 2002). Due to their extensive use and the wide range of applications, phthalates are distributed along the aquatic environmental compartments being reported in water (Penalver et al., 2000), wastewater, sludge (Roslev et al., 2007), and less commonly in sediment (Chaler et al., 2004).

In PCPs phthalates play many roles. One such role is as a solubilizer, which helps both essential oils and fragrance oils to attach to various types cosmetics such as shampoos, creams and gels. Phthalates are also commonly used as humectants, which help to retain moisture, soften the skin, and skin penetration enhancers (Paulsen, 2015).

3.2.10 Siloxanes

Siloxanes are a relatively new group of PCPs, consisting of a polymeric organic silicone that comprises a backbone of alternating silicon-oxygen units with an organic chain attached to every silicon atom, conferring them a low surface tension, physiologic inertness, high thermal stability, and a smooth texture (Liu et al., 2014). Siloxanes are used in a broad range of consumer products (antiperspirants, skin-care creams, hair conditioners, and color cosmetics) and in industrial ones, such as automotive polishes, fuel additives, and antifoaming agents. They are considered high production volume chemicals, having annual productions for some of them of 45–227 thousand tons worldwide; however, recent reports raise concern about the potential toxic effects of cyclic siloxanes (Horii and Kannan, 2008). Siloxanes are likely to be discharged into sewage systems through the use of "rinse-off" products and partially adsorbed onto sludge in WWTPs due to their high Kow and released to the aquatic environment through wastewater discharges (Sparham et al., 2008; Richardson, 2008; Sanchís et al., 2013), having also been found in sewage sludge (Liu et al., 2014; Sanchís et al., 2013) and sediment (Sanchís et al., 2013; Sparham et al., 2011). The siloxane family includes octamethylcyclotetrasiloxane (D4), decamethylcyclopentasiloxane (D5), dodecamethylcyclohexasiloxane (D6), and tetradecamethylcycloheptasiloxane (D7) (Richardson, 2008).

To date very little data exists pertaining to acute toxicity of DEET to aquatic organisms. Costanzo et al. (2007) summarizes all data published through 2006 and since its publication no additional information has been reported (Costanzo et al., 2007). Data indicates DEET is only slightly toxic to aquatic organisms (Table 3.2) (Michael and Grant, 1974). Although DEET is relatively resistant to breakdown and commonly found in surface water, no known studies exist that have examined chronic toxicity of DEET exposure to aquatic organisms. DEET has been observed to inhibit cholinesterase in rats (Chaney et al., 2000), and it is possible similar effects could be observed in fish even though no research has been conducted. Additionally no studies have been conducted to examine potential endocrine effects on aquatic organisms although studies have been conducted in rats. These studies indicated that DEET has no effect on sperm count, morphology, or viability, in male rats after 9 week exposure (Lebowitz et al., 1983), thus indicating little potential effects in aquatic species. Based on available information, Costanzo et al. (2007) performed a preliminary risk assessment and concluded DEET is not likely to produce biological effects at environmentally relevant concentrations in aquatic ecosystems; however, due to lack of information on chronic toxicity a definitive assessment could not be made. Similar conclusions are still applicable today as chronic toxicity of DEET to aquatic organisms remains undetermined (Costanzo et al., 2007)

Similar to DEET a preliminary risk assessment has also been conducted on the moth repellant 1,4-dichlorobenzene. Invertebrates, specifically *D. magna*, appear the most sensitive from short-term exposure (Table 3.2) whereas fish appear to be most sensitive to long-term exposures (Table 3.3) (Boutonnet et al., 2004). Based on observed environmental concentrations, it is unlikely acute or chronic effects are occurring to freshwater and marine organisms. Additionally there is little potential for bioaccumulation of 1,4-dichlorobenzene, and to date there is no indication that it can cause endocrine effects.

3.3 Pathways of personal care products in the environment

Based on study of Ellis (2006), PCPs along with pharmaceuticals can enter the environment through several pathways (Fig. 3.1). Untreated household effluent and treated effluents from industries and hospital services containing some partially degraded and refractory PCPs may directly discharge into various receiving water bodies without improper treatment. The presence of PCPs in the aquatic environments such as ground water, surface water, drinking water, and wastewater was reported recently by many researchers (de García et al., 2013, 2014; Chen et al., 2012; Ebele et al., 2017). Domestic sewage is the primary source for PCPs discharged into the aquatic environments. PCPs, including shampoos, body washes, toothpastes, sunscreens, cosmetics, and hand lotions, can be discharged into sewerage systems and surface water through the daily washing activities of human beings. Additionally sludge processing from wastewater treatment plants can be a way for environmental pollution.

3.4 Fates of personal care products in wastewater treatment plants

Personal care products are mainly discharged into aquatic environments through sewage treatment plants (STPs) before they reach the receiving soil, surface

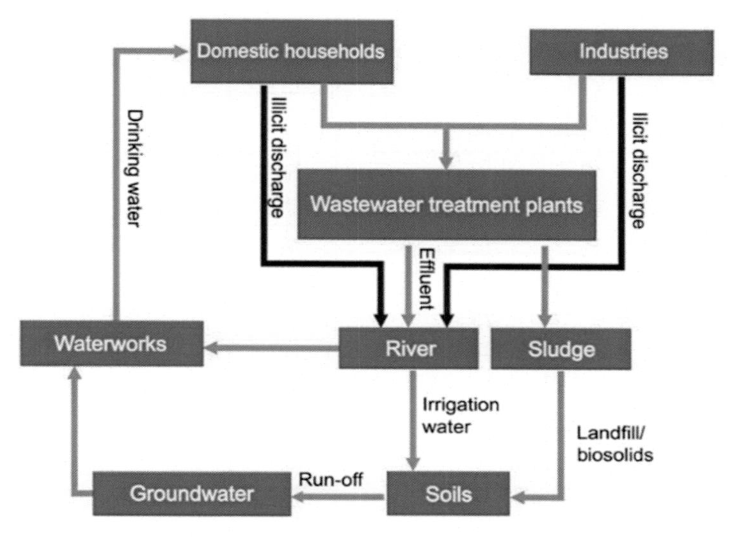

FIGURE 3.1

Sources and pathways of PCPs (Montes-Grajales et al., 2017). *PCPs*, Personal care products.

From Montes-Grajales, D., Fennix-Agudelo, M., Miranda-Castro, W., 2017. Occurrence of personal care products as emerging chemicals of concern in water resources: a review. Sci. Total Environ. 595, 601–614. Originally adapted from Ellis, J.B., 2006. Pharmaceutical and personal care products (PPCPs) in urban receiving waters, Environ. Pollut., 144, 184–189.

water, sediment, and groundwater. They are frequently detected at various concentrations in influent, effluent, reclaimed water, and receiving water bodies. The potential fates of PCPs in STPs (e.g., biodegradation/biotransformation, retention of solid/sludge, and release into receiving water bodies) are dependent on their original chemical structures and the associated metabolites and/or transformation products. Typical PCP removal processes include ASP, tertiary treatment with nutrient removal, membrane bioreactors, and advanced oxidation processes (AOPs) (Yang et al., 2017).

Most work has focused on investigating the occurrences and fates of PCPs in sewage and STPs, and their elimination efficiency. However, indepth studies of the mass balance and removal mechanisms of pharmaceutical and personal care products (PPCPs) (e.g., biotransformation, sedimentation, adsorption, biodegradation, volatilization, and hydrolysis, etc.) in STPs, and their inhibitory effects on biological processes (e.g., ASP and nutrient removal) have not been fully established. Therefore it is necessary to further evaluate the effects of PCPs on the performance of different treatment methods for different purposes (e.g., ultimate discharge or water reuse). Tables 3.4 and 3.5 summarize the influent and effluent concentrations of common PCPs detected in STPs in different countries (Yang et al., 2017).

Table 3.4 The concentrations and removal (%) of selected PCPs in conventional STPs in different countries (Yang et al., 2017).

Order	Representative compounds	Influent (ng/L)	Final effluent (ng/L)	Overall removal (%)	Sludge (ng/kg)	Location	References
A. Bactericides/disinfectants							
1a	TCS	892	202	77	645	India (two states)	Subedi et al. (2015a)
		2300	48	>90		USA, California	Yu et al. (2013)
		547	112	79		Korea, Ulsan	Behera et al. (2011)
		300	NA	55		USA	Blair et al. (2015)
1b	TCC	1150	49	>80	5570	India (two states)	Subedi et al. (2015a)
		540	NA	11		USA	Blair et al. (2015)
B. Fragrances							
2a	Calaxilid fragrance (HHCB)	2560–4520	NA	61 ≥ 99		Korea, Busan	Lee et al. (2010)
2b	Toxalide fragrance (AHTN)	550–1210		NA		Korea, Busan	Lee et al. (2010)
C. Insect repellents							
3a	DEET	600–1200	60–624	69 ± 21		China, Beijing	Sui et al. (2010)
		66	40	40		China, Shanghai	Wang et al. (2014)
D. Preservatives							
4a	BuP	15–27	3	>80		China, Guangzhou	Yu et al. (2011)
		160–170	1	>99		China, Guangzhou	Yu et al. (2011)
4b	MeP	290–10,000	6–50	>90		Spain (northwest)	González-Mariño et al. (2011)
		334	11	96		Spain, Valencia	Carmona et al. (2014)
		36.8; 97.9	0.14; 0.14	99.7; 99	41.6; 58.5	New York, USA (two STPs)	Wang and Kannan (2016)
4c	PrP	520–2800	2–210	>90		Spain (northwest)	González-Mariño et al. (2011)
		1630	<5	99		Spain, Valencia	Carmona et al. (2014)

(Continued)

Table 3.4 The concentrations and removal (%) of selected PCPs in conventional STPs in different countries (Yang et al., 2017). *Continued*

Order	Representative compounds	Influent (ng/L)	Final effluent (ng/L)	Overall removal (%)	Sludge (ng/kg)	Location	References
E. Sunscreen UV filters							
5a	4MBC	169	43	12 ($n = 60$)	49	Hong Kong (five regions)	Tsui et al. (2014)
5b	EHMC	462	150	93 ($n = 60$)	68	Hong Kong (five regions)	Tsui et al. (2014)
		309	126	59		Hong Kong (five regions)	Tsui et al. (2014)
		601	347	42		Hong Kong (five regions)	Tsui et al. (2014)
5c	BMDM	289	147	49		Hong Kong (five regions)	Tsui et al. (2014)
5d	EHS	93	8	91		Hong Kong (five regions)	Tsui et al. (2014)
5e	HMS	151	31	79		Hong Kong (five regions)	Tsui et al. (2014)
5f	IAMC	43	24	44		Hong Kong (five regions)	Tsui et al. (2014)
5g	ODPABA	138	56	17		Hong Kong (five regions)	Tsui et al. (2014)
5h	OC	8	0	>99			Tsui et al. (2014)
5i	OXB	ND	41.2	1.53		India (two states)	Subedi et al. (2015a)

BMDM, *Butyl methoxydibenzoylmethane;* BuP, *butylparaben;* EHMC, *2-ethyl-hexyl-4-trimethoxycinnamate;* EHS, *ethylhexyl salicylate;* HMS, *homosalate;* IAMC, *isoamyl p-methoxycinnamate;* 4MBC, *4-methyl-benzilidine-camphor;* MeP, *methylparaben;* OC, *octocrylene;* ODPABA, *octyl dimethyl-p-aminobenzoic acid;* OXB, *oxycodone;* PCP, *Personal care products;* STPs, *sewage treatment plants;* TCS, *triclosan;* TCC, *triclocarban.*

With permission.

Table 3.5 The concentrations and removal (%) of selected PCPs in conventional WTPs in different countries (Yang et al., 2017).

Order	Representative compounds	Raw water (ng/L)	Treated water (ng/L)	Overall removal (%)	Mineral waters (ng/L)	Tap waters (ng/L)	Location	References
A Parabens								
1a	Butylparaben				36	28	Spain, Valencia	Carmona et al. (2014)
1b	Ethylparaben						Spain, Valencia	Carmona et al. (2014)
1c	Methylparaben				40	12	Spain, Valencia	Spain, Valencia
1d	Propylparaben							Carmona et al. (2014)
1e	Methyl triclosan	74	ND	>99			China (polit)	Zhao et al. (2014)
B Bactericides/disinfectants								
2a	TCS				4	ND	Spain, Valencia	Carmona et al. (2014)
		ND				ND	USA, New York	Subedi et al. (2015b)
		3	ND	>99			USA (17 WTPs)	Benotti et al. (2008)
		74–102	<0.1	>99			Spain (one city in south-eastern)	Azzouz and Ballesteros (2013)
		1230	100	92			Israel and Palestine	Dotan et al. (2016)
2b	TCC				12	13	Spain, Valencia	Carmona et al. (2014)
		7.18				5.4	USA, New York	Subedi et al. (2015b)
C. Insect repellents								
3a	DEET	85	49	42			USA (17 WTPs)	Benotti et al. (2008)
		19.8–78.4	ND	>99			China, Taihu	Lin et al. (2016)
4a	Oxybenzone	19.4				1.39	USA, New York	Subedi et al. (2015b)

DEET, N,N-*diethyl-metatoluamide;* NA, *not available;* ND, *not detectable;* PCP, *personal care products;* TCC, *triclocarban;* TCS, *triclosan;* WTPs, *wastewater treatment plant.*
With permission.

3.5 Health effects of personal care products on biota and humans

The widespread presence of PCPs in receiving water resources is a growing concern because of its impact on environmental and human health. PCPs detected widely in wastewater, surface water, and groundwater. The occurrence of PCPs has been also reported in the groundwater resources of Spain (Cabeza et al., 2012) Germany (Reh et al., 2013), UK (Lapworth et al., 2012), USA (Barnes et al., 2008), Switzerland (Kahle et al., 2009), Australia (Liu et al., 2011), and China (Peng et al., 2014).

They can become harmful to human and animal health because it is possible that their residues enter and accumulate in the food chain through wastewater effluent release and the reuse of treated sewage and sludge for agricultural practices (Rajapaksha et al., 2014; Vithanage et al., 2014; Yang et al., 2017). Also PCPs may be partially metabolized or incompletely biodegraded in water matrices. Thus the excreted metabolites can become secondary pollutants and be further modified in receiving water bodies (Yang et al., 2017).

The main concern of these compounds related to their ability as endocrine disruptors is interfering with the reproductive system and the normal development of living organisms. There is limited data available about chronic and subchronic effects of PCPs in biota. For instance, UV filters such as benzophenone 3 (BP3), benzophenone 4 (BP4), and ethylhexyl methoxycinnamate (EHMC) lead to alteration of the gene transcription related to the sexual hormones production, while it has been observed that octocrylene (OC) interferes with blood flow, blood vessel formation, and organ development in adult and embryo zebrafish (Bluthgen et al., 2012; Zucchi et al., 2011a,b).

Remarkably high concentrations were found in human fat (28–189 g/kg for HHCB, 8–33 g/kg for AHTN), human blood (average 0.77 g/L for HHCB and 0.27 g/L for AHTN), and breast milk (16–108 g/kg HHCB, 11–58 g/kg AHTN). These findings reflect the high bioaccumulation tendency of HHCB and AHTN in accordance to the water–octanol partition coefficients (log Kow) of 5.9 and 5.7 for HHCB and AHTN, respectively. After usage the compounds reach the aquatic environment via domestic wastewater, where they can interfere with fish and other organisms. According to REACH, HHCB is classified as dangerous for the environment and very toxic for aquatic organisms, with the warning H410 (very toxic to aquatic life). The predicted no-effect concentration (PNEC) of HHCB according to REACH is 4.4 g/L. The aquatic toxicity is stated as no observed effect concentration (NOEC) of 0.093 mg/L for fish (Lange et al., 2015).

Studying the dietary impact of TCC in rats, for instance, at concentration higher than 25 mg/kg body weight per day had some effect on anemia and body, liver, and spleen weights in rats fed for 2 years. Butyl and propylparaben were able to influence the sperm quality of juvenile rats. Spongiform myelinopathy has been reported in the brainstem of rats exposed to near-lethal doses of DEET

(Verschoyle et al., 1992). Data about possible risks to human health on PCPs exposure are even scarce. Nevertheless humans have a continuous and close contact to PCPs, and the effects of such an exposure are mostly difficult to predict (Cruz and Barceló, 2015). PCPs have been reported to be present in diverse human samples. For instance, fragrances have been reported to be at ng/g lipids in human breast milk (Schlumpf et al., 2010; Yin et al., 2012) and human adipose tissue (Kannan et al., 2005); TCS has been reported at the ng/mL level in urine (Asimakopoulos et al., 2014; Frederiksen et al., 2013) and at ng/g lipids in adipose, liver, and brain tissue (Geens et al., 2012); and parabens have been found at the same concentrations in urine (Asimakopoulos et al., 2014), breast tissue (Darbre et al., 2004), and human milk (Schlumpf et al., 2010). Similarly UV filters have been determined in urine (Asimakopoulos et al., 2014), human milk (Schlumpf et al., 2010), and semen. TCS is degraded to dioxins and is toxic to aquatic bacteria at levels found in the environment (Ricart et al., 2010). There is also a general concern about the capabilities of TCS regarding the generation of antibiotic resistance. It is suggested that TCS and other antimicrobial compounds could cause bacterial resistance against antibiotics (Aiello et al., 2007) and may be related to allergic sensitisation in children (Bertelsen et al., 2013). TCC may be able to induce the production of methemoglobin (Fe^{+3}-based protein complex, similar to hemoglobin but unable to carry oxygen) through the transformation by heat into a primary amine in the bloodstream (Johnson et al., 1963). Exposure to fragrances has been associated with a wide range of health effects, such as allergic contact dermatitis, asthma, headaches, and mucosal symptoms (Elberling et al., 2005; Bridges, 2002). Although humans metabolize phthalates, easily excreting them in 24–48 h through urine (Hauser and Calafat, 2005), the continuous exposure to it seems to be able to interact with a nuclear receptor (peroxisome proliferator-activated receptors) that has an important role in adipogenesis and lipid storage, disrupting homeostasis, and increasing the risk for obesity and, thus, increasing diabetes risk (Svensson et al., 2011), as well as immune and asthma responses (Kimber and Dearman, 2010). Extensive topical application of DEET has resulted in poisonings (with symptoms like tremor, restlessness, slurred speech, seizures, impaired cognitive functions, and coma) including deaths and being linked to possible neurotoxic effects (Abou-Donia, 1996). Phthalates have been linked to asthma and allergies and behavior changes (Kim et al., 2009; Bornehag et al., 2004). In addition, some compounds generate a significant concern due to their carcinogenic potential. One study has tried to correlate low levels of parabens with breast cancer tissue (Darbre et al., 2004), and phthalates have been related to hepatic and pancreatic cancer in mice and rats (Ito et al., 2007), and a survey in Mexico reported a positive correlation between phthalate concentrations in urine and the risk of developing breast cancer (López-Carrillo et al., 2009). It seems clear that there is growing concerns about the potential carcinogenicity of these compounds.

3.6 Control strategies for personal care product contamination

Both WWTPs and WTPs are designed to remove organic and inorganic suspended materials, flocculated matter, and pathogens from wastewater or drinking water sources; however, neither drinking nor wastewater treatment processes are specifically designed to remove PCPs from water. Owing to their high chemical stability in a reaction chemical and low biodegradability, most PCPs are impossible to be completely eliminated by conventional treatment processes. In recent years several technologies were developed to provide more efficiency removal, including advanced oxidation processes (AOPs) and membrane filtration, activated carbon (Liang et al., 2014; Nakada et al., 2017).

3.6.1 Membrane filtration

Membrane filtration processes, such as nanofiltration (NF) and reverse osmosis (RO), are promising alternatives for the elimination of PCPs from wastewater (Nghiem et al., 2004; Yoon et al., 2006, 2010). Ultrafiltration (UF) and microfiltration (MF) have been proven to remove PCPs. However, their removal performances are relatively poor because membrane pore sizes are considerably larger than PCP molecules. For comparison, pressure-driven membrane processes, NF and RO, were applied to the drinking water treatment (Watkinson et al., 2007). These processes generally show significant PPCP removal efficiencies; however these membranes are still slightly permeable to some relatively small pollutants (Schäfer et al., 2011).

The removal capabilities of two different types of submerged NF flat sheet modules for removal of pharmaceuticals from STPs were investigated (Röhricht et al., 2009). Approximately 60% of diclofenac and naproxen were retained by both types of membranes, whereas only a small proportion of carbamazepine was removed. Hence diclofenac and naproxen may be obstructed by the negatively charged membrane surface, whereas carbamazepine may not (Nghiem et al., 2005). However, these removal efficiencies may not be sufficient to justify the use of such a system as an additional treatment step in STPs. For more polar compounds, the NF membrane showed higher removal efficiencies than the UF membrane. The removal of selected PPCPs by NF and RO has also been compared in previous studies (Yangali-Quintanilla et al., 2011). The average retention efficiency of NF is 82% for neutral pollutants and 97% for ionic contaminants, whereas RO can achieve 85%–99%. Real et al. (2012) compared the efficiencies of different system configurations in the elimination of PCPs from selected water sources. When ozonation was combined with NF, the removal efficiency was significantly affected by such variables as ozone dose and treatment sequences. For instance NF followed by ozonation removed >97% of pollutants from natural water, with an ozone dose of 2.25 mg/L and >90% from secondary effluent, with an ozone dose of 3.75 mg/L (Real et al., 2012). In contrast, a high removal efficiency (>70% in the permeate stream) was achieved by ozonation with initial dose of 2.25 mg/L followed by NF in natural waters (Watkinson et al., 2007). Although NF and RO processes exhibit efficient PPCP removal, pollutants in a highly concentrated form remaining in the retentate require further treatment.

3.6.2 Granular activated carbon

Granular activated carbon (GAC) and powdered activated carbon (PAC) were investigated for the sorptive removal of PCPs (Yang et al., 2011; Boehler et al., 2012; Margot et al., 2013). GAC is typically used in rapid filters, whereas PAC is an efficient method in removing seasonally occurring taste and odor in WTPs (Scheurer et al., 2010; Zoschke et al., 2011). In this review we focus on GAC because it has been used widely in drinking water treatment and tertiary treatment in STPs. Stackelberg et al. (2007) found that GAC facilities in a conventional WTP accounted for 53% removal of the tested PPCPs, whereas disinfection and sedimentation accounted for 32% and 15%, respectively (Stackelberg et al., 2007). In a study by Hernández-Leal et al. (2011), the removal efficiencies for tonalide and nonylphenol ranged from 50% to >90% (galaxolide). Contact time was found to markedly affect the extent of carbon adsorption. Short contact times resulted in low removal efficiencies (Hernández-Leal et al., 2011). Correspondingly long contact times increase surface loading and the number of accessible adsorption sites (Bolong et al., 2009; Meinel et al., 2015). In general adsorption by activated carbon has greater potential for removal of antibiotics than coagulation and flocculation processes (Choi et al., 2008).

Activated carbon has also demonstrated as an effective advanced treatment process in removing PPCP residues from treated effluents. Ek et al. (2014) conducted a pilot-scale study to evaluate the performance of activated carbon in removing pharmaceutical residues from treated wastewater (Ek et al., 2014). The results suggested that activated carbon beds with 90%–98% PCP removal may be a competitive alternative to treatment with ozone. Similar conclusions were drawn by Grover et al. (2011), who studied the removal of pharmaceuticals from sewage effluent in a full-scale STP. 43%–64% of steroidal estrogens were successfully removed by GAC (Grover et al., 2011).

GAC treatment removes PCP compounds by physical adsorption onto the GAC bed and to a lesser extent through biodegradation, thus avoiding generation of harmful reaction by-products. Moreover a high removal of metals is expected. After a certain time in use, the bed needs to be regenerated or replaced. Hence the key influencing parameters are the amount of fresh GAC required for the treatment and the number of regeneration cycles before the bed needs to be replaced (Tarpani and Azapagic, 2018).

The key operating parameters for GAC—the amount of fresh GAC and the number of regeneration cycles before the bed has to be replaced—were estimated based on two criteria commonly considered in the design of GAC: empty-bed contact time (EBCT) and the bed service time. The initial amount of fresh granular activated carbon was estimated for different EBCTs, as follows (Tarpani and Azapagic, 2018):

$$V_{GAC} = EBCT \times Q_{inf} \tag{3.1}$$

Where V_{GAC} is volume of granular activated carbon in the bed (m^3), EBCT is empty-bed contact time (min), and Q_{inf} is influent flow to be treated (m^3/min).

The maximum number of bed regenerations was fixed at 10 to ensure the bed's initial characteristics were maintained, with ~50% of the bed regenerated less than five times. It was suggested 7–12 refills for an efficient operation of the GAC treatment. Consequently the total number of bed replacements (N_{BR}) over the lifespan of the unit (60 years) is defined as follows (Tarpani and Azapagic, 2018):

$$N_{BR} = T_{treatment}/(n_{max}\ t_{GAC}) \tag{3.2}$$

Where N_{BR} is total number of bed replacement over the lifespan of 60 years (–), $T_{treatment}$ is treatment time (days), n_{max} is maximum number of bed regeneration before replacement (10), and t_{GAC} is bed service time (days).

The amount of fresh and regeneration GAC was then calculated by adapting the following method (Tarpani and Azapagic, 2018):

$$F_{GAC} = m_{GAC}[1 + N_{BR} + m_{loss}(n_r + N_{BR}\ n_{max})] \quad (kg) \tag{3.3}$$

$$F_{GAC} = m_{GAC}(n_r + N_{BR}\ n_{max}) \tag{3.4}$$

Where F_{GAC} is amount of fresh GAC needed for the treatment (kg), m_{GAC} is amount of granular activated carbon in the bed (kg), m_{loss} is percentage of GAC lost during regeneration (%), n_r is number of regeneration after the previous bed replacement (–), and R_{GAC} is amount of regenerated GAC (kg).

Table 3.6 shows the operating parameters for GAC cited in Eqs. (3.1)–(3.4) per 1000 m^3 of wastewater.

3.6.3 Advanced oxidation processes

AOPs, such as ozonation, UV, photocatalysis, and Fenton reaction, have been used for drinking water treatment (e.g., odor/taste control and disinfection) and to lesser extent in wastewater disinfection (Huber et al., 2003; Klavarioti et al., 2009; Gerrity et al., 2010). AOPs may change the polarity and functional groups of the target PCPs (McMonagle, 2013; Papageorgiou et al., 2014). Thus AOPs are suitable for water reuse purposes that involve direct human contact, such as household wastewater reuse applications (Hernández-Leal et al., 2011). It has been reported that WTPs equipped with AOPs further eliminated PCPs.

Table 3.6 the operating parameters for GAC cited in Eqs. (3.1)–(3.4) per 1000 m^3 of wastewater (Tarpani and Azapagic, 2018).

	EBCT (min)	Q_{inf} (m^3/h)	V_{GAC} (m^3)	m_{GAC} (kg)	t_{GAC} (days)	N_{BR}	n_r	Fresh GAC (kg)	Regenerated GAC (kg)
GAC	20	2,667	889	501,396	330	6	6	6,818,986	33,092,136
	30		1,334	752,094	220	9	9	14,966,671	74,457,306
	40		1,778	1,002,792	110	19	9	40,011,401	199,555,608

EBCT, *Empty-bed contact time;* GAC, *granular activated carbon.*
With permission.

Compounds such as caffeine, indomethacin, and sulfamethoxazole were removed at efficiencies of 89.5%, 84.2%, and 92.2%, respectively (Lin et al., 2016).

A study on pilot-scale experiments in a WTP was conducted by (Borikar et al., 2014). The results indicated that conventional WTPs equipped with either ozone/ H_2O_2 or UV/H_2O_2 greatly improved PPCP removal from 26% to 97% or 92%, respectively. Among the tested PCPs, carbamazepine, fluoxetine, naproxen, gemfibrozil, and TCS showed near complete removal. Diclofenac and ibuprofen were also removed by up to 97% and 98%, respectively. However, pharmaceuticals demonstrated some resistance in that, the highest removal of atorvastatin was only 88%. Fast (2015) conducted a holistic analysis, including a ranking system, to determine the performance of several AOPs. The findings indicated that H_2O_2/ozone presented the highest average ranking in reducing PCPs. In addition, performance improved significantly when oxidation was combined with other unit processes (Fast, 2015). Česen et al. (2015) demonstrated that removal rates of 99% for CP and 94% for IF were be achieved using a UV/O_3/H_2O_2 system with 5 g/L of H_2O_2 for 120 min. By coupling this AOP with a biological treatment, the removal rates of CP and IF could be further enhanced >99% (Česen et al., 2015). Real et al. (2012) demonstrated that a combined process using UV radiation (254 nm; for 30 min) and NF was very effective, with removal rates of >80% in the majority of the experiments. However, some recent reports have expressed substantial concern regarding the application of AOPs for PPCP removal (Real et al., 2012). Yang et al. (2016) evaluated the performance of UV/chlorine and UV/H_2O_2 processes in water purification to degrade PCP residues after sand filtration. The results showed that UV/chlorine exhibited superior PCP removal and disinfection by-products (DBPs) were formed after chlorination (Yang et al., 2016).

3.7 Conclusion

Nowadays, PCPs in aquatic environments have been recognized as an important environmental problem. Numerous studies have conducted on the presence and fates of PCPs during treatment in STPs and WTPs. Discharge of wastewater effluents into the receiving water is a reason for the PCPs occurrence in surface water, ground water, and drinking water. Therefore high PCPs concentration in wastewater effluent and sludge could be ultimately led to their presence into food chain and aquatic environments. The ability of bioaccumulation and endocrine disruption effects are concern related to PCPs. PCP removal, while limited with conventional treatment processes, can be enhanced with recent treatment processes such as advanced oxidation, activated carbon, or the use of membrane filtration such as RO and NF. Although AOPs are effective at removing PCPs, these processes usually have high treatment cost per unit volume, and further studies on inexpensive methods to remove PCPs compounds were still recommended. For this reason, further research is required to understand the effects of PCPs on treatment efficiency, microbial community structure in biological processes, and process stability in STPs and WTPs. The results could provide a theoretical base for improving management and the protection of receiving water bodies and increasing the use of reclaimed water.

References

Abou-Donia, M.B., 1996. Neurotoxicity resulting from coexposure to pyridostigmine bromide, DEET, and permethrin: implications of Gulf War chemical exposures. J. Toxicol. Environ. Health Part. A 48, 35–56.

Amin, M.M., Tabatabaeian, M., Chavoshani, A., Amjadi, E., Hashemi, M., Ebrahimpour, K., et al., 2019. Paraben content in adjacent normal-malignant breast tissues from women with breast cancer. Biomed. Environ. Scie. 32, 893–904.

Aiello, A.E., Larson, E.L., Levy, S.B., 2007. Consumer antibacterial soaps: effective or just risky? Clin. Infect. Dis. 45, S137–S147.

Antwi, F.B., Shama, L.M., Peterson, R.K., 2008. Risk assessments for the insect repellents DEET and picaridin. Regul. Toxicol. Pharmacol. 51, 31–36.

Api, A., 2001. Toxicological profile of diethyl phthalate: a vehicle for fragrance and cosmetic ingredients. Food Chem. Toxicol. 39, 97–108.

Asimakopoulos, A.G., Wang, L., Thomaidis, N.S., Kannan, K., 2013. Benzotriazoles and benzothiazoles in human urine from several countries: a perspective on occurrence, biotransformation, and human exposure. Environ. Int. 59, 274–281.

Asimakopoulos, A.G., Thomaidis, N.S., Kannan, K., 2014. Widespread occurrence of bisphenol A diglycidyl ethers, p-hydroxybenzoic acid esters (parabens), benzophenone type-UV filters, triclosan, and triclocarban in human urine from Athens, Greece. Sci. Total. Environ. 470, 1243–1249.

Azzouz, A., Ballesteros, E., 2013. Influence of seasonal climate differences on the pharmaceutical, hormone and personal care product removal efficiency of a drinking water treatment plant. Chemosphere 93, 2046–2054.

Baena-Nogueras, R.M., González-Mazo, E., Lara-Martín, P.A., 2013. Determination and occurrence of secondary alkane sulfonates (SAS) in aquatic environments. Environ. Pollut. 176, 151–157.

Balk, F., Ford, R.A., 1999. Environmental risk assessment for the polycyclic musks AHTN and HHCB in the EU: I. Fate and exposure assessment. Toxicol. Lett. 111, 57–79.

Balmer, M.E., Buser, H.-R., Muller, M.D., Poiger, T., 2005. Occurrence of some organic UV filters in wastewater, in surface waters, and in fish from Swiss lakes. Environ. Sci. Technol. 39, 953–962.

Barnes, K.K., Kolpin, D.W., Furlong, E.T., Zaugg, S.D., Meyer, M.T., Barber, L.B., 2008. A national reconnaissance of pharmaceuticals and other organic wastewater contaminants in the United States—I) Groundwater. Sci. Total. Environ. 402, 192–200.

Bazin, I., Gadal, A., Touraud, E., Roig, B., 2010. Hydroxy Benzoate Preservatives (Parabens) in the Environment: Data for Environmental Toxicity Assessment. Xenobiotics in the Urban Water Cycle. Springer.

Behera, S.K., Kim, H.W., Oh, J.-E., Park, H.-S., 2011. Occurrence and removal of antibiotics, hormones and several other pharmaceuticals in wastewater treatment plants of the largest industrial city of Korea. Sci. Total. Environ. 409, 4351–4360.

Benijts, T., Lambert, W., De Leenheer, A., 2004. Analysis of multiple endocrine disruptors in environmental waters via wide-spectrum solid-phase extraction and dual-polarity ionization LC-ion trap-MS/MS. Anal. Chem. 76, 704–711.

Benotti, M.J., Trenholm, R.A., Vanderford, B.J., Holady, J.C., Stanford, B.D., Snyder, S. A., 2008. Pharmaceuticals and endocrine disrupting compounds in US drinking water. Environ. Sci. Technol. 43, 597–603.

Bertelsen, R.J., Longnecker, M.P., LøVik, M., Calafat, A.M., Carlsen, K.H., London, S.J., et al., 2013. Triclosan exposure and allergic sensitization in Norwegian children. Allergy 68, 84–91.

Bester, K., 2009. Analysis of musk fragrances in environmental samples. J. Chromatogr. A 1216, 470–480.

Bina, B., Mohammadi, F., Amin, M.M., Pourzamani, H.R., Yavari, Z., 2018. Determination of 4-nonylphenol and 4-tert-octylphenol compounds in various types of wastewater and their removal rates in different treatment processes in nine wastewater treatment plants of Iran. Chin. J. Chem. Eng. 26, 183–190.

Bjerregaard, P., Hansen, P.R., Larsen, K.J., Erratico, C., Korsgaard, B., Holbech, H., 2008. Vitellogenin as a biomarker for estrogenic effects in brown trout, Salmo trutta: laboratory and field investigations. Environ. Toxicol. Chemistry: An. Int. J. 27, 2387–2396.

Blair, B., Nikolaus, A., Hedman, C., Klaper, R., Grundl, T., 2015. Evaluating the degradation, sorption, and negative mass balances of pharmaceuticals and personal care products during wastewater treatment. Chemosphere 134, 395–401.

Bluthgen, N., Zucchi, S., Fent, K., 2012. Effects of the UV filter benzophenone-3 (oxybenzone) at low concentrations in zebrafish (*Danio rerio*). Toxicol. Appl. Pharmacol. 263, 184–194.

Boehler, M., Zwickenpflug, B., Hollender, J., Ternes, T., Joss, A., Siegrist, H., 2012. Removal of micropollutants in municipal wastewater treatment plants by powder-activated carbon. Water Sci. Technol. 66, 2115–2121.

Bolong, N., Ismail, A., Salim, M.R., Matsuura, T., 2009. A review of the effects of emerging contaminants in wastewater and options for their removal. Desalination 239, 229–246.

Borikar, D., Mohseni, M., Jasim, S., 2014. Evaluations of conventional, ozone and UV/H2O2 for removal of emerging contaminants and THM-FPs. Water Qual. Res. J. Can. 50, 140–151.

Bornehag, C.-G., Sundell, J., Weschler, C.J., Sigsgaard, T., Lundgren, B., Hasselgren, M., et al., 2004. The association between asthma and allergic symptoms in children and phthalates in house dust: a nested case–control study. Environ. Health Perspect. 112, 1393–1397.

Boutonnet, J.-C., Thompson, R.S., De Rooij, C., Garny, V., Lecloux, A., Van Wijk, D., 2004. 1, 4-Dichlorobenzene marine risk assessment with special reference to the OSPARCOM region: North Sea. Environ. Monit. Assess. 97, 103–117.

Brausch, J.M., Rand, G.M., 2011. A review of personal care products in the aquatic environment: environmental concentrations and toxicity. Chemosphere 82, 1518–1532.

Bridges, B., 2002. Fragrance: emerging health and environmental concerns. Flavour. Fragr. J. 17, 361–371.

Cabeza, Y., Candela, L., Ronen, D., Teijon, G., 2012. Monitoring the occurrence of emerging contaminants in treated wastewater and groundwater between 2008 and 2010. The Baix Llobregat (Barcelona, Spain). J. Hazard. Mater. 239, 32–39.

Carlsson, G., Norrgren, L., 2004. Synthetic musk toxicity to early life stages of zebrafish (*Danio rerio*). Arch. Environ. Contam. Toxicol. 46, 102–105.

Carmona, E., Andreu, V., Picó, Y., 2014. Occurrence of acidic pharmaceuticals and personal care products in Turia River Basin: from waste to drinking water. Sci. Total. Environ. 484, 53–63.

Česen, M., Kosjek, T., Laimou-Geraniou, M., Kompare, B., Šlrok, B., Lambropolou, D., et al., 2015. Occurrence of cyclophosphamide and ifosfamide in aqueous environment and their removal by biological and abiotic wastewater treatment processes. Sci. Total. Environ. 527, 465–473.

Chaler, R., Cantón, L., Vaquero, M., Grimalt, J.O., 2004. Identification and quantification of n-octyl esters of alkanoic and hexanedioic acids and phthalates as urban wastewater markers in biota and sediments from estuarine areas. J. Chromatogr. A 1046, 203–210.

Chaney, L.A., Wineman, R.W., Rockhold, R.W., Hume, A.S., 2000. Acute effects of an insect repellent, N, N-diethyl-m-toluamide, on cholinesterase inhibition induced by pyridostigmine bromide in rats. Toxicol. Appl. Pharmacol. 165, 107–114.

Chen, H., Li, X., Zhu, S., 2012. Occurrence and distribution of selected pharmaceuticals and personal care products in aquatic environments: a comparative study of regions in China with different urbanization levels. Environ. Sci. Pollut. Res. 19, 2381–2389.

Choi, K.-J., Kim, S.-G., Kim, S.-H., 2008. Removal of antibiotics by coagulation and granular activated carbon filtration. J. Hazard. Mater. 151, 38–43.

Chou, Y.-J., Dietrich, D.R., 1999. Toxicity of nitromusks in early lifestages of South African clawed frog (*Xenopus laevis*) and zebrafish (*Danio rerio*). Toxicol. Lett. 111, 17–25.

Coronado, M., De Haro, H., Deng, X., Rempel, M.A., Lavado, R., Schlenk, D., 2008. Estrogenic activity and reproductive effects of the UV-filter oxybenzone (2-hydroxy-4-methoxyphenyl-methanone) in fish. Aquat. Toxicol. 90, 182–187.

Costanzo, S., Watkinson, A., Murby, E., Kolpin, D.W., Sandstrom, M.W., 2007. Is there a risk associated with the insect repellent DEET (N, N-diethyl-m-toluamide) commonly found in aquatic environments? Sci. Total. Environ. 384, 214–220.

Cruz, S., Barceló, D., 2015. Personal Care Products in the Aquatic Environment. Springer.

da Silva, S.S., Chiavone-Filho, O., De Barros Neto, E.L., Mota, A.L., Foletto, E.L., Nascimento, C.A., 2014. Photodegradation of non-ionic surfactant with different ethoxy groups in aqueous effluents by the photo-Fenton process. Environ. Technol. 35, 1556–1564.

Darbre, P., Byford, J., Shaw, L., Horton, R., Pope, G., Sauer, M., 2002. Oestrogenic activity of isobutylparaben in vitro and in vivo. J. Appl. Toxicol.: An. Int. J. 22, 219–226.

Darbre, P., Aljarrah, A., Miller, W., Coldham, N., Sauer, M., Pope, G., 2004. Concentrations of parabens in human breast tumours. J. Appl. Toxicol.: An. Int. J. 24, 5–13.

Daughton, C.G., Ternes, T.A., 1999. Pharmaceuticals and personal care products in the environment: agents of subtle change? Environ. Health Perspect. 107, 907–938.

Davis, E.E., 1985. Insect repellents: concepts of their mode of action relative to potential sensory mechanisms in mosquitoes (Diptera: Culicidae). J. Med. Entomol. 22, 237–243.

de García, S.O., Pinto, G.P., Encina, P.G., Mata, R.I., 2013. Consumption and occurrence of pharmaceutical and personal care products in the aquatic environment in Spain. Sci. Total. Environ. 444, 451–465.

De García, S.A.O., Pinto, G.P., García-Encina, P.A., Irusta-Mata, R., 2014. Ecotoxicity and environmental risk assessment of pharmaceuticals and personal care products in aquatic environments and wastewater treatment plants. Ecotoxicology 23, 1517–1533.

Díaz-Cruz, M.S., García-galán, M.J., Guerra, P., Jelic, A., Postigo, C., Eljarrat, E., et al., 2009. Analysis of selected emerging contaminants in sewage sludge. TrAC. Trends Anal. Chem. 28, 1263–1275.

Dietrich, D.R., Chou, Y.-J., 2001. Ecotoxicology of musks. ACS Symposium Series. ACS Publications, pp. 156–167.

Dietrich, D.R., Hitzfeld, B.C., 2004. Bioaccumulation and Ecotoxicity of Synthetic Musks in the Aquatic Environment. Series Anthropogenic Compounds. Springer.

Dobbins, L.L., Usenko, S., Brain, R.A., Brooks, B.W., 2009. Probabilistic ecological hazard assessment of parabens using *Daphnia magna* and Pimephales promelas. Environ. Toxicol. Chem. 28, 2744–2753.

Domínguez, C., Reyes-Contreras, C., Bayona, J.M., 2012. Determination of benzothiazoles and benzotriazoles by using ionic liquid stationary phases in gas chromatography mass spectrometry. Application to their characterization in wastewaters. J. Chromatogr. A 1230, 117–122.

Dotan, P., Godinger, T., Odeh, W., Groisman, L., Al-Khateeb, N., Rabbo, A.A., et al., 2016. Occurrence and fate of endocrine disrupting compounds in wastewater treatment plants in Israel and the Palestinian West Bank. Chemosphere 155, 86–93.

Dymicky, M., Huhtanen, C., 1979. Inhibition of Clostridium botulinum by p-hydroxybenzoic acid n-alkyl esters. Antimicrob. Agents Chemother. 15, 798–801.

Ebele, A.J., Abdallah, M.A.-E., Harrad, S., 2017. Pharmaceuticals and personal care products (PPCPs) in the freshwater aquatic environment. Emerg. Contam. 3, 1–16.

Ek, M., Baresel, C., Magner, J., Bergström, R., Harding, M., 2014. Activated carbon for the removal of pharmaceutical residues from treated wastewater. Water Sci. Technol. 69, 2372–2380.

Eklund, T., 1980. Inhibition of growth and uptake processes in bacteria by some chemical food preservatives. J. Appl. Bacteriol. 48, 423–432.

Elberling, J., Linneberg, A., Dirksen, A., Johansen, J., FRøLund, L., Madsen, F., et al., 2005. Mucosal symptoms elicited by fragrance products in a population-based sample in relation to atopy and bronchial hyper-reactivity. Clin. Exp. Allergy 35, 75–81.

Ellis, J.B., 2006. Pharmaceutical and personal care products (PPCPs) in urban receiving waters. Environ. Pollut. 144, 184–189.

Eriksson, E., Auffarth, K., Eilersen, A.M., Henze, M., Ledin, A., 2003. Household chemicals and personal care products as sources for xenobiotic organic compounds in grey wastewater. Water Sa 29, 135–146.

Fast, S.A., 2015. Holistic Analysis of Emerging Contaminant Removal Using Advanced Oxidation Processes. Mississippi State University.

Fent, K., Kunz, P.Y., Gomez, E., 2008. UV filters in the aquatic environment induce hormonal effects and affect fertility and reproduction in fish. CHIMIA Int. J. Chem. 62, 368–375.

Frederiksen, H., Aksglaede, L., Sorensen, K., Nielsen, O., Main, K.M., Skakkebaek, N.E., et al., 2013. Bisphenol A and other phenols in urine from Danish children and adolescents analyzed by isotope diluted TurboFlow-LC–MS/MS. Int. J. Hyg. Environ. Health 216, 710–720.

Fromme, H., Otto, T., Pilz, K., 2001. Polycyclic musk fragrances in different environmental compartments in Berlin (Germany). Water Res. 35, 121–128.

Gago-Ferrero, P., Diaz-Cruz, M.S., Barceló, D., 2012. An overview of UV-absorbing compounds (organic UV filters) in aquatic biota. Anal. Bioanal. Chem. 404, 2597–2610.

Gatermann, R., Biselli, S., Huhnerfuss, H., Rimkus, G., Hecker, M., Karbe, L., 2002. Synthetic musks in the environment. Part 1: species-dependent bioaccumulation of polycyclic and nitro musk fragrances in freshwater fish and mussels. Arch. Environ. Contam. Toxicol. 42, 437–446.

Geens, T., Neels, H., Covaci, A., 2012. Distribution of bisphenol-A, triclosan and n-nonylphenol in human adipose tissue, liver and brain. Chemosphere 87, 796–802.

Gerrity, D., Stanford, B.D., Trenholm, R.A., Snyder, S.A., 2010. An evaluation of a pilot-scale nonthermal plasma advanced oxidation process for trace organic compound degradation. Water Res. 44, 493–504.

Geyer, H., Rimkus, G., Wolf, M., Attar, A., Steinberg, C., Kettrup, A., 1994. Synthetic nitro musk fragrances and bromocyclen—new environmental chemicals in fish and mussels as well as in breast milk and human lipids. Z. Umweltchem. Okotox 6, 9–17.

Glassmeyer, S.T., Furlong, E.T., Kolpin, D.W., Cahill, J.D., Zaugg, S.D., Werner, S.L., et al., 2005. Transport of chemical and microbial compounds from known wastewater discharges: potential for use as indicators of human fecal contamination. Environ. Sci. Technol. 39, 5157–5169.

Golden, R., Gandy, J., Vollmer, G., 2005. A review of the endocrine activity of parabens and implications for potential risks to human health. Crit. Rev. Toxicol. 35, 435–458.

González-Mariño, I., Quintana, J.B., Rodríguez, I., Cela, R., 2011. Evaluation of the occurrence and biodegradation of parabens and halogenated by-products in wastewater by accurate-mass liquid chromatography-quadrupole-time-of-flight-mass spectrometry (LC-QTOF-MS). Water Res. 45, 6770–6780.

Grover, D., Zhou, J., Frickers, P., Readman, J., 2011. Improved removal of estrogenic and pharmaceutical compounds in sewage effluent by full scale granular activated carbon: impact on receiving river water. J. Hazard. Mater. 185, 1005–1011.

Halden, R.U., Paull, D.H., 2005. Co-occurrence of triclocarban and triclosan in US water resources. Environ. Sci. Technol. 39, 1420–1426.

Hany, J., Nagel, R., 1995. Detection of sunscreen agents in human breast-milk. Dtsch. Lebensmittel-Rundschau 91, 341–345.

Hauser, R., Calafat, A., 2005. Phthalates and human health. Occup. Env. Med. 62, 806–818. Find this article online.

Heberer, T., Gramer, S., Stan, H.J., 1999. Occurrence and distribution of organic contaminants in the aquatic system in Berlin. Part III: determination of synthetic musks in Berlin surface water applying solid-phase microextraction (SPME) and gas chromatography-mass spectrometry (GC-MS). Acta Hydrochimica et. Hydrobiologica 27, 150–156.

Hernández-Leal, L., Temmink, H., Zeeman, G., Buisman, C., 2011. Removal of micropollutants from aerobically treated grey water via ozone and activated carbon. Water Res. 45, 2887–2896.

Horii, Y., Kannan, K., 2008. Survey of organosilicone compounds, including cyclic and linear siloxanes, in personal-care and household products. Arch. Environ. Contam. Toxicol. 55, 701.

Houlihan, J., Brody, C., Schwan, B., Meunick, R., Doyle, M.B., Patterson, J., et al., 2002. Not too pretty: phthalates, beauty products & the FDA.

Huber, M.M., Canonica, S., Park, G.-Y., Von Gunten, U., 2003. Oxidation of pharmaceuticals during ozonation and advanced oxidation processes. Environ. Sci. Technol. 37, 1016–1024.

Hubinger, J.C., Havery, D.C., 2006. Analysis of consumer cosmetic products for phthalate esters. J. Cosmet. Sci. 57, 127–137.

Inoue, K., Kawaguchi, M., Okada, F., Takai, N., Yoshimura, Y., Horie, M., et al., 2003. Measurement of 4-nonylphenol and 4-tert-octylphenol in human urine by column-switching liquid chromatography–mass spectrometry. Anal. Chim. Acta 486, 41–50.

Inui, M., Adachi, T., Takenaka, S., Inui, H., Nakazawa, M., Ueda, M., et al., 2003. Effect of UV screens and preservatives on vitellogenin and choriogenin production in male medaka (*Oryzias latipes*). Toxicology 194, 43–50.

Ito, Y., Yamanoshita, O., Asaeda, N., Tagawa, Y., Lee, C.-H., Aoyama, T., et al., 2007. Di (2-ethylhexyl) phthalate induces hepatic tumorigenesis through a peroxisome proliferator-activated receptor α-independent pathway. J. Occup. Health 49, 172–182.

Jeong, H., Kim, J., Kim, Y., 2017. Identification of linkages between EDCs in personal care products and breast cancer through data integration combined with gene network analysis. Int. J. Environ. Res. Public. Health 14, 1158.

Johnson, R.R., Navone, R., Larson, E.L., 1963. An unusual epidemic of methemoglobinemia. Pediatrics 31, 222–225.

Jonkers, N., Sousa, A., Galante-Oliveira, S., Barroso, C.M., Kohler, H.-P.E., Giger, W., 2010. Occurrence and sources of selected phenolic endocrine disruptors in Ria de Aveiro, Portugal. Environ. Sci. Pollut. Res. 17, 834–843.

Käfferlein, H.U., Göen, T., Angerer, J., 1998. Musk xylene: analysis, occurrence, kinetics, and toxicology. Crit. Rev. Toxicol. 28, 431–476.

Kahle, M., Buerge, I.J., Muller, M.D., Poiger, T., 2009. Hydrophilic anthropogenic markers for quantification of wastewater contamination in ground-and surface WATERS. Environ. Toxicol. Chem. 28, 2528–2536.

Kannan, K., Reiner, J.L., Yun, S.H., Perrotta, E.E., Tao, L., Johnson-Restrepo, B., et al., 2005. Polycyclic musk compounds in higher trophic level aquatic organisms and humans from the United States. Chemosphere 61, 693–700.

Kasprzyk-Hordern, B., Dinsdale, R.M., Guwy, A.J., 2008. The occurrence of pharmaceuticals, personal care products, endocrine disruptors and illicit drugs in surface water in South Wales, UK. Water Res. 42, 3498–3518.

Kim, B.-N., Cho, S.-C., Kim, Y., Shin, M.-S., Yoo, H.-J., Kim, J.-W., et al., 2009. Phthalates exposure and attention-deficit/hyperactivity disorder in school-age children. Biol. Psychiatry 66, 958–963.

Kimber, I., Dearman, R.J., 2010. An assessment of the ability of phthalates to influence immune and allergic responses. Toxicology 271, 73–82.

Klavarioti, M., Mantzavinos, D., Kassinos, D., 2009. Removal of residual pharmaceuticals from aqueous systems by advanced oxidation processes. Environ. Int. 35, 402–417.

Kolpin, D.W., Furlong, E.T., Meyer, M.T., Thurman, E.M., Zaugg, S.D., Barber, L.B., et al., 2002. Pharmaceuticals, hormones, and other organic wastewater contaminants in US streams, 1999– 2000: a national reconnaissance. Environ. Sci. Technol. 36, 1202–1211.

Kunz, P.Y., Fent, K., 2006. Multiple hormonal activities of UV filters and comparison of in vivo and in vitro estrogenic activity of ethyl-4-aminobenzoate in fish. Aquat. Toxicol. 79, 305–324.

Kunz, P.Y., Gries, T., Fent, K., 2006. The ultraviolet filter 3-benzylidene camphor adversely affects reproduction in fathead minnow (Pimephales promelas). Toxicol. Sci. 93, 311–321.

Lange, C., Kuch, B., Metzger, J.W., 2015. Occurrence and fate of synthetic musk fragrances in a small German river. J. Hazard. Mater. 282, 34–40.

Lapworth, D., Baran, N., Stuart, M., Ward, R., 2012. Emerging organic contaminants in groundwater: a review of sources, fate and occurrence. Environ. Pollut. 163, 287–303.

Lebowitz, H., Young, R., Kidwell, J., Mcgowan, J., Langloss, J., Brusick, D., 1983. DEET (N, N-diethyltoluamide) does not affect sperm number, viability and head morphology in male rats treated dermally. Drug. Chem. Toxicol. 6, 379–395.

Lee, H.-B., Peart, T.E., Svoboda, M.L., 2005. Determination of endocrine-disrupting phenols, acidic pharmaceuticals, and personal-care products in sewage by solid-phase extraction and gas chromatography–mass spectrometry. J. Chromatogr. A 1094, 122–129.

Lee, I.-S., Lee, S.-H., Oh, J.-E., 2010. Occurrence and fate of synthetic musk compounds in water environment. Water Res. 44, 214–222.

Liang, R., Hu, A., Hatat-Fraile, M., Zhou, N., 2014. Fundamentals on Adsorption, Membrane Filtration, and Advanced Oxidation Processes for Water Treatment. Nanotechnology for Water Treatment and Purification. Springer.

Lin, T., Yu, S., Chen, W., 2016. Occurrence, removal and risk assessment of pharmaceutical and personal care products (PPCPs) in an advanced drinking water treatment plant (ADWTP) around Taihu Lake in China. Chemosphere 152, 1–9.

Liu, Y.-S., Ying, G.-G., Shareef, A., Kookana, R.S., 2011. Simultaneous determination of benzotriazoles and ultraviolet filters in ground water, effluent and biosolid samples using gas chromatography–tandem mass spectrometry. J. Chromatogr. A 1218, 5328–5335.

Liu, N., Shi, Y., Li, W., Xu, L., Cai, Y., 2014. Concentrations and distribution of synthetic musks and siloxanes in sewage sludge of wastewater treatment plants in China. Sci. Total. Environ. 476, 65–72.

López-Carrillo, L., Hernández-ramírez, R.U., Calafat, A.M., Torres-Sánchez, L., Galván-Portillo, M., Needham, L.L., et al., 2009. Exposure to phthalates and breast cancer risk in northern Mexico. Environ. Health Perspect. 118, 539–544.

Loraine, G.A., Pettigrove, M.E., 2006. Seasonal variations in concentrations of pharmaceuticals and personal care products in drinking water and reclaimed wastewater in southern California. Environ. Sci. Technol. 40, 687–695.

Lv, M., Sun, Q., Xu, H., Lin, L., Chen, M., Yu, C.-P., 2014. Occurrence and fate of triclosan and triclocarban in a subtropical river and its estuary. Mar. Pollut. Bull. 88, 383–388.

Madsen, T., Boyd, H.B., Nylen, D., Pedersen, A.R., Petersen, G.I., Simonsen, F., 2001. Environmental and health assessment of substances in household detergents and cosmetic detergent products. Environ. Proj. 615, 221.

Margot, J., Kienle, C., Magnet, A., Weil, M., Rossi, L., De Alencastro, L.F., et al., 2013. Treatment of micropollutants in municipal wastewater: ozone or powdered activated carbon? Sci. Total. Environ. 461, 480–498.

McAvoy, D.C., Schatowitz, B., Jacob, M., Hauk, A., Eckhoff, W.S., 2002. Measurement of triclosan in wastewater treatment systems. Environ. Toxicol. Chem.: An. Int. J. 21, 1323–1329.

McMonagle, H., 2013. Evaluation of the Role of Advanced Wastewater Treatment in Removal of Priority Pollutants From Municipal Point-Discharges. Dublin City University. Oscail.

Meinel, F., Ruhl, A., Sperlich, A., Zietzschmann, F., Jekel, M., 2015. Pilot-scale investigation of micropollutant removal with granular and powdered activated carbon. Water, Air, Soil. Pollut. 226, 2260.

Michael, A., Grant, G., 1974. Toxicity of the repellent deet (N, N-diethyl-meta-toluamide) to Gaziibusia affinis (Baird and Girard).

Moldovan, Z., 2006. Occurrences of pharmaceutical and personal care products as micropollutants in rivers from Romania. Chemosphere 64, 1808–1817.

Montes-Grajales, D., Fennix-Agudelo, M., Miranda-Castro, W., 2017. Occurrence of personal care products as emerging chemicals of concern in water resources: a review. Sci. Total. Environ. 595, 601–614.

Murphy, M.E., Montemarano, A.D., Debboun, M., Gupta, R., 2000. The effect of sunscreen on the efficacy of insect repellent: a clinical trial. J. Am. Acad. Dermatol. 43, 219–222.

Murray, K.E., Thomas, S.M., Bodour, A.A., 2010. Prioritizing research for trace pollutants and emerging contaminants in the freshwater environment. Environ. Pollut. 158, 3462–3471.

Nagtegaal, M., 1997. Detection of sunscreen agents in water and fish of the meerfelder maar, the eifel, Germany. Umweltwissenschaften und Schadstoff-Forschung 9, 79–86.

Nakada, N., Hanamoto, S., Jurgens, M.D., Johnson, A.C., Bowes, M.J., Tanaka, H., 2017. Assessing the population equivalent and performance of wastewater treatment through the ratios of pharmaceuticals and personal care products present in a river basin: application to the River Thames basin, UK. Sci. Total. Environ. 575, 1100–1108.

Negreira, N., Rodríguez, I., Rodil, R., Cela, R., 2012. Assessment of benzophenone-4 reactivity with free chlorine by liquid chromatography quadrupole time-of-flight mass spectrometry. Anal. Chim. Acta 743, 101–110.

Nghiem, L.D., Schäfer, A.I., Elimelech, M., 2004. Removal of natural hormones by nanofiltration membranes: measurement, modeling, and mechanisms. Environ. Sci. Technol. 38, 1888–1896.

Nghiem, L.D., Schäfer, A.I., Elimelech, M., 2005. Pharmaceutical retention mechanisms by nanofiltration membranes. Environ. Sci. Technol. 39, 7698–7705.

Olmez-Hanci, T., Arslan-Alaton, I., Basar, G., 2011. Multivariate analysis of anionic, cationic and nonionic textile surfactant degradation with the H_2O_2/UV-C process by using the capabilities of response surface methodology. J. Hazard. Mater. 185, 193–203.

Papageorgiou, A., Voutsa, D., Papadakis, N., 2014. Occurrence and fate of ozonation by-products at a full-scale drinking water treatment plant. Sci. Total. Environ. 481, 392–400.

Paulsen, L., 2015. The health risks of chemicals in personal care products and their fate in the environment.

Peck, A.M., 2006. Analytical methods for the determination of persistent ingredients of personal care products in environmental matrices. Anal. Bioanal. Chem. 386, 907–939.

Pedersen, S., Selck, H., Salvito, D., Forbes, V., 2009. Effects of the polycyclic musk HHCB on individual-and population-level endpoints in *Potamopyrgus antipodarum*. Ecotoxicol. Environ. Saf. 72, 1190–1199.

Penalver, A., Pocurull, E., Borrull, F., Marce, R., 2000. Determination of phthalate esters in water samples by solid-phase microextraction and gas chromatography with mass spectrometric detection. J. Chromatogr. A 872, 191–201.

Peng, X., Ou, W., Wang, C., Wang, Z., Huang, Q., Jin, J., et al., 2014. Occurrence and ecological potential of pharmaceuticals and personal care products in groundwater and reservoirs in the vicinity of municipal landfills in China. Sci. Total. Environ. 490, 889–898.

Peng, X., Xiong, S., Ou, W., Wang, Z., Tan, J., Jin, J., et al., 2017. Persistence, temporal and spatial profiles of ultraviolet absorbents and phenolic personal care products in riverine and estuarine sediment of the Pearl River catchment, China. J. Hazard. Mater. 323, 139–146.

Petrovic, M., Barceló, D., Diaz, A., Ventura, F., 2003. Low nanogram per liter determination of halogenated nonylphenols, nonylphenol carboxylates, and their non-halogenated precursors in water and sludge by liquid chromatography electrospray tandem mass spectrometry. J. Am. Soc. Mass. Spectr. 14, 516–527.

Poiger, T., Buser, H.-R., Balmer, M.E., Bergqvist, P.-A., Muller, M.D., 2004. Occurrence of UV filter compounds from sunscreens in surface waters: regional mass balance in two Swiss lakes. Chemosphere 55, 951–963.

Pryor, S.W., Hay, A.G., Walker, L.P., 2002. Nonylphenol in anaerobically digested sewage sludge from New York State. Environ. Sci. Technol. 36, 3678–3682.

Quednow, K., Puttmann, W., 2009. Temporal concentration changes of DEET, TCEP, terbutryn, and nonylphenols in freshwater streams of Hesse, Germany: possible influence of mandatory regulations and voluntary environmental agreements. Environ. Sci. Pollut. Res. 16, 630–640.

Rajapaksha, A.U., Vithanage, M., Lim, J.E., Ahmed, M.B.M., Zhang, M., Lee, S.S., et al., 2014. Invasive plant-derived biochar inhibits sulfamethazine uptake by lettuce in soil. Chemosphere 111, 500–504.

Ramskov, T., Selck, H., Salvitod, D., Forbes, V.E., 2009. Individual-and population-level effects of the synthetic musk, hhcb, on the deposit-feeding polychaete, *Capitella sp.* I. Environ. Toxicol. Chem. 28, 2695–2705.

Real, F.J., Benitez, F.J., Acero, J.L., Roldan, G., 2012. Combined chemical oxidation and membrane filtration techniques applied to the removal of some selected pharmaceuticals from water systems. J. Environ. Sci. Health, Part. A 47, 522–533.

Reemtsma, T., Miehe, U., Duennbier, U., Jekel, M., 2010. Polar pollutants in municipal wastewater and the water cycle: occurrence and removal of benzotriazoles. Water Res. 44, 596–604.

Reh, R., Licha, T., Geyer, T., Nödler, K., Sauter, M., 2013. Occurrence and spatial distribution of organic micro-pollutants in a complex hydrogeological karst system during low flow and high flow periods, results of a two-year study. Sci. Total. Environ. 443, 438–445.

Ricart, M., Guasch, H., Alberch, M., Barceló, D., Bonnineau, C., Geiszinger, A., et al., 2010. Triclosan persistence through wastewater treatment plants and its potential toxic effects on river biofilms. Aquat. Toxicol. 100, 346–353.

Richardson, S.D., 2008. Environmental mass spectrometry: emerging contaminants and current issues. Anal. Chem. 80, 4373–4402.

Rodil, R., Moeder, M., 2008. Stir bar sorptive extraction coupled to thermodesorption–gas chromatography–mass spectrometry for the determination of insect repelling substances in water samples. J. Chromatogr. A 1178, 9–16.

Rodil, R., Quintana, J.B., López-Mahía, P., Muniategui-Lorenzo, S., Prada-Rodríguez, D., 2008. Multiclass determination of sunscreen chemicals in water samples by liquid chromatography – tandem mass spectrometry. Anal. Chem. 80, 1307–1315.

Röhricht, M., Krisam, J., Weise, U., Kraus, U.R., During, R.A., 2009. Elimination of carbamazepine, diclofenac and naproxen from treated wastewater by nanofiltration. Clean–Soil Air Water 37, 638–641.

Roslev, P., Vorkamp, K., Aarup, J., Frederiksen, K., Nielsen, P.H., 2007. Degradation of phthalate esters in an activated sludge wastewater treatment plant. Water Res. 41, 969–976.

Routledge, E.J., Parker, J., Odum, J., Ashby, J., Sumpter, J.P., 1998. Some alkyl hydroxy benzoate preservatives (parabens) are estrogenic. Toxicol. Appl. Pharmacol. 153, 12–19.

Sanchís, J., Martínez, E., Ginebreda, A., Farre, M., Barceló, D., 2013. Occurrence of linear and cyclic volatile methylsiloxanes in wastewater, surface water and sediments from Catalonia. Sci. Total. Environ. 443, 530–538.

Schäfer, A.I., Akanyeti, I., Semião, A.J., 2011. Micropollutant sorption to membrane polymers: a review of mechanisms for estrogens. Adv. Colloid Interface Sci. 164, 100–117.

Scheurer, M., Storck, F.R., Brauch, H.-J., Lange, F.T., 2010. Performance of conventional multi-barrier drinking water treatment plants for the removal of four artificial sweeteners. Water Res. 44, 3573–3584.

Schlumpf, M., Lichtensteiger, W., 2001. "In Vitro and in Vivo Estrogenicity of UV Screens": Response. Environ. Health Perspect. A359–A361.

Schlumpf, M., Schmid, P., Durrer, S., Conscience, M., Maerkel, K., Henseler, M., et al., 2004. Endocrine activity and developmental toxicity of cosmetic UV filters—an update. Toxicology 205, 113–122.

Schlumpf, M., Kypke, K., Wittassek, M., Angerer, J., Mascher, H., Mascher, D., et al., 2010. Exposure patterns of UV filters, fragrances, parabens, phthalates, organochlor pesticides, PBDEs, and PCBs in human milk: correlation of UV filters with use of cosmetics. Chemosphere 81, 1171–1183.

Schmitt, C., Oetken, M., Dittberner, O., Wagner, M., Oehlmann, J., 2008. Endocrine modulation and toxic effects of two commonly used UV screens on the aquatic invertebrates *Potamopyrgus antipodarum* and *Lumbriculus variegatus*. Environ. Pollut. 152, 322–329.

Schramm, K.-W., Kaune, A., Beck, B., Thumm, W., Behechti, A., Kettrup, A., et al., 1996. Acute toxicities of five nitromusk compounds in Daphnia, algae and photoluminescent bacteria. Water Res. 30, 2247–2250.

Schreurs, R., Lanser, P., Seinen, W., Van Der Burg, B., 2002. Estrogenic activity of UV filters determined by an in vitro reporter gene assay and an in vivo transgenic zebrafish assay. Arch. Toxicol. 76, 257–261.

Shang, D.Y., Macdonald, R.W., Ikonomou, M.G., 1999. Persistence of nonylphenol ethoxylate surfactants and their primary degradation products in sediments from near a municipal outfall in the Strait of Georgia, British Columbia, Canada. Environ. Sci. Technol. 33, 1366–1372.

Soni, M., Carabin, I., Burdock, G., 2005. Safety assessment of esters of p-hydroxybenzoic acid (parabens). Food Chem. Toxicol. 43, 985–1015.

Sparham, C., Van Egmond, R., O'connor, S., Hastie, C., Whelan, M., Kanda, R., et al., 2008. Determination of decamethylcyclopentasiloxane in river water and final effluent by headspace gas chromatography/mass spectrometry. J. Chromatogr. A 1212, 124–129.

Sparham, C., Van Egmond, R., Hastie, C., O'connor, S., Gore, D., Chowdhury, N., 2011. Determination of decamethylcyclopentasiloxane in river and estuarine sediments in the UK. J. Chromatogr. A 1218, 817–823.

Stackelberg, P.E., Gibs, J., Furlong, E.T., Meyer, M.T., Zaugg, S.D., Lippincott, R.L., 2007. Efficiency of conventional drinking-water-treatment processes in removal of pharmaceuticals and other organic compounds. Sci. Total. Environ. 377, 255–272.

Subedi, B., Balakrishna, K., Sinha, R.K., Yamashita, N., Balasubramanian, V.G., Kannan, K., 2015a. Mass loading and removal of pharmaceuticals and personal care products, including psychoactive and illicit drugs and artificial sweeteners, in five sewage treatment plants in India. J. Environ. Chem. Eng. 3, 2882–2891.

Subedi, B., Codru, N., Dziewulski, D.M., Wilson, L.R., Xue, J., Yun, S., et al., 2015b. A pilot study on the assessment of trace organic contaminants including pharmaceuticals and personal care products from on-site wastewater treatment systems along Skaneateles Lake in New York State, USA. Water Res. 72, 28–39.

Sui, Q., Huang, J., Deng, S., Yu, G., Fan, Q., 2010. Occurrence and removal of pharmaceuticals, caffeine and DEET in wastewater treatment plants of Beijing, China. Water Res. 44, 417–426.

Svensson, K., Hernández-Ramírez, R.U., Burguete-García, A., Cebrián, M.E., Calafat, A. M., Needham, L.L., et al., 2011. Phthalate exposure associated with self-reported diabetes among Mexican women. Environ. Res. 111, 792–796.

Tanwar, S., Di Carro, M., Ianni, C., Magi, E., 2014. Occurrence of PCPs in Natural Waters From Europe. Personal Care Products in the Aquatic Environment. Springer.

Tarpani, R.R.Z., Azapagic, A., 2018. Life cycle environmental impacts of advanced wastewater treatment techniques for removal of pharmaceuticals and personal care products (PPCPs). J. Environ. Manag. 215, 258–272.

Tay, K.S., Rahman, N.A., Abas, M.R.B., 2009. Degradation of DEET by ozonation in aqueous solution. Chemosphere 76, 1296–1302.

Terasaki, M., Makino, M., Tatarazako, N., 2009. Acute toxicity of parabens and their chlorinated by-products with *Daphnia magna* and *Vibrio fischeri* bioassays. J. Appl. Toxicol. 29, 242–247.

Ternes, T.A., Joss, A., Siegrist, H., 2004. Peer Reviewed: Scrutinizing Pharmaceuticals and Personal Care Products in Wastewater Treatment. ACS Publ.

Tolls, J., Berger, H., Klenk, A., Meyberg, M., Muller, R., Rettinger, K., et al., 2009. Environmental safety aspects of personal care products-a European perspective. Environ. Toxicol. Chem. 28, 2485.

Tsui, M.M., Leung, H., Lam, P.K., Murphy, M.B., 2014. Seasonal occurrence, removal efficiencies and preliminary risk assessment of multiple classes of organic UV filters in wastewater treatment plants. Water Res. 53, 58–67.

Verschoyle, R., Brown, A., Nolan, C., Ray, D., Lister, T., 1992. A comparison of the acute toxicity, neuropathology, and electrophysiology of N, N-diethyl-m-toluamide and N, N-dimethyl-2, 2-diphenylacetamide in rats. Fundamental Appl. Toxicol. 18, 79–88.

Vithanage, M., Rajapaksha, A.U., Tang, X., Thiele-Bruhn, S., Kim, K.H., Lee, S.-E., et al., 2014. Sorption and transport of sulfamethazine in agricultural soils amended with invasive-plant-derived biochar. J. Environ. Manag. 141, 95–103.

Wang, W., Kannan, K., 2016. Fate of parabens and their metabolites in two wastewater treatment plants in New York State, United States. Environ. Sci. Technol. 50, 1174–1181.

Wang, D., Sui, Q., Lu, S.-G., Zhao, W.-T., Qiu, Z.-F., Miao, Z.-W., et al., 2014. Occurrence and removal of six pharmaceuticals and personal care products in a wastewater treatment plant employing anaerobic/anoxic/aerobic and UV processes in Shanghai, China. Environ. Sci. Pollut. Res. 21, 4276–4285.

Watkinson, A., Murby, E., Costanzo, S., 2007. Removal of antibiotics in conventional and advanced wastewater treatment: implications for environmental discharge and wastewater recycling. Water Res. 41, 4164–4176.

Weiss, S., Reemtsma, T., 2005. Determination of benzotriazole corrosion inhibitors from aqueous environmental samples by liquid chromatography-electrospray ionization-tandem mass spectrometry. Anal. Chem. 77, 7415–7420.

Weiss, S., Jakobs, J., Reemtsma, T., 2006. Discharge of three benzotriazole corrosion inhibitors with municipal wastewater and improvements by membrane bioreactor treatment and ozonation. Environ. Sci. Technol. 40, 7193–7199.

Winkler, M., Kopf, G., Hauptvogel, C., Neu, T., 1998. Fate of artificial musk fragrances associated with suspended particulate matter (SPM) from the River Elbe (Germany) in comparison to other organic contaminants. Chemosphere 37, 1139–1156.

Wollenberger, L., Breitholtz, M., Kusk, K.O., Bengtsson, B.-E., 2003. Inhibition of larval development of the marine copepod *Acartia tonsa* by four synthetic musk substances. Sci. Total. Environ. 305, 53–64.

Yamagishi, T., Miyazaki, T., Horii, S., Akiyama, K., 1983. Synthetic musk residues in biota and water from Tama River and Tokyo Bay (Japan). Arch. Environ. Contam. Toxicol. 12, 83–89.

Yang, X., Flowers, R.C., Weinberg, H.S., Singer, P.C., 2011. Occurrence and removal of pharmaceuticals and personal care products (PPCPs) in an advanced wastewater reclamation plant. Water Res. 45, 5218–5228.

Yang, X., Sun, J., Fu, W., Shang, C., Li, Y., Chen, Y., et al., 2016. PPCP degradation by UV/chlorine treatment and its impact on DBP formation potential in real waters. Water Res. 98, 309–318.

Yang, Y., Ok, Y.S., Kim, K.-H., Kwon, E.E., Tsang, Y.F., 2017. Occurrences and removal of pharmaceuticals and personal care products (PPCPs) in drinking water and water/sewage treatment plants: a review. Sci. Total. Environ. 596, 303–320.

Yangali-Quintanilla, V., Maeng, S.K., Fujioka, T., Kennedy, M., Li, Z., Amy, G., 2011. Nanofiltration vs. reverse osmosis for the removal of emerging organic contaminants in water reuse. Desalin. Water Treat. 34, 50–56.

Yin, J., Wang, H., Zhang, J., Zhou, N., Gao, F., Wu, Y., et al., 2012. The occurrence of synthetic musks in human breast milk in Sichuan, China. Chemosphere 87, 1018–1023.

Yoon, Y., Westerhoff, P., Snyder, S.A., Wert, E.C., 2006. Nanofiltration and ultrafiltration of endocrine disrupting compounds, pharmaceuticals and personal care products. J. Membr. Sci. 270, 88–100.

Yoon, Y., Ryu, J., Oh, J., Choi, B.-G., Snyder, S.A., 2010. Occurrence of endocrine disrupting compounds, pharmaceuticals, and personal care products in the Han River (Seoul, South Korea). Sci. Total. Environ. 408, 636–643.

Yu, Y., Huang, Q., Cui, J., Zhang, K., Tang, C., Peng, X., 2011. Determination of pharmaceuticals, steroid hormones, and endocrine-disrupting personal care products in sewage sludge by ultra-high-performance liquid chromatography–tandem mass spectrometry. Anal. Bioanal. Chem. 399, 891–902.

Yu, Y., Wu, L., Chang, A.C., 2013. Seasonal variation of endocrine disrupting compounds, pharmaceuticals and personal care products in wastewater treatment plants. Sci. Total. Environ. 442, 310–316.

Zhang, N.-S., Liu, Y.-S., Van Den Brink, P.J., Price, O.R., Ying, G.-G., 2015. Ecological risks of home and personal care products in the riverine environment of a rural region in South China without domestic wastewater treatment facilities. Ecotoxicol. Environ. Saf. 122, 417–425.

Zhao, X., Chen, Z.-L., Wang, X.-C., Shen, J.-M., Xu, H., 2014. PPCPs removal by aerobic granular sludge membrane bioreactor. Appl. Microbiol. Biotechnol. 98, 9843–9848.

Zoschke, K., Engel, C., Börnick, H., Worch, E., 2011. Adsorption of geosmin and 2-methylisoborneol onto powdered activated carbon at non-equilibrium conditions: influence of NOM and process modelling. Water Res. 45, 4544–4550.

Zucchi, S., Bluthgen, N., Ieronimo, A., Fent, K., 2011a. The UV-absorber benzophenone-4 alters transcripts of genes involved in hormonal pathways in zebrafish (*Danio rerio*) eleuthero-embryos and adult males. Toxicol. Appl. Pharmacol. 250, 137–146.

Zucchi, S., Oggier, D.M., Fent, K., 2011b. Global gene expression profile induced by the UV-filter 2-ethyl-hexyl-4-trimethoxycinnamate (EHMC) in zebrafish (*Danio rerio*). Environ. Pollut. 159, 3086–3096.

CHAPTER

Surfactants: an emerging face of pollution

4

Jayesh Bhatt[1], Avinash Kumar Rai[1], Meghavi Gupta[1], Shubhang Vyas[1], Rakshit Ameta[2], Suresh C. Ameta[1], Afsane Chavoshani[3] and Majid Hashemi[4,5]

[1]*Department of Chemistry, PAHER University, Udaipur, India*

[2]*Department of Chemistry, J. R. N. Rajasthan Vidyapeeth (Deemed to be University), Udaipur, India*

[3]*Department of Environmental Health Engineering, School of Health, Isfahan University of Medical Sciences, Isfahan, Iran*

[4]*Environmental Health Engineering, School of Public Health, Kerman University of Medical Sciences, Kerman, Iran*

[5]*Environmental Health Engineering Research Center, Kerman University of Medical Sciences, Kerman, Iran*

4.1 Introduction

The production of soap-like materials is known from around 2800 BC in ancient Babylon. At that time soap consisted of water, alkali, and cassia oil. The word soap is derived from the word "sapo," and it was manufactured from tallow and ashes. Soap may be called detergent, similar to soap, which was manufactured in ancient China from vegetation and herbs. Industrially bar soaps were manufactured first in the late 18th century. William Sheppard patented a liquid version of soap in 1865. The process of cleaning things such as clothing, floors, and bathrooms become much easier with soap. Later on liquid soap was found to be more effective than flake soap as a detergent, and it leaves fewer residues on skin, clothes, and surfaces of wash basins.

The term surfactant was derived from surface active agent. It is a molecule, which lowers down surface tension. The word amphiphile was also coined for such molecules by Paul Winsor. It is a combination of two Greek words. [amphi means "double," (from both sides), and philos means friendship or affinity]. It has two parts: a polar portion, which exhibits a strong affinity for polar solvents such as water, and is commonly called hydrophilic part or hydrophile. The other part is nonpolar, which is called hydrophobic or lipophile derived from Greek terms phobos (fear) and lipos (grease).

Micropollutants and Challenges. DOI: https://doi.org/10.1016/B978-0-12-818612-1.00004-0

Surfactants are known to exhibit other properties apart from lowering surface tension, and therefore they find their main use as soap, detergent, dispersion, emulsifier, wetting agent, foaming agent, corrosion inhibitor, antistatic agent, bactericide, etc.

Surfactants are the main components in household detergent which includes:

- laundry detergent like washing powder, laundry soap, laundry detergent, washing paste, and laundry tablets;
- home cleaning supplies such as detergent, floor cleaner, toilet fine and clean appliances cleaning; and
- personal toiletries like shampoo, shower gel, hand liquid, and cleanser.

Silicone surfactants, fluorocarbon surfactants, gemini surfactants, nonionic types of surfactants having branched chain fatty acid alcohol, fatty amines, etc. are the newer developments in the field of surfactant. Ecofriendliness is the most desired portion of surfactant formulations, whether it is used either for domestic or industrial purpose. There is an urgent necessity to develop biodegradable surfactants so as to save our environment from this emerging face of pollution: detergent or surfactant pollution.

One of the major disadvantages of using soap is that it does not give lather in hard water; thus limiting its scope in cleaning and washing. Surfactants were developed as an alternative to soap because it can be used in hard water. As a result detergents have almost replaced the conventional soap from the market, leaving aside only bathing soap.

Surfactants are compounds that lower the surface tension (or interfacial tension) between two liquids, between a gas and a liquid, or between a liquid and a solid. A surfactant contains both a water-insoluble (or oil-soluble) component and a water-soluble component.

The annual production of surfactants in the world was about 13 million tonnes in 2008, but the world market for surfactants reached a volume of more than 33 billion $ in 2014. Annual revenues are expected to increase by 2.5% every year, reaching 40.4 billion $ in 2022. The commercially most significant type of surfactants is the anionic surfactant, linear alkylbenzene sulfonate (LAS). It is widely used in cleaners and detergents.

The soap is biodegradable as it is sodium salt of carboxylic acid, but the detergent is sodium salt of sulfuric acid, and therefore soaps are degraded by microbes, but detergents are toxic/harmful to microbial systems. Therefore they get accumulated in water and will survive for a long term, causing a threat to environment.

Surfactants play an important role as cleaning, wetting, dispersing, emulsifying, foaming, and antifoaming agents in many practical applications and products, including detergents, fabric softeners, emulsions, soaps, paints, adhesives, inks, antifogs, ski waxes, snowboard wax, deinking of recycled papers, laxatives, etc. and agrochemical formulations such as herbicides, insecticides, biocides (sanitizers), and spermicides (nonoxynol-9), personal care products including cosmetics, shampoos, shower gel, hair conditioners (after shampoo), toothpastes, etc. A large amount of surfactants in different forms are discharged into the environment through effluents, which are resulted in harming aquatic life, polluting the water, and endangering human health. Therefore it is of utmost importance to monitor and control release of various types of surfactants in environmental water. The surfactants are used on a large scale basis from everyday household use to industrial cleaning and textile manufacturing.

As surfactants are hazardous for people and environment, detection/determination and removal/degradation of surfactants/detergent from wastewater are urgently needed. Several methods have been attempted for removing or decreasing detergent (surfactant) concentration including coagulation, adsorption, biodegradation, photocatalytic degradation, etc.

4.2 Classification of surfactants

The most accepted and scientifically sound classification of surfactants is based on their dissociation in water. Surfactants are classified into four broad categories: cationic, nonionic, anionic, and amphoteric (Fig. 4.1).

4.2.1 Cationic surfactants

These are surfactants, which give rise to a positively charged surfactant ion (cationic) and an anionic as counter ion on dissolving in water. These surfactants became particularly important, when the commercial potential of their bacteriostatic properties was established in early 1930s. These surfactants are relatively more expensive than anionic because of the high pressure hydrogenation reaction to be carried out during their preparation. These are dissolved in water to generate the surface active positive ions. They have sufficient surface activity in an acidic medium, but may precipitate and lose their activity in alkaline medium. Cationic surfactants (CS) are normally classified in following categories: open-chain CS, heterocyclic group CS, and bonded intermediate connection CS depending on their chain structure. These are widely used for sterilization, rust, corrosion, breaking, corrosion, and mineral flotation.

Some commonly used CS are cetrimonium bromide, cetrimonium chloride, quaternary ammonium compound, cetyltrimethylammonium bromide (CTMAB), benzalkonium chloride (BAC), etc.

4.2.1.1 Detection/determination

Li and Zhao (2004) developed a novel spectrophotometric method for the determination of CS by using a new reagent benzothiaxolyldiazoaminoazobenzene (BTDAB). CTMAB and cetylpyrdinium chloride (CPC) in 0.06–0.10 M sodium hydroxide react with BTDAB to form a violet–red 1:2 (CS:BTDAB) ion association complex in the presence of triton X-100. It was reported that Beer's law was

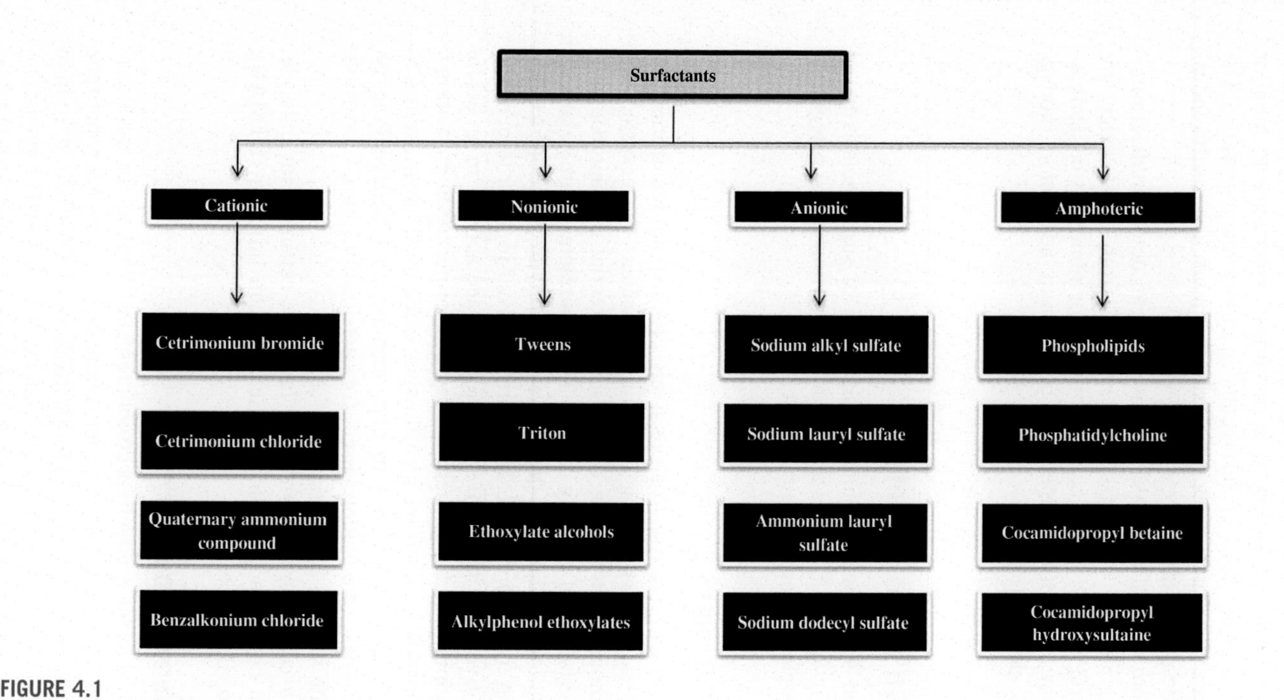

FIGURE 4.1

Classification diagram of Surfactants.

obeyed for CTMAB or CPC in the concentration range of 0–100 g 25 mL^{-1} of solution. The present method is based on a color reaction, and therefore it is simple and rapid. Apart from it there is no use of toxic organic solvents. This method was applied in determination of trace CS in industrial wastewater with satisfactory results.

Tsai and Ding (2004) analyzed the contents of alkyltrimethylammonium compounds (ATMACs) in commercial hair conditioners and fabric softeners. They used gas chromatography–mass spectrometry (GC–MS) with electron impact (EI) and low-pressure positive-ion chemical ionization (PICI) modes. The proposed method involves mixed diluted samples (pH adjusted to 10.0) with potassium iodide so as to enhance the extraction of iodide–$ATMA^+$ ion pairs by direct liquid–liquid extraction. These pairs were then demethylated to give corresponding nonionic alkyldimethylamines (ADMAs) by thermal decomposition. It was reported that contents of total measured ATMAC ranged from 0.4% to 6.9% and 3.3% to 4.6% for hair conditioners and fabric softeners, respectively.

A new sensitive potentiometric surfactant sensor was prepared by Maduniccacic et al. (2008), which was based on a highly lipophilic 1,3-didecyl-2-methyl-imidazolium cation and a tetraphenylborate antagonist ion. This sensor responded fast to the presence of CPC, hexadecyltrimethylammonium bromide (CTAB), and hyamine with slope 59.8, 58.6, and 56.8 mV/decade, respectively. This sensor served as an end-point detector in ion-pair surfactant potentiometric titrations, where sodium tetraphenylborate is used as titrant. Several commercial CS and disinfectant products were estimated by this method. Proposed sensor showed satisfactory analytical performances in the pH range of 2–11, and it also exhibited excellent selectivity performance for CPC as compared to other organic compounds.

Devi and Chattopadhyaya (2011) prepared a new membrane electrode using CPC based Sn(IV) phosphate (CPC-SnP) as the electroactive material. This electrode exhibited a linear response for the surfactant cetylpyridinium chloride in the concentration range of 5.0×10^{-3}–5.0×10^{-6} mol/dm^3. The working pH range and the response time for electrode were found to be 2–6 and 30 s, respectively. They determined CPC in mouth wash with good results as compared to two phase titration method. Present electrode could be utilized as an indicator electrode in the potentiometric titration of cationic surfactant CPC as well as in its direct determination.

Parham et al. (2011) developed a new, simple, and sensitive flotation-spectrophotometric method for the determination of cetylpyridinium chloride, which is based on the formation of an ion-associate between CPC and orange II (OR). It floats in the interface of aqueous phase and *n*-hexane by vigorous shaking. The aqueous solution was removed, and the adsorbed ion-associate present on to the wall of a separating funnel was dissolved in a small volume of methanol. The absorbance was measured at 480 nm, where a linear graph was observed in the concentration range of 15–800 ng/mL of CPC. The limit of detection (LOD) was reported to be 10.8 ng/mL. Proposed method was successfully applied for the determination of CPC in a commercial mouth wash product.

Wang et al. (2011) developed an octanethiolate gold particle modified boron-doped diamond (OT-Au-BDD) electrode for the determination of cetylpyridinium bromide (CPB) using $K_3Fe(CN)_6$ as a probe. They first deposited gold nanoparticles on the hydrogen-terminated BDD surface by an electrochemical method, thus producing gold particle modified BDD (Au-BDD). The octanethiol was subsequently self-assembled onto the Au-BDD surface to prepare OT-Au-BDD electrode. As-prepared electrode gave a linear response in the range of $5 \times 10^{-7}-10^{-4}$ M with detection limit of 0.2 μM to CPB. It was reported that better reproducibility and stability was observed as compared to self-assembled octanethiol monolayer modified gold (OT-Au) electrode because of outstanding electrochemical properties of BDD.

A new, highly selective and sensitive fluorescent detection method was developed by Huang et al. (2014) for determination of quaternary pyridinium salts cationic surfactant in an aqueous solution of polyoxyethylene-23-lauryl ether (Brij 35). This method is based on the emission of a fluorescent dye, disodium-4,4′-bis-(4-anilino-6-morpholino-*s*-triazin-2-ylamino)-2,2′-stilbenedisulfonate (CXT), which was quenched by CPB. The fluorescence intensity of CXT showed distinct changes in presence of CPB with excellent selectivity and sensitivity at pH 8.0, but fluorescence of the CXT/CPB complex was found to be not affected by other common surfactants in environmental samples. Fluorescent intensity of CXT (5.0 μM) shows a linear response toward the concentration of CPB in the range from 0 to 20 μM, and the limit of detection was calculated to be 65 nM. This method can be successfully used in the analysis of CPB in real environmental samples.

A rapid and novel fluorescence-sensing system has been proposed by Li et al. (2015) using a complex of acridine orange (AO) and polystyrene sulfonate (PSS) to determine CTAB in aqueous solutions. It was revealed that AO interacts with PSS, forming a complex via electrostatic attraction and hydrophobic interaction. They reported that fluorescence of AO was greatly quenched in the presence of PSS. This proposed method can be applied for the discrimination and detection of surfactants with different hydrocarbon chain lengths due to their different binding affinity towards PSS. The detection limit for CTAB was found to be as low as 0.2 μg/mL, and the linear range was observed from 0.5 to 3.5 μg/mL. The sensor was successfully applied in detection of CTAB in water samples.

A simple, rapid, and sensitive colorimetric assay was reported by Shrivas et al. (2015) for the quantitation of CS^+ in domestic effluent, municipal waste, and surface water samples. Proposed method is based on the aggregation of tartrate-capped gold nanoparticles (AuNPs) through both; electrostatic and hydrophobic interactions occurring between CS^+ and AuNPs. It was observed that aggregation resulted in a color change from pink to blue. Cetylpyridinium chloride was selected as a model compound, where optimal conditions were found to be in pH 9.0, reaction time 5 min, and concentration of AuNPs 25 nM. Quantitative determination of CPC, CTAB, dodecyltrimethylammonium bromide (DTAB) could be made by this method in the range of 10–500, 10–200, and 10–300 ng/mL with limit of detection 3 ng/mL for CPC, CTAB, and DTAB, respectively.

Two selective electrodes of the types carbon paste (CPE) and screen-printed sensors (SPE) have been constructed by Ali et al. (2015) based on incorporation of zeolite ionophore. The proposed electrodes showed Nernstian behavior with a linear concentration range of 4.61×10^{-7}–1.0×10^{-2} and 1.26×10^{-7}–1.0×10^{-2} mol/L and lower limit of detection of 4.61×10^{-7} and 1.26×10^{-7} mol/L for modified CPE and SPE sensors, respectively. It was reported that electrodes display good selectivity for CTAB with respect to a number of other inorganic and organic species. The response was found to be not affected by variation in pH between 2.0–8.5 and 2.0–9.0 for modified CPE and SPE, respectively. These sensors were successfully tried for determination of CTAB in both pure solution and in different spiked real water samples.

Mohammad and Mobin (2015) found a most efficient method for the separation of ternary mixture of surfactants [alkyldimethylbenzylammonium chloride (ADBAC) + CTAB + Triton X-100]. They used a thin-layer chromatographic system using silica gel as stationary phase and a mixture of methanol, 0.1% aqueous sodium thiocyanate, acetone, and ethyl acetate in 7:3:2:3 ratio as eco-favourable mobile phase. Chromatographic parameters have also been calculated. The effect of the presence of different foreign substances like cations, anions, amino acids, and vitamins as impurities has been examined. The detection limits for ADBAC, CTAB, and Triton X-100 have been determined, and the lowest possible detectable amounts (μg/spot) of ADBAC, cetrimide, CPB, CTAB, and triton X-100 were found to be 1.4, 0.1, 0.1, 0.1, and 1.6, respectively, with a reasonably good sensitivity. This method has been successfully applied for the identification of ADBAC in a household cleaning product.

Based on quick, easy, cheap, effective, rugged, and safe (QuEChERS) extraction, liquid chromatography electrospray tandem mass spectrometry was developed by Díez et al. (2015) for analysis of didecyldimethylammonium chloride and four BAC homologue residues in fruits and vegetables. Experiment was conducted using three independent series in orange, lettuce, and avocado composite samples spiked at three levels (10, 100, and 500 μg/kg). They used five quantification strategies to avoid matrix effects on quantitative results:

- solvent calibration,
- matrix-matched calibration prepared on samples,
- on sample extract aliquots,
- standard addition prepared on samples, and
- on sample extract aliquots.

The proposed validated method was used to monitor 108 fruit and vegetable commercial samples.

The fluorescence quenching of cucurbit[7]uril-palmatine (CB[7]-PAL) complexes by cetylpyridinium chloride was studied in aqueous solutions by Jian-Hong et al. (2016). It was observed that the rate of the fluorescence quenching was proportional to the concentration of CPC. This fluorescent probe method was developed for the determination of CPC in pharmaceutical preparations and human urine, which is simple, accurate, and high sensitive utilizing supramolecular complex formation with CB[7]. The linear range of the method was found to be from 0.019 to 3.556 μg/mL with the detection limit of 0.007 μg/mL.

A method for the determination of CS in soil samples was developed by Idkowiak et al. (2017). Five different CS were selected such as BAC, 1-dodecyl-3-methylimidazolium bromide, didecyldimethylammonium bromide, trihexyl(tetradecyl)phosphonium bromide, and trihexyl(tetradecyl)phosphonium chloride. The developed method was used for analysis of soil samples with methanol. The samples were subjected to analysis as disulphine blue active substances spectrophotometrically. The limits of detection for this method ranged from 2 to 27 μg/g.

Hassouna et al. (2017) developed and validated a simple, specific, precise, accurate, and stability-indicating RP-HPLC method for the determination of azelastine hydrochloride (AZH) and BAC in eye drops formulations. This method was performed on the Thermo CPS CN column (150×4.6 mm, 5 μm particle size) using buffer solution of pH 4.5 containing 50 mM potassium dihydrogen phosphate and 5.7 mM hexane sulfonate:acetonitrile (50:50 v/v) as the mobile phase at a flow rate of 1.5 mL/min, UV detection at 212 nm, and run time of 7.5 min. This method was validated on the basis of accuracy, precision, selectivity, LOD, LOQ, robustness, ruggedness, linearity and range. A linear relationship was obtained in the ranges of 6.25–50 μg/mL and 5.0–50 μg/mL for AZH and BAC, respectively.

4.2.1.2 Degradation/removal

Rao and Dube (1996) studied photocatalytic degradation of binary and ternary mixtures of three surfactants, viz., dodecylbenzene sulfonic acid sodium salt (DBS), cetylpyridinium chloride (CPC), and triton X-100 (TX-100). TiO_2 powder and Degussa P-25 TiO_2 were used as photocatalysts. These studies were also employed for degradation of some commercial soap/detergent formulations. It was observed that surfactants DBS, CPC, and TX-100 were adsorbed on both P-25 TiO_2 and TiO_2 using their ionic/polar functional groups to about 25%–45% of their initial concentration of these surfactants in dark. Photodegraded surfactant solutions showed gradual disappearance base on UV-Vis spectra, due to rupture of aromatic rings of DBS, CPC, and TX-100.

The photocatalytic degradation of cetylpyridinium chloride (CPC) was also investigated by Singhal et al. (1997) over titanium dioxide photocatalyst. The effects of various parameters like pH, concentration of surfactant, amount of semiconductor, particle size, light intensity, etc., were observed on rate of removal. This photocatalytic degradation followed pseudofirst order kinetics.

Tada et al. (2000) studied the effect of SiO_x monolayer coverage on the rate of TiO_2 photocatalytic oxidation of CPB in aqueous solutions. It was found that rate of CPB removal from the solution ($5 < pH < 7$) increases with the surface modification at concentrations below 4.5×10^{-4} M. It was also observed that its promoting effect was enhanced with the decreasing concentration. The acceleration of the reaction with the SiO_x monolayer coverage has been attributed to the increase in the rate of adsorption due to the electrostatic attraction of cetylpyridinium ion; however, suppression of Br^- adsorption has also been suggested as a minor contribution.

Eng et al. (2006) opined that high oxidation power of ferrate(VI) ($Fe^{VI}O_4^{2-}$) can be utilized in developing cleaner or greener technology for removal of organic contaminants. The unique property of ferrate(VI) was to degrade cetylpyridinium chloride (CPC) almost completely. It was reported that ferrate(VI) oxidizes CPC within a minute and molar consumption of ferrate(VI) was almost equal to the oxidized CPC. A decrease in total organic carbon (TOC) from CPC was found to be more than 95% at pH 9.2, suggesting that CPC has mineralized to carbon dioxide. Other product of the oxidation was detected as ammonium ion.

Bassey and Grigson (2010) were able to degrade a quaternary ammonium surfactant benzyldimethylhexadecylammonium chloride (BDHAC) in a minimal salt medium using two strains of bacteria, isolated from marine sediments. These bacteria were determined to be *Bacillus niabensis* and *Thalassospira* sp., and these were capable of degrading BDHAC when it was present in the range of 2–4 mg/mL. 90% BDHAC was degraded within 7 days, but limited growth of the strain was observed at 2 and 4 mg/mL BDHAC. The presence of a potential metabolite *N,N*-dimethylbenzylamine suggests that the cleavage of the C-alkyl-N bond is a major step in catabolism of BDHAC.

Duman and Ayranci (2010) used activated carbon cloth (ACC) as adsorbent for the removal of eight CS and these are: benzyltrimethylammonium chloride (BTMACl), benzyldimethyldecylammonium chloride (BDMDACl), benzyltriethylammonium chloride (BTEACl), benzyldimethylhexadecylammonium chloride (BDMHDACl), benzyltributylammonium chloride (BTBACl), benzyldimethyltetradecyl ammonium chloride (BDMTDACl), *N*-dodecylpyridinium chloride (*N*-DPCl), and cetylpyridinium chloride from their aqueous solutions. Adsorption process was monitored via in situ UV spectroscopic technique. The best fit for degradation was found with the pseudosecond order model, and isotherm data were analyzed with Freundlich and the Langmuir models, but Freundlich model fitted well with experimental data than the Langmuir model. It was also discussed in terms of the pH, the nature of CS (e.g., functional groups, size, and hydrophobicity) and the nature of the ACC (e.g., surface charge and pore size).

Ismail et al. (2010) evaluated sorptive behavior of four quaternary ammonium compounds (QACs): hexadecyl trimethyl ammonium chloride (C_{16}TMA), dodecyl trimethyl ammonium chloride (C_{12}TMA), hexadecyl benzyl dimethyl ammonium chloride (C_{16}BDMA), and dodecyl benzyl dimethyl ammonium chloride (C_{12}BDMA) in municipal primary, waste activated, mesophilic digested, and thermophilic digested sludges. An equilibrium was reached within 4 h. At the QAC concentration of 300 mg/L and a sludge volatile solids concentration of 1 g/L, the extent of adsorption was found to be 13%, 88%, 67%, and 89% for the C_{12}TMA, C_{16}TMA, C_{12}BDMA, and C_{16}BDMA, respectively. About 40% of the sludge-C_{12}TMA desorbed, whereas less than 5% of the sludge-C_{16}BDMA was desorbed in 10 days at pH 7. However, effect of pH was almost negligible on the desorption extent of C_{12}TMA at a pH range 4–10 even over 10 days. While more than 50% desorption of C_{16}BDMA was observed on increasing the solution pH to 10.

Ren et al. (2011) investigated performance of activated sludge in the removal of tetradecyl benzyl dimethyl ammonium chloride (C_{14}BDMA) by adsorption from aqueous solution. The effect of different pH, contact time, ionic strength, and temperature was observed on adsorption. It was observed that equilibrium was attained within 2 h, and adsorption capacity increases on increasing solution pH, reaching a constant value above pH 9. The ionic strength was found to show negative effect on removal of C_{14}BDMA. Kinetics data were best described with the pseudosecond order model. It was indicated that adsorption of C_{14}BDMA onto activated sludge was a feasible, spontaneous, and exothermic process in nature in the temperature range of 15°C–35°C.

Sonochemical degradation of a cationic surfactant, laurylpyridinium chloride (LPC) was studied in water by Singla et al. (2011) at concentrations of 0.1–0.6 mM, much below its critical micelle concentration 15 mM. A broad range of decomposition products, hydrocarbon gases, and water-soluble species are produced as evident from GC, EMI, and HPLC, where propionamide and acetamide were identified as two major degradation intermediates. These are probably formed as the result of the opening of the pyridinium ring by the attack of radical.

Bae et al. (2012) investigated adsorption of cetylpyridinium chloride on pyrite surface in its suspension. It was reported that maximum adsorption capacity of pyrite for CPC was 357 mmol/kg in pyrite suspension (4 g/L) at pH 7 equilibrated with CPC (0.1–1.8 mmol/L). Three different isoelectric points are at pH 7.0, 8.0, and 9.0. It was found that adsorption isotherms in suspension at pH 5 and 7 with NaCl (0.01 and 0.1 M) showed a decreasing pattern in CPC adsorption capacity.

Eng et al. (2012) carried out photocatalytic oxidation of anionic, dodecylbenzenesulfonate (DBS), and cationic, cetylpyridinium chloride (CPC) surfactants using a swirl-flow monolithic reactor. The kinetics of degradation in aqueous TiO_2 suspensions was investigated by varying catalyst loading, light intensity, initial surfactant concentration, dissolved oxygen concentration, and pH. It was reported that optimal catalyst required for degradation of DBS and CPC was 0.75 and 1.5 g/L, respectively, at neutral pH. The photocatalytic oxidation of DBS and CPC showed decrease in rate on increasing pH. It was found that removal of DBS was more efficient in acidic condition, while CPC showed efficacy in basic pH range. A decrease in TOC levels of surfactant during oxidation was found to be more effective at a neutral pH as compared to acidic or basic pH.

López et al. (2012) presented photocatalytic degradation of BAC, which is a bactericide widely used in pharmaceutical and cosmetic formulations, and it is risky for the biota. BAC is rapidly degraded but its mineralization is relatively low, which indicates that intermediate compounds formed are more resistant to the photocatalytic treatment. However, a combination of this with a biological treatment improved the extent of treatment. It was revealed that photocatalysis as a pretreatment improved the biodegradability of wastewaters. It was also observed that once the level of chemical oxygen demand (COD) was reduced, then photocatalysis can be used for the total elimination of BKC from the solution.

Moradidoost et al. (2012) investigated photocatalytic degradation of cetylpyridinium chloride in aqueous phase using various semiconductors such as TiO_2, ZnO, and SnO_2. The effect of different conditions including the amount of photocatalyst, pH, initial concentration, and presence of anions on rate of degradation was studied. It was indicated that maximum degradation (87%) of surfactant was achieved when ZnO was used as catalyst. It was revealed that efficiency was enhanced on increasing amount of photocatalyst, but it decreases with the increase in the initial concentration of cetylpyridinium chloride. Photodegradation rate of CPC was found to obey pseudofirst order kinetics represented by the Langmuir-Hinshelwood model.

Ferrate(VI) is commonly used as an efficient multifunctional water treatment reagent because of its interesting properties, such as strong oxidation, absorption, flocculation, disinfection, and deodorization. Yang et al. (2013) studied removal of cationic surfactant CPB on ferrate (K_2FeO_4). The influence of different operating variables on the mineralization efficiency was observed as a function of ferrate dosage, initial pH, and reaction time to achieve optimum conditions. Under optimal treatment conditions, $pH > 5$ and ferrate dosage = 1.5 times, CPB could be degraded in 5 min. Removal efficiency of CPB above 99% and TOC removal of about 91.3% could be achieved in a minute.

Adsorption of cetylpyridinium chloride (CPC) was investigated by Obeid et al. (2014) using magnetic alginate beads (MagAlgbeads). They proposed magnetic adsorbent (magsorbent) by encapsulation of magnetic functionalized nanoparticles in an alginate gel. The influence of several parameters such as contact time, pH, and initial surfactant concentration was studied on CPC adsorption. It was revealed that adsorption was accompanied by shrinking of the beads corresponding to 45% reduction of the volume. It was indicated that equilibrium time was strongly dependent on the concentration of surfactant. Adsorption also depends on the pH solution as pH is likely to affect the ionization state of adsorption sites. The maximum adsorption was obtained in a large pH range (3.2–12.0) It was suggested that magnetic alginate beads could be used as a new efficient magsorbent for hydrophobic pollutants due to the formation of micelle-like surfactants that aggregate in them.

Ourari et al. (2014) investigated removal of CPB from aqueous solutions using a mineral adsorbent called Maghnite (a bentonite). It was observed that CPB is adsorbed on Maghnite with an adsorption capacity of 0.438 mmol/g. It was indicated that CPB is efficient for the deinking of newspaper solutions, particularly when the natural Maghnite is associated to CPB. CPB shows a double action here; one is to protect and second is to optimize aquatic environment.

Photocatalytic degradation of detergent like cetylpyridinium chloride was studied by Asthana et al. (2014) in a batch process using zinc oxide nanoparticles. The effect of operational parameters such as UV light, pH, catalyst loading, and oxidants was studied on degradation. It was demonstrated that CPC is rapidly degraded in aqueous solution within relatively short time of about 60 min, and optimum conditions were observed at pH = 8.0, ZnO NPs = 40 mg/100 mL, [CPC] = 9.0×10^{-5} M, and [H_2O_2] = 8×10^{-5} M. A synergistic effect was observed on the addition of oxidants like H_2O_2 and $K_2S_2O_8$ into illuminated ZnO NPs suspension, leading to an enhancement of the process; however, excess amount of H_2O_2 and $K_2S_2O_8$ resulted in an adverse effect on rate of reaction.

Bergero and Lucchesi (2015) encapsulated *Pseudomonas putida* A (ATCC 12633) in Ca-alginate and investigated their ability to degrade the quaternary ammonium compounds (QACs) such as tetradecylbenzyldimethylammonium chloride (C_{14}BDMA), hexadecylbenzyldimethylammonium chloride (C_{16}BDMA), and BAC (a mixture 1:1 of C_{14}BDMA and C_{16}BDMA). This degradation was carried out at pH 7.4 buffered medium at 30°C keeping agitation at 100 rpm. It was observed that incubation immobilized cells degraded 90% of 35–315 mg/L of C_{14}BDMA, C_{16}BDMA, and BAC in 24 h while free cells were unable to degrade at concentrations above 35 mg/L. It was found that beads could be reused up to four cycles without any significant change in the degradation efficiency.

Ibrahim et al. (2016) carried out batch adsorption experiments for the removal of cetyldimethylbenzylammonium chloride (BAC) from aqueous solutions using mesoporous silica as adsorbent. The silica was prepared using sodium silicate as a precursor, and CTAB was used as template using sol–gel method. The effects of operating variables on the efficiency of the process were investigated such as dosage of mesoporous silica (0.05–0.1, 0.2, and 0.3 g) and temperature (298K–328K). The equilibrium data fitted well to Freundlich isotherm model. It was indicated that reaction was better represented by pseudosecond order and the intraparticle diffusion model, which suggests that the adsorption process proceeds by surface sorption and intraparticle diffusion. It was observed that these systems were spontaneous and exothermic in nature.

The removal of CPC was investigated in water solutions by Flilissa et al. (2016) using electrocoagulation, phosphate-assisted electrocoagulation, and adsorption on electrogenerated adsorbents. Electrocoagulation studies were carried out with aluminum electrodes in CPC synthetic solutions. After 2 h of electrolysis in 0.1 M NaCl solutions, it was observed that CPC was mainly removed by electroreduction upto 28% and 24% for starting concentrations of CPC at 0.5 and 1.0 mM, respectively. It was reported that a change of the cathode from aluminum to carbon or steel showed no reasonable change in the removal efficiency of electrolysis in 0.1 M NaCl solution. However, after 2 h, its electrolysis in 0.1 M NaCl in the presence of 0.1 M phosphate buffer, 80% removal was achieved for 0.5 or 1 mM CPC solutions.

A combination of UV irradiation and chlorine (UV/chlorine), a newly interested advanced oxidation process, was used by Huang et al. (2017) to degrade dodecylbenzyldimethylammonium chloride (DDBAC). It was observed that UV/chlorine showed synergistic effects on DDBAC degradation compared to UV irradiation or chlorination alone. Radical quenching experiments indicated that this degradation involved both UV photolysis and radical species oxidation, accounting about 48.4% and 51.6%, respectively. It was revealed that chlorine dosage and pH were main factors affecting the efficiency of UV/chlorine. The degradation of DDBAC decreased from 81.4% to 56.6% at pH 3.6–9.5 in. 12 min. They concluded that complete detoxification of DDBAC solution could be achieved by UV/chlorine treatment for 120 min, which is relatively more effective than UV irradiation or chlorination alone.

Advanced oxidation of a mixture of surfactants namely BACs, benzyl dimethyl dodecyl ammonium chloride (BDDA), and benzyl dimethyl tetradecyl ammonium chloride (BDTA) was investigated by Khan et al. (2017) using O_3 and H_2O_2 at elevated pH. It was reported that about 1.28 g/h O_3 and 200 mg/L of H_2O_2 degraded 90% of the initial BACs (BDDA:BDTA at 50:25 mg/L) at pH 11 within 30 min. They also assessed toxicity of BACs in the treated water using two fresh water algae species (*Chlorella vulgaris* and *Chlamydomonas reinhardtii*), which were completely removed by O_3/H_2O_2 oxidation. It was reported that AOP improved the biodegradability of BACs as evident from the ratio of the 5-day biological oxygen demand (BOD_5) to COD.

The treatment of a disinfectant wastewater containing BAC was investigated by Hong et al. (2017) by $S_2O_8^{2-}/Fe^{2+}$ process. A microbial fuel cell (MFC) system was applied as biosensor for the toxicity test of degradation products. The optimal condition was achieved for the degradation of BAC, pH = 5.0 $[S_2O_8^{2-}]/[Fe^{2+}] = 1:2$, where the removal rate of BAC reaches about 91.45%, and the COD removal efficiency reaches to 52.46% within 60 min. It was observed that voltage output and maximal power density were increased from 25.42 to 37.80 mV and 1.01 to 12.13 mW/m^2, respectively, when 100 mg/L BAC was treated by $S_2O_8^{2-}/Fe^{2+}$ oxidation process.

Xu et al. (2017) investigated gamma degradation of toxic nonoxidizing biocide dodecyl dimethyl benzyl ammonium chloride (DDBAC). 70%–100% of degradation was achieved depending on the initial concentration and the absorbed dose; however, dissolved organic carbon was removed only by 10%–33%. DDBAC degradation rate constant ratios of $^{\cdot}OH$, $H^{\cdot}$, and e_{aq}^{-} was calculated to be 7.4:1.4:1. It was reported that acute toxicity of 10 mg/L DDBAC was removed by 60% at absorbed doses of 0.5–3.0 kGy. It was proved by them that gamma irradiation was found effective in removal of DDBAC and its toxicity.

Adsorption of these cationic surfactants (QACs) such as Br-tetradecyltrimethylammonium (TTAB), Cl-tetradecylbenzyldimethylammonium (C_{14}BDMA), and Cl-hexadecylbenzyldimethylammonium (C_{16}BDMA) to activated sludge was investigated by Bergero and Lucchesi (2018) in wastewater treatment plant. When adsorption equilibrium was reached after 2 h, it was revealed that at initial concentration of 200 mg/L about 81%, 90%, and 98% of TTAB, C_{14}BDMA, and C_{16}BDMA were adsorbed, respectively. After every six successive desorption cycles, about 21% and 12.7% of TTAB and C_{14}BDMA were desorbed from the sludge. It was found that the more is the hydrophobic compound, the lesser will be its extent of desorption. In wastewater samples with supplementing activated sludge from wastewater with TTAB 200 mg/L and Ca-alginate beads containing the QACs-degrading microorganisms *Pseudomonas putida* A (ATCC 12633) and *Aeromonas hydrophila* MFB03, then 10 mg/L of TTAB was detected in the liquid phase and, 6–8 mg/L was adsorbed to the sludge after 24 h. Total TTAB amount (phase solid and liquid) did not change without Ca-alginate beads or with empty beads.

Dodecyltrimethylammonium chloride (DTAC) is a quaternary ammonium compound (QAC) that pollutes the environment. A synergistic effect with UV irradiation and persulfate (UV/PS) was observed by Lee et al. (2018) to degrade it. The removal of DTAC was about 91% with the PS dosage of 75.6 μM (UV/PS) with UV (870 mJ/cm^2). The contributions of HO and $SO^{\cdot}_4{}^-$ in DTAC degradation were found to be about 30% and 62% at pH 7, respectively. It was reported that $SO_4{}^-$ and $^{\cdot}OH$ were not influenced significantly in acidic medium (pH 3– 7), but they were affected in basic medium (pH 7– 11). It was observed that wastewater matrices of $HCO_3{}^-$, Cl^-, and humic acid inhibited the DTAC elimination while there was no significant impact on its elimination in presence of $NO_3{}^-$ and $SO_4{}^{2-}$. The rate constant in the reverse osmosis influent (ROI) and reverse osmosis concentrate (ROC) was observed to be 0.04–0.1 min^{-1} and 0.02–0.05 min^{-1}, respectively as the PS dosage was increased from 18.9 to 113.4 μM. It was also revealed that inhibitive effects of matrix in ROI and ROC were 70% and 81%, respectively.

4.2.2 Nonionic surfactants

These are characterized by the presence of hydrophilic head groups that do not ionize appreciably in water. These include polyoxyethylenated alkylphenols, alcohol ethoxylates, alkylphenol ethoxylates, alkanolamides, etc. Nonionic surfactants are next to anionic surfactant with around 45% of the overall industrial production. Most of these nonionic surfactants are either in liquid or slurry form, and their solubility decrease in water with the increasing temperature. Nonionic surfactants have some different physicochemical properties as compared to anionic or CS due to their structural features. Here hydrophilic groups are divided into four different categories such as polyethylene glycol, polyhydric alcohols, polyether type, and glycosidic type. Nonionic surfactants have covalently bonded oxygen-containing hydrophilic groups, which are bonded to hydrophobic parent structures. The water solubility of the oxygen groups is the result of hydrogen bonding. Hydrogen bonding decreases with the increasing temperature, and therefore water solubility of nonionic surfactants decreases at higher temperatures. These are less sensitive to water hardness as compared to anionic surfactants, and they foam less strongly. Even some nonionic surfactants can be nonfoaming or low foaming. This property makes them a good choice as an ingredient in low-foaming detergents. These surfactants are widely used in the textile, food, paper, medicines, plastic, fiber, pesticides, glass, dyes, industries, etc. Comparatively they have a better performance than ionic surfactants.

Tweens, triton, ethoxylate alcohols, alkylphenol ethoxylates (APEs), fatty alcohol ethoxylates, polyethylene glycol dodecyl ethers, spans, polyethylene oxides, etc. are some good examples.

4.2.2.1 Detection/determination

Petrović and Barceló (2000) developed reversed-phase liquid chromatography and mass spectrometric method for the simultaneous determination of anionic and nonionic surfactants. They were able to determine less polar compounds: alcohol ethoxylates (AEOs), nonylphenol ethoxylates (NPEOs), coconut diethanol amides, poly(ethylene glycol)s, and phthalate esters by this technique in positive ionization mode, while more polar compounds such as nonylphenolcarboxylates, nonylphenol (NP), octylphenol, and bisphenol A were analyzed by ion-pair LC-ESI-MS under negative ionization conditions. It was successfully applied to the trace determination of anionic and nonionic surfactants in sewage sludge collected from different sewage treatment plants. It was observed that polyethoxylates (AEOs and NPEOs) can be found in all samples at ppm levels (in the range of 10 − 190 mg/kg AEOs and 2 − 135 mg/kg NPEOs, respectively).

The development of a quantitative method for simultaneous determination of three surfactants, amphoteric (cocoamphoacetate, CAA), anionic (sodium laureth sulfate, SLES), and nonionic (alcohol ethoxylate, AE) using a reversed-phase C18 HPLC coupled with an ESI ion-trap mass spectrometer (MS) was reported by Levine et al. (2005) The method enables rapid determinations in small sample volumes containing inorganic salts (up to 3.5 g/L) and multiple classes of surfactants with high specificity by applying surfactant specific tandem mass spectrometric strategies. It was reported that dynamic linear ranges of 2–60, 1.5–40, and 0.8–56 mg/L (10 μL injection) were observed for CAA, SLES, and AE, respectively.

Dodecanol ethoxylates ($C_{12}EO_x$) represent Alcohol ethoxylates (AEs) from both; renewable and petrochemical sources. Zembrzuska et al. (2016) developed LC-MS/MS method for AEs monitoring of water from the river Warta in Poznan (Poland). The concentration of $C_{12}EO_x$ having 2–9 oxyethylene subunits was determined. The proposed method provides LOD of between 1 and 9 ng/L, and it is suitable for the monitoring of nonionic surfactants fingerprints in river water. The lowest determined concentration was 17 ± 1 ng/L, while the highest was 2.6 ± 0.14 μg/L. The total concentration of $C_{12}EO_2$–$C_{12}EO_9$ homologs varied between 1.4 and 11.2 μg/L.

4.2.2.2 Degradation/removal

Ang and Abdul (1992) investigated biodegradation of a nonionic alcohol ethoxylate surfactant by native microbes from a contaminated site. They carried out three sets of experiments consisting of 13 microcosms to evaluate the rate of biodegradation and also effect of nutrients and supplementary oxygen on the process of degradation. It was revealed that initial concentrations (1000, 650, 250, and 180 mg/L) in the presence of ground water, sterilized soil, and surfactant solutions were reduced to less than 5 mg/L in 36, 20, 17, and 17 days, respectively. It was reported that biodegradation rate in microcosms with added nutrients was even more than twice the rate in the reactor without nutrients. Various ratios of carbon, nitrogen, and phosphorus nutrients were tried, but it was observed that a ratio of carbon: nitrogen: phosphorus (10:2:1) was optimum for biodegradation of surfactant under the microcosm conditions. It was observed that addition of 5 mg/L of oxygen (in the form of hydrogen peroxide) increased the degradation rate of surfactant by about 30%.

The sonochemical degradation of aqueous solutions of triton X-100 was performed by Destaillats et al. (2000) using ultrasonic frequency of 358 kHz with an applied power of 50 W. It was revealed that hydrophobic alkyl chain was the preferential site for oxidation. Sonochemical degradation of the corresponding alkylphenols (e.g., tertoctylphenol) was carried out under the same conditions. It was revealed that there was a substantial increase in the rate constant for the degradation of triton X-100 below its CMC. It indicates that micelle formation helps in effectively isolating the free surfactant monomers from the water–air interface of the oscillating cavitation bubbles, thus resulting in a decrease in overall efficiency of the sonochemical process.

Adsorption of anionic surfactant [sodium dodecylbenzenesulfonate (SDBS)] and nonionic surfactant (an alcohol ethoxylates with 12 carbons and 9 oxyethyl groups, A12E9) mixtures was studied by Rao and He (2006). They observed effects of order of addition and mixing ratios of two surfactants, pH, and ionic strength on adsorption. It was found that saturated adsorption amount of SDBS and A12E9 on soils decreased, when A12E9 was added into soils first compared to that of second, possibly resulting from the screening of A12E9 to part adsorption sites on soils and also the hydrocarbon chain–chain interactions between SDBS and A12E9. The adsorption of SDBS on soils decreased with the increase of molar fraction of A12E9 in mixed surfactant solutions. Adsorption amount of SDBS and A12E9 on soils decreased on increasing pH of mixed surfactant solutions. The reduction of ionic strength in soils resulted in a decrease of adsorption amount of SDBS and A12E9 on soils.

Nine-mole nonylphenol ethoxylates (NPE9) containing wastewater was discharged to the highly anoxic environment of a 4500 L septic tank, before it is distributed to the oxic subsurface via 100 m of leach line. Periodic soil pore water and/or ground water samples were collected and analyzed for nonylphenol ethoxylates (NPEs), nonylphenol ether carboxylates, and nonylphenol after 180 days of injecting detergent to the septic system. It was observed by Huntsman et al. (2006) that NPE9 and degradation intermediates were found to be reduced by about 99.99%. Similarly 18% reduction in molar concentration within the septic tank was also observed. Further 96.7% reduction of molar concentration was reported within the leach lines. It was indicated that degradation of the surfactant occurred within the anoxic portion of the disposal system with a continued rapid biodegradation in the oxic unsaturated zone.

Karthikeyan and Ranjith (2007) carried out degradation of anionic and nonionic surfactants in aqueous solution by ozonation. They also studied the effect of ozonation and its operational variables such as ozone concentration, ozonation time, and ionic nature of surfactants. Two types of surfactants were taken; sodium dodecyl sulfate (SDS) as an anionic surfactant, and Tween 80 (T80) as a nonionic surfactant. The degradation was recorded in terms of their critical micille concentration (CMC) values, COD, and absorbance. It was observed that CMC value of the residual surfactant concentrations after ozonation was about 4 mg/L in the case of SDS while it was 1.25 mg/L for T80. The degradation efficiency was determined to be about 86.67% and 97.14% for SDS and T80 while residual concentrations were found to be 4 and 2 mg/L, respectively. A reduction in COD was measured as 73% and 77% for T80 and SDS, respectively.

Yoshikawa et al. (2008) carried out photocatalytic oxidative degradation of a nonionic surfactant, tween 80 (polyoxyethylene sorbitan monooleate) using TiO_2 photocatalyst. They used UV irradiation type slurry bubble column as a photoreactor. The rate constant of degradation was found to be proportional to surface area TiO_2 and the square root of UV light intensity; however, it was independent on the concentration of tween 80.

The sonolytic degradation of the nonionic surfactant, octaethylene glycol monododecyl ether was studied by Singla et al. (2009) at different initial concentrations, lower and more than its critical micelle concentration. It was reported that the degradation rate increases with increase in concentration of the surfactant till its CMC but above CMC, a saturation like behavior was observed, which suggests that the surfactant in its monomer form is better involved in the process of degradation. It was revealed that its degradation process involved two distinct primary processes occurring at the bubble/solution interface: (1) Hydroxylation/oxidation of the surfactant and (2) pyrolytic fragmentation of the surfactant. Hydroxylation of the ethoxy chain may produce various short-chain carboxyalkyl-polyethylene glycol intermediates. Polyethylene glycol chain was formed due to the scission of the C(12)E(8) molecule, which undergoes rapid hydroxylation/oxidation to yield simple organic components undergoing further degradation. However, it was also indicated that the sonolytic mineralization of the surfactant is little bit difficult to achieve at reasonable rates, may be due to the relatively low surface activity of the degradation products formed during sonolysis.

Esmaeilzadeh et al. (2011) have studied the adsorption of different types of surfactants, cationic (dodecyl trimethylammonium bromide, DTAB), anionic (sodium dodecyl sulfate, SDS), and nonanionic (lauryl alcohol-7 mole ethoxylate, LA7) on carbonate rock in presence of ZrO_2 spherical nanoparticles. ZrO_2 nanoparticles with tetrahedral structure (spherical in shape with size 17–19 nm) showed significant effect on adsorption of surfactants on the carbonate rock. It was reported that conductivity of DTAB in aqueous solution containing calcite powder decreases more as compared to other surfactants in presence of ZrO_2 nanoparticles.

The oxidative degradation of nonionic surfactants by the photoFenton process has been investigated by Ono et al. (2012). They also investigated kinetics of mixtures of nonionic surfactant and other type surfactants such as mixtures of nonionic and ionic surfactants, which are normally used to utilize their synergistic effects in many applications. Effects of operating parameters on the degradation of commercial nonionic surfactant Sannonic SS-90 (polyoxyethylene alkyl ether) were evaluated such as dosages of Fenton reagents (iron and hydrogen peroxide) and UV light intensity. It was observed that the increasing the dosages of the Fenton reagents increased the degradation rate but after a particular dose, further addition of the reagents does not enhance rate of degradation. It was revealed that the existence of the anionic surfactant such as sodium dodecylbenzenesulfonate inhibited the degradation of the nonionic surfactant due to the formation of complex with Fe ion, but the presence of cationic surfactant such as C_{12}TMA affected it insignificantly.

Arslan-Alaton et al. (2013) made use of sulfate radical based photochemical oxidation process to degrade octylphenol polyethoxylate (OPPE) Triton X-45. They investigated the effect of initial $S_2O^{2-}{}_8$ (0–5.0 mM) and OPPE (10–100 mg/L) concentrations on its removal along at pH 6.5. It was indicated that very rapid OPPE degradation (100%) was accompanied with high TOC abatement rates (90%) and it could be achieved for 10 and 20 mg/L aqueous OPPE at elevated $S_2O^{2-}{}_8$ concentrations ($\geq$2.5 mM). $S_2O^{2-}{}_8$/UV-C treatment was completely removed OPPE up to an initial concentration of 40 mg/L in the presence of 2.5 mM $S_2O^{2-}{}_8$. While TOC removal efficiencies was lowered down to only 40%. This was also compared with the H_2O_2/UV-C oxidation process, and found that performance was comparable for degradation and mineralization of OPPE. However, the pseudofirst order rate coefficient for OPPE degradation in $S_2O^{2-}{}_8$/UV-C oxidation (0.044 min^{-1}) was found to be significantly lower than that calculated for phenol (0.397 min^{-1}). It was also revealed that the selectivity of $SO_4^{-\cdot}$ was significantly higher than that of $^{\cdot}OH$ radical.

Gao et al. (2014) measured concentrations of nonylphenol ethoxylates (NPnEO, $n = 1$ to 2) and nonylphenol (NP) in water and sludge samples from sewage treatment plant (STP) with an anaerobic/oxic (A/O) and a biological aerated filter (BAF) process. It was observed A/O process exhibited significant improvement in performance as compared to the BAF process. It was found that mean values of NP, NP1EO, and NP2EO concentrations in influents from the STP were almost similar with the range of $1.8-2.0 \times 10^3$ ng/L. The removal efficiency of NP, NP1EO, and NP2EO from the aqueous phase was 78%, 84%, and 89%, respectively, while it was relatively lower for the BAF process, as 55%, 76%, and 79%, respectively. Their high concentrations were detected in the sludge samples with a maximum 2.7 μg/g d.w., indicating that such an improvement in their elimination may be due to adsorption by the sludge. They reported that the quantity of NP in the effluent from the oxic unit was increased by 32%, which indicates that NP1EO and NP2EO undergo degradation in the oxic conditions.

The efficiency of decomposition of nonionic surfactant was studied by Kos et al. (2014) using Fenton method in the presence of iron nanocompounds, and it was compared with the classical Fenton method. Water solutions of the surfactant tergitol TMN-10 were subjected to treatment by the classical Fenton method and in the presence of iron nanocompounds. They determined COD and total organic carbon (TOC) in samples of liquid solutions. The optimum conditions were achieved for this treatment, by varying the doses of iron and nanoiron, hydrogen peroxide, and pH of the solution. It was observed that iron oxide nanopowder was found to catalyze the process of detergent decomposition, increasing its efficiency and degree of mineralization. It was revealed that the efficiency of the decomposition of surfactant in the process with the use of iron nanocompounds was about 10%–30% higher than that in traditional method.

The removal and degradation of the nonionic surfactant linear alcohol ethoxylate (LAE) Genapol C-100 was evaluated by Motteran et al. (2014) in an anaerobic fluidized bed reactor. They observed high removal efficiencies of the COD (88%) and LAE (98%) even at high surfactant concentrations during the 492 days of operation. A higher removal of LAE (98%) was obtained for the 97.9 mg LAE L^{-1} influent in the absence of the cosubstrate. A greater diversity of volatile fatty acid was also observed. At the end of the reactor operation, 2.05 mg of LAE was found to be adsorbed in the biomass, and 98.5% was biodegraded.

TiO_2–WO_3 heterostructures were synthesized by Anandan and Wu (2014) at room temperature and ambient pressure, using a sonochemical approach and that too in short reaction time. As-prepared TiO_2–WO_3 heterostructures consisted of globular agglomerates (~250 nm in. diameter), which were composed of very small (<5 nm) dense particles (WO_3) dispersed inside the globules as evident from TEM and EDX. It was indicated by XRD data that there may be some phase transitions, occurring due to the formation of intimate bond between TiO_2 and WO_3. The catalytic activity of these TiO_2–WO_3 heterostructures was tested for the degradation of wastewater pollutant containing tergitol (NP-9) by its combination with ozonation, and two-fold degradation rate was observed as compared with ozone process alone.

Aqueous solutions of Brij 30 were photocatalytically and photochemically degraded by Kabdaşli et al. (2015) via TiO_2/UV-A, H_2O_2/UV-C, and PS/UV-C processes. They reported high rates of removals of Brij 30 during TiO_2/UV-A treatment (10 min), but a longer period of about 4–6 h was required to obtain significant total organic carbon (TOC) removals. A positive effect on treatment efficiencies was observed on increasing the TiO_2 dosage. An improvement in Brij 30 removal from 64% to 79% within 10 min was achieved at pH 3.0 on increasing the TiO_2 dosage from 1.0 to 1.5 g/L whereas 68% and 88% TOC removal was observed after 6 h. The removal of Brij 30 was very fast and complete with both H_2O_2/UV-C and PS/UV-C treatments, which was accompanied by a significant mineralization rates ranging between 74% and 80%.

Four different oxidation processes were attempted by da Silva et al. (2015) such as direct photolysis (DP) and three advanced oxidation processes (heterogeneous photocatalysis—HP, electrochemical oxidation—EO, and photoassisted electrochemical oxidation—PEO) for the treatment of wastewater containing nonylphenol ethoxylate (NPnEO). These processes were compared in terms of UV/

Vis spectrum, mineralization (total organic carbon), reaction kinetics, energy efficiency, and phytotoxicity. They used solution containing NPnEO as a surrogate of the degreasing wastewater. It was observed that photoassisted processes ware able to degrade the surfactant, and it produces biodegradable intermediates in the reaction. The best results could be achieved in terms of degradation, mineralization, reaction kinetics, and energy consumption, when PEO was carried out with a 250 W lamp and a current density of 10 mA/cm^2.

Menzies et al. (2017) studied biodegradation of four classes of high volume surfactants, such as nonionic alkyl ethoxylates (AE), and anionic alkyl ethoxysulfates (AES), alkyl sulfate (AS), and LAS. More than 97% loss of parent was reported along with CO_2 production (16%–94%) after incubation for 96–100 h in. raw domestic wastewater. It was revealed that the first order biodegradation rate of $C_{12}E_3$, $C_{14}E_3$, and $C_{16}E_3$ was affected by alkyl chain length with rates ranging from 6.8 to 0.49 h^{-1}. However, a comparison of $C_{14}E_1$, $C_{14}E_3$, and $C_{14}E_9$ indicated that the number of ethoxy units did not show any effect on biodegradation rate. It was observed that LAS did not exhibit first order decay kinetics and that its primary degradation was slow.

Aguilar et al. (2018) used nanoparticles of silver deposited on TiO_2 as catalysts for the removal of three organic molecules. They prepared catalysts by the photodeposit method. Photocatalysis is a known efficient treatment for the degradation of various organic compounds; however, its efficiency may be increased and it can be made visible-active on modifying the catalyst by the addition of silver. It was observed that the amount of the dopant does not affect band gap of catalyst, but it influences the surface area of the catalyst. The removal of acetaminophen and nonylphenol polyethylene glycol (NPG) was increased on decreasing the concentration of dopant. It was revealed that removal capacity also depends on molecule to be treated. It was also reported that removal percentages for the NPG were higher in presence of visible light as compared to UV radiation.

4.2.3 Anionic surfactants

These are surfactant that gives rise to a negatively charged surfactant ion (anionic) and a cation as counter ion, when dissolved in water. Sulfonic acid salts, alkylbenzene sulfonates, alcohol sulfates, phosphoric acid esters, carboxylic acid salts, etc. are good examples of anionic surfactants. These account for about half of the world production of surfactants. Anionic surfactants can be further divided into five peptide condensates depending upon their hydrophilic groups and these are: carboxylic acid salt type, sulfonate, sulfate salt type, phosphate ester, and fatty acid salts. Anionic surfactants are the earliest known surfactants. These are widely used as detergents, dispersants, foaming agents, stabilizers, emulsifiers, antistatic agents in most of the families and industries. Some commonly used anionic surfactants are sodium alkyl sulfate, sodium lauryl sulfate, Ammonium lauryl sulfate, sodium dodecyl sulfate, etc.

4.2.3.1 Detection/determination

Sablayrolles et al. (2009) developed a complete analytical method for LAS trace determination. They applied this method successfully to LAS (C10–C13) uptake in carrot plants. The developed methodology provides determination of LAS at low detection limits (5 g dry matter) for carrot sample (2 g dry matter) with reasonable good recoveries rate ($>90\%$). LAS is generally found in the carrot leaves, and its percentage transfer remains very low (0.02%).

A sensor for SDS has been developed by Abounassif et al. (2015). Here sensing membranes contain 1-dodecanthiol as an electroactive material (ionophore). As-prepared sensor showed a rapid, fixed, and nonNernstian behavior over a range of SDS (1×10^{-3} to 6.0×10^{-6} M) with detection limits of 5×10^{-6} M in pH range of 3.0–5.0. A direct determination of 100 μg/mL of SDS showed moderated recovery (98.1%) using this DS-PVC sensor. This DS-PVC membrane sensor could be successfully used for assay of SDS in mouth tooth wash.

A comparative study was made by Ramcharan and Bissessu (2016) for the quantification of LAS based on traditional liquid–liquid extraction (LLE) and HPLC–UV methods. In LLE, LAS formed complexes via ion association with methylene blue (MB)(AS–MB) complex. This complex was extracted with chloroform at λmax of 653 nm. Optimized conditions for quantification of a LAS peak using HPLC–UV were also achieved by isocratic elution on a C_{18} column with a 95% acetonitrile and 5% 0.7 M acetic acid mobile phase. It was reported that both these methods displayed percentage recoveries of more than 90%. LAS concentration in laundry samples was determined in the range of 116 and 454 mg/L

Sakac et al. (2017) prepared a new solid state sensor for potentiometric determination of surfactants with a layer of multiwalled carbon nanotubes. 1,3-Didecyl-2-methylimidazolium–tetraphenylborate ion-pair was used as a sensing material. It was reported that detection limits for SLS and dodecyl benzene sulfonate (DBS) were 2×10^{-7} and 3×10^{-7} M, respectively. This sensor was applied for the end-point determination in potentiometric titrations of anionic surfactants. The operational life-time of the proposed sensor was prolonged upto more than 3 months.

A new procedure has been described by Acevedo et al. (2018) for detection of milk adulterants. This method is based on simultaneous protein precipitation and liquid–liquid microextraction (LLME) of the ion-pair formed between the analytes and methylene blue. Sodium dodecyl sulfate (100 μL) was used as a model surfactant, along with 25 μg of methylene blue, and 2.9 mg of EDTA. A linear response was achieved from the range of 10–50 mg/L with detection limit (99.7% confidence level). This proposed method is a simple, rapid, cost-effective, and environmentally friendly approach for detection of milk adulterations by anionic surfactants.

A novel molecularly imprinted polymer (MIP) sensor was developed by Motia et al. (2018) for the electrochemical detection of sodium lauryl sulfate (SLS).

A screen-printed gold electrode (Au-SPE) based on MIP was prepared by an electrochemical polymerization of 2-aminothiophenol (2-ATP) in the presence of SLS as target molecule. They also constructed non-imprinted polymer (NIP) sensor omitting the SLS template for control. The selectivity of this proposed MIP sensor was observed using ethylenediaminetetraacetic acid (EDTA), tween 80, and urea, with satisfactory selectivity toward SLS. Different experimental parameters such as number of cycle for the electropolymerization, incubation time of 2-ATP, elution and SLS incubation time, were varied to achieve optimum condition to improve performance of the sensor. Proposed electrochemical sensor exhibited a logarithmic working range from 0.1 to 1000 pg/mL and a detection limit of 0.18 pg/mL under optimal conditions. This sensor has remarkable properties over others such as higher sensitivity and selectivity, good reproducibility, wider logarithmic range, lower detection limit, and long term stability. As-developed MIP sensor was used to determine SLS contents in environmental waters and also cosmetic products samples.

Haq et al. (2018) developed a high performance thin-layer chromatographic (HPTLC) process for estimation of SLS in bulk form and in marketed products. Silica gel was used as stationary phase. The mobile phase used in this study was a mixture of same environment-friendly solvents such as butanol, hexane, and glacial acetic acid in 7:2:1 ratio. This method was validated for various parameters like accuracy, linearity, precision, robustness, solution stability, recovery, and specificity. They also calculated LOD as well as LOQ which were found to be 31 and 92 ng per band, respectively. It was claimed that developed "green" HPTLC process is an appropriate, reproducible, and selective method for the quantification of SLS in SLS in commercially available toothpaste.

4.2.3.2 Degradation/removal

The photocatalytic degradation of sodium lauryl sulfate over TiO_2 has been carried out by Ameta et al. (2000) The effects of different parameters such as pH, concentration of surfactant, amount of semiconductor, effect of particle size, light intensity, etc. have been observed, and a tentative mechanism was proposed by them.

Adsorption characteristics of sodium dodecyl sulfate (SDS) were studied by Adak et al. (2005) on neutral alumina. Alumina was found to be good adsorbent for SDS and it could be used for the removal of anionic surfactant from wastewater, particularly when it is present in high concentrations. The SDS wastewater from actual laundary was treated by both; batch and continuous mode. Initial concentration of SDS in wastewater was 8068 ppm. It was reported that optimal adsorbent dose and equilibrium time were 120 g/L and 1 h, respectively, and under these conditions, the removal efficiency was found to be 94%. After use alumina could be efficiently regenerated by aqueous NaOH.

Photocatalytic degradation of anionic surfactant linear alkylbenzene sulfonates (LABS) was studied by Giahi et al. (2012) using ZnO nanoparticles (diameter

20 nm) in a batch process on irradiation with UV light. The effect of different experimental parameters such as surfactant concentration and pH was evaluated.

ZnO nanocatalyst was synthesized by Samadi et al. (2014) and they examined its use for UV-induced removal of sodium dodecyl sulfate from aquatic solutions. This anionic surfactant was selected because of its toxicity, wide use in industrial laundry, and as a major pollutant in municipal wastewater systems. The ZnO nanocrystals were synthesized using the precipitation method using $ZnSO_4 \cdot 7H_2O$ as a precursor and NaOH as precipitant. The effect of various operating conditions such as ZnO suspension concentration, initial surfactant concentration, and pH of solution were investigated. It was revealed that 98% of surfactant was removed in 40 min, and its removal efficiency was increased with increasing pH up to 9.0 and a decrease was obtained on further increase in pH.

The photocatalytic efficiency of titanium dioxide was observed by Malakootian et al. (2015) in removing anionic surfactant from the sewage water. The performance of photocatalysis for the removal of anionic surfactant was evaluated with operating variables such as, nanoTiO_2 (0.25–1.5 g/L), pH (5-7-9), surfactant concentration, and UV irradiation time. The removal of surfactant was monitored spectrophotometrically. The maximum removal efficiency could be achieved (95%) with synthetic solution under the optimal conditions such as pH = 5, nanoTiO_2 0.75 g/L, concentration of surfactant 100 mg/L and 60 min exposure time. It was revealed that removal process with real solutions (wastewater beverage chime) lowers down to 74% under the optimal conditions.

Malakootian et al. (2016) investigated efficiency of photoFenton process in removal of SDS from aqueous solutions. The effect of different operating parameters was investigated on removal such as pH, concentrations of H_2O_2 and Fe^{2+}, UV radiation, concentration of SDS, and contact time. Optimum conditions were found to be pH = 3, H_2O_2 = 0.5 mmol, Fe^{2+} = 40 mg/L, UV-C = 45 watt, initial concentration of SDS = 25 mg/L and contact time = 30 min. It was observed that maximum removal of 94.36% and 71% and be achieved for synthetic solutions and soft drink wastewater, respectively. It was also revealed that proposed method can remove SDS from real sample of water with efficiency of 71% within 30 min.

Nguyen et al. (2016) investigated degradation of sodium dodecyl sulfate by photoelectrochemical (PEC) and electrochemical in dark (EC-dark) processes, both. It was reported that only 1% SDS was removed by physical absorption on α-Fe_2O_3 surface. It was found that the PEC method was more efficient as compared to the EC-dark process. It was revealed that simple PEC process can remove the sulfate group in SDS completely and reduce about 90% of TOC. Residual organics contains hydroxyl and carboxylic groups, which are less toxic than SDS.

Wahyuni et al. (2016) have studied surfactant in a laundry wastewater using photodegradation under UV/TiO_2/H_2O_2 (photoFenton-like) and UV/Fe^{2+}/H_2O_2 (photoFenton) processes. Concentration of surfactant left in the wastewater was determined by UV/Visible spectrophotometry using methylene blue. It was observed that the photodegradation effectiveness depends on TiO_2 dose, Fe(II) and H_2O_2 concentrations, pH, and time. It was proved that UV/Fe^{2+}/H_2O_2 (photoFenton) process was found more effective in the photodegradation of surfactant than UV/TiO_2/H_2O_2 (photoFenton-like) process.

SDS is an anionic surfactant, which is quite commonly used all over the world, as an important foaming component of shampoos, toothpaste, and detergents. Adekanmbi and Usinola (2017) isolated bacteria capable of utilizing SDS from sediment and wastewater samples of a detergent manufacturing plant and laundry. They employed SDS utilizing bacteria in the degradation in a batch culture for 10 days on a rotary shaker at 150 rpm. Two of the selected eight SDS-degrading bacteria; *Staphylococcus aureus* WAW1 and *Bacillus cereus* WAW2 were selected for biodegradation setup based on their growth consistency. It was reported that *Staphylococcus aureus* WAW1 was able to degrade about 36.8% of SDS while *Bacillus cereus* WAW2 was able to degrade more SDS upto 51.4%.

Takayanagi et al. (2017) quantified contributions of four removal mechanisms of sodium dodecyl benzene sulfonate (SDBS), that is, reductive degradation, oxidative degradation, adsorption, and precipitation. The effective removal of SDBS by zero valent iron (ZVI) was attributed to the adsorption capability of iron oxides/hydroxides on ZVI surface at nearly neutral pH. The highest removal rate of SDBS and the maximum TOC (total organic carbon) removal were obtained at pH 6.0 and it was 77.8%. However, contributions of degradation, precipitation, and adsorption to TOC removal were only 4.6%, 14.9%, and 58.3%, respectively. It was also reported that in oxidative degradation by the Fenton reaction at pH 3.0. TOC removal was only 9.8% while contributions of degradation, precipitation, and adsorption to TOC removal were 2.3%, 4.6%, and 2.9%, respectively.

Nurdin et al. (2018) modified titanium dioxide photocatalyst with nickel and nitrogen using titanium tetraisopropoxide (TTIP) as a precursor via a microwave-assisted method. It was reported that Ni and N dopants led to lowering of TiO_2 band gap, so as make it efficient in visible light. The crystalline size of Ni–N–TiO_2 was 13.275 nm (anatase) as evident for XRD. It was observed that TiO_2 modified by Ni and N exhibited band gap of 1.95 eV. Photocatalytic degradation of sodium lauryl sulfate under visible irradiation was found to be 93.75%.

An ultrasonic bath with frequency of 130 kHz was used by Dehghani et al. (2019) to investigate the effects of different operational parameters on degradation of anionic surfactant such as sonication time, initial concentration, and power. LABS solutions were selected as a model system using methylene blue. Experiments were carried out with initial concentrations of 0.2, 0.5, 0.8, and 1.0 mg/L, frequency of 130 kHz, acoustic powers value of 400 and 500 W, and at pH 6.8–7.0. It was reported that LABS degradation rate was found to increase with increasing sonication time and power both, but it decreased on increasing concentration of LABS.

4.2.4 Amphoteric or zwitterionic surfactants

Wherever a single surfactant molecule show both properties, anionic and cationic, then it is called amphoteric or zwitterionic surfactant. Amphoteric surfactants can also be divided into different types such as imidazoline, betaine, lecithin, and amino acid types depending on type of anion. The toxicity of amphoteric surfactants is comparatively low. It is quite gentle to the skin and also accompanied by good biodegradability. Amphoteric surfactants have wide applications in the personal care products shampoos, shower gels, cosmetics, etc. and can also be used in industrial softeners and antistatic agents. Phospholipids, phosphatidylcholine, cocamidopropyl hydroxysultaine, cocamidopropyl betaine, phosphatidylserine, phosphatidylethanolamine, etc. are good examples of amphoteric surfactants.

4.2.4.1 Detection/determination

Wang and Zhou (2016) developed a rapid and sensitive ultrahigh performance liquid chromatography-tandem mass spectrometry (UHPLC–MS/MS) method to quantify dimethylaminopropylamine (DMAPA) and lauramidopropyldimethylamine (LAPDMA) in cosmetic products. It was reported that recoveries were observed at three spiking levels of low, medium, and high concentrations ranging from 98.4% to 112%.

Steuer et al. (2016) reported a method for the rapid and simultaneous measurement of closely related phosphatidylcholine-derived metabolites. Plasma, serum and urine samples were simply deproteinized and separated by hydrophilic interaction liquid chromatography (HILIC). Detection and quantification were performed using LC–MS/MS with electrospray ionization in positive mode.

Cocamidopropyl betaine (CAPB) is a zwitterionic surfactant which is synthesized using coconut oil and normally supplied in form of an aqueous solution with 25%–37% w/w. Gholami et al. (2018) developed a novel method based on UV–visible spectroscopy for determination of CAPB. Eriochrome black T (EBT) was used as a specific color indicator, where a red shift and color change were observed. CAPB was detected in commercial real samples under optimal condition.

4.2.4.2 Degradation/removal

CAPB (300 mg/L) degradation was observed by Merkova et al. (2018) with the application of two isolated strains. It was revealed that almost complete mineralization of the CAPB could be achieved during 4 days when mineral medium containing ammonium salt was used as a nitrogen source, but same process required more than 29 days of incubation in nitrogen free conditions. Degradation with and without the source of nitrogen showed that *Pseudomonas sp.* FV played an important role in degradation of CAPB, which is capable of utilizing the alkyl chains of the surfactant. The other strain used *Rhizobium sp.* FM was also able to degrade intermediates originating from the primary biodegradation stage.

4.3 Adverse effects of surfactants

4.3.1 On aquatic plants

The degree of damage by surfactants to aquatic plants depends on its concentration in wastewater. If the concentration of surfactants is too high in the water, it will affect the growth of algae and other microorganisms in water. It resulted in a decrease in primary productivity of water bodies. Thus the food chain of aquatic organisms in water bodies is undermined. Surfactants can cause acute poisoning, which can lead to increase in permeability of membrane, resulting into gradually disintegration of material exosmose and cell structure. As a result, contents of superoxide dismutase, catalase, peroxidase activity, and chlorophyll also decrease. Adverse effects of surfactants have been shown in Fig. 4.2. The accumulation of these surfactants is increasing day by day. Depending on relationship between chemical structure of surfactants and the toxicity of water to aquatic organisms can be summarized in these three points:

- the greater is the hydrophobicity (HLB value is smaller) of surfactants, the greater will be its aquatic toxicity,

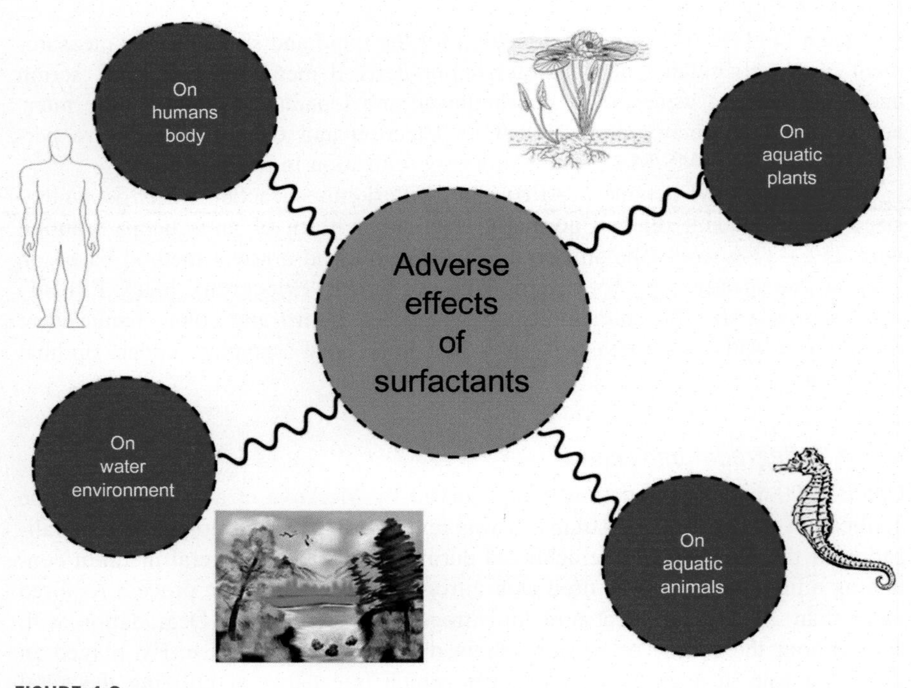

FIGURE 4.2

Environmental adverse effects of Surfactants.

- toxicity of aquatic organisms will be lower with more ethoxylate group, and
- the toxicity of anionic surfactants decreases as compared with nonionic surfactants.

4.3.2 On aquatic animals

A certain toxicity of these surfactants will also pass into the animal. It may be through either animal feeding or skin penetration way. When concentration of surfactant present in water is too high, then surfactants can enter different parts of body such as gills, blood, kidney, pancreas, gall bladder, and liver. Fish very easily absorbs surfactants by the surface of body and gills, then these are distributed to different body tissues and organs with the blood circulation. When a fish is exposed to surfactants, serum transaminases and alkaline acid phosphatase activity increase, producing adverse effects on fish. Contamination enters the body of fish through the food chain, and it produces inhibition to various enzymes in the human body resulting in loss of immunity of body.

4.3.3 On water environment

If wastewater containing surfactants is discharged into the environment, it can cause serious water pollution problems. If the concentration of the surfactant reaches to 0.1 mg/L, then water may give persistent foams. Majority of bubbles disappear easily in the water and form an insulating layer of foam. Such insulating layer resulted in making exchange between the water body and gas atmosphere weak, so as to reduce dissolved oxygen. Quite a large number of microorganisms are died because of hypoxia and water bodies are deteriorated. As surfactants' concentration is increased, but below its critical micelle concentration (CMC), surface tension decreases rapidly. But when the surfactant concentration exceeds CMC in the water, it can also increase the concentration of insoluble and soluble-water pollutants in the water. These surfactants dissolve materials, which have no original adsorption energy into layered material. Such a solubilization behavior can lead to pollution indirectly changing the properties of water adversely. These surfactants can also kill microorganisms present in the environment and thus inhibiting the material degradation of many toxic substances.

4.3.4 On humans body

The toxic effects of surfactants on the human body may be divided into their effects on the skin and within the body. Long term use of detergents or surfactants can cause skin irritation, and it may also lead to some degree of damage. When surfactant enters into the human body, they damage the activity of various enzymes, and as a result, normal physiological function of body is disrupted or adversely affected. Surfactants have some toxicity, and they may accumulate in the human body, and therefore, it will be difficult to degrade.

In general nonionic surfactants are not electrically charged and therefore, do not combine with various proteins. As a result they have minimal irritation to the skin while toxicity of CS is more, while toxicity of anionic surfactants is somewhere in between that of nonionic surfactants and CS. There are some reports that sodium dodecyl benzene sulfonate is absorbed through the skin. They may damage liver and sometimes cause other chronic symptoms. They may be teratogenic and carcinogenic also in certain cases.

Since most of the detergent contain a large amount of polyphosphate as net agent and therefore large amounts of phosphorus will be there in wastewater. It could easily lead to eutrophication. When the concentration of surfactants exceeds a certain limit of concentration in sewage treatment of wastewater, it affects aeration, sedimentation, sludge nitrification, and many other important or useful processes. It will increase the difficulty in wastewater treatment. The presence of surfactants will promote emulsification and dispersion in water-insoluble oil and polychlorinated organics, and hence, efficiency of pollutant treatment is also reduced.

4.4 Conclusion

The use of surfactant is increasing day by day because of its common utility in shampoos, detergent cakes, toothpastes, mouth washes, cleaning of textiles, and floor, etc. Out of all the four categories, CS are occupied about half the production of surfactants industries and little less than half is nonionic surfactants. Only a little share is of anionic and zwitterionic (amphoteric) surfactants. Their abundant use is going to create a havoc, because most of these are not biodegradable. It is emerging as a new face of water and soil pollution. Therefore, there is an urgent demand either to cut short the production of these surfactants (which is difficult to execute because of developing pressure of the society towards cleanliness) or to develop some newer surfactants, which are biodegradable. Till then, it is very important to detect the presence and determination of quantity of surfactants in wastewater or sludge. One should also find rapid, simple, and low cost methods for the degradation of these toxic contaminants. Otherwise the time is not far off, when the whole world will be in the cancerous grip of surfactant/detergent pollution.

References

Abounassif, M.A., Hefnawy, M.M., Al-Robian, H., Mostafa, G.A.E., 2015. Dodecanthiol as novel sensing material for potentiometric determination of sodium dodecyl sulphate anionic surfactant. Int. J. Electrochem. Sci. 10, 8668–8679.

Acevedo, M.S.M.S.F., Lima, M.J.A., Nascimento, C.F., Rocha, F.R.P., 2018. A green and cost-effective procedure for determination of anionic surfactants in milk with liquid-liquid microextraction and smartphone-based photometric detection. Microchem. J. 143, 259–263.

Adak, A., Bandyopadhyay, M., Pal, A., 2005. Removal of anionic surfactant from wastewater by alumina: a case study. Colloids Surf. A: Physicochem. Eng. Asp 254, 165–171.

Adekanmbi, A.O., Usinola, I.M., 2017. Biodegradation of sodium dodecyl sulphate (SDS) by two bacteria isolated from wastewater generated by a detergent manufacturing plant in Nigeria. Jordan J. Biol. Sci. 10, 251–255.

Aguilar, C.A., Montalvo, C., Zermeño, B.B., Cerón, R.M., Cerón, J.G., Anguebes, F., et al., 2018. Photocatalytic degradation of acetaminophen, tergitol and nonylphenol with catalysts TiO_2/Ag under UV and vis light. Int. J. Environ. Sci. Technol. 16 (2), 843–852.

Ali, T.A., Soliman, M.H., Mohamed, G.G., Farag, A.B., Samah, M.K., 2015. Development of a new modified screen-printed and carbon paste electrodes for selective determination of cetyltrimethylammonium bromide in different water samples. Int. J. Electrochem. Sci. 10, 3192–3206.

Ameta, S., Punjabi, P.B., Rao, P., Singhal, B., 2000. Use of titanium dioxide as photocatalyst for photodegradation of sodium lauryl sulfate. J. Indian Chem. Soc. 77 (3), 157–160.

Anandan, S., Wu, J.J., 2014. Ultrasound assisted synthesis of TiO_2-WO_3 heterostructures for the catalytic degradation of Tergitol (NP-9) in water. Ultrason. Sonochem. 21 (4), 1284–1288.

Ang, C.C., Abdul, A.S., 1992. A laboratory study of the biodegradation of an alcohol ethoxylate surfactant by native soil microbes. J. Hydrol. 138 (10), 191–209.

Arslan-Alaton, I., Olmez-Hanci, T., Genç, B., Dursun, D., 2013. Advanced oxidation of the commercial non-ionic surfactant octylphenol polyethoxylate Triton™ X-45 by the persulfate/UV-C process: effect of operating parameters and kinetic evaluation. Front. Chem. 1. Available from: https://doi.org/10.3389/fchem.2013.00004.

Asthana, S., Pal, S.K., Sharma, M., 2014. Solar light assisted nano ZnO photo catalytic mineralization – the green technique for the degradation of detergents. Int. J. Chem. Pharm. Anal. 1 (3), 141–147.

Bae, S., Mannan, M.B., Lee, W., 2012. Adsorption of cationic cetylpyridinium chloride on pyrite surface. J. Ind. Eng. Chem. 18 (4), 1482–1488.

Bassey, D.E., Grigson, S.J.W., 2010. Degradation of benzyldimethyl hexadecylammonium chloride by *Bacillus niabensis* and *Thalassospira sp.* isolated from marine sediments. Toxicol. Environ. Chem. 93 (1), 44–56.

Bergero, M.F., Lucchesi, G.I., 2015. Immobilization of *Pseudomonas putida* a (ATCC 12633) cells: a promising tool for effective degradation of quaternary ammonium compounds in industrial effluents. Int. Biodeterior. Biodegrad. 100, 38–43.

Bergero, M.F., Lucchesi, G.I., 2018. Degradation of cationic surfactants using immobilized bacteria: its effect on adsorption to activated sludge. J. Biotechnol. 272-273, 1–6.

da Silva, S.W., Klauck, C.R., Siqueira, M.A., Bernardes, A.M., 2015. Degradation of the commercial surfactant nonylphenol ethoxylate by advanced oxidation processes. J. Hazard. Mater. 282, 241–248.

Dehghani, M.H., Zarei, A., Yousefi, M., 2019. Efficiency of ultrasound for degradation of an anionic surfactant from water: surfactant determination using methylene blue active substances method. Methods X 6, 805–814.

Destaillats, H., Hung, H.M., Hoffmann, M.R., 2000. Degradation of alkylphenol ethoxylate surfactants in water with ultrasonic irradiation. Environ. Sci. Technol. 34 (2). Available from: https://doi.org/10.1021/es990384x.

Devi, S., Chattopadhyaya, M.C., 2011. A new PVC membrane ion selective electrode for determination of cationic surfactant in mouthwash. J. Surfactants Deterg. 15 (3), 387–391.

Díez, C., Feinberg, M., Spörri, A.S., Cognard, E., Ortelli, D., Edder, P., et al., 2015. Evaluation of quantification methods to compensate for matrix effects in the analysis of benzalkonium chloride and didecyldimethylammonium chloride in fruits and vegetables by LC-ESI-MS/MS. Food Anal. Methods 9 (2), 485–499.

Duman, O., Ayranci, E., 2010. Adsorptive removal of cationic surfactants from aqueous solutions onto high-area activated carbon cloth monitored by in situ UV spectroscopy. J. Hazard. Mater. 174 (1-3), 359–367.

Eng, Y.Y., Sharma, V.K., Ray, A.K., 2006. Ferrate(VI): green chemistry oxidant for degradation of cationic surfactant. Chemosphere 63 (10), 1785–1790.

Eng, Y.Y., Sharma, V.K., Ray, A.K., 2012. Degradation of anionic and cationic surfactants in a monolithic swirl-flow photoreactor. Separ. Purific. Technol 92, 43–49.

Esmaeilzadeh, P., Bahramian, A., Fakhroueian, Z., 2011. Adsorption of anionic, cationic and non-ionic surfactants on carbonate rock in presence of ZrO_2 nanoparticles. Physi. Proc 22, 63–67.

Flilissa, A., Méléard, P., Darchen, A., 2016. Cetylpyridinium removal using phosphate-assisted electrocoagulation, electroreduction and adsorption on electrogenerated sorbents. Chem. Eng. J. 284, 823–830.

Gao, D., Li, Z., Guan, J., Li, Y., Ren, N., 2014. Removal of surfactants nonylphenol ethoxylates from municipal sewage-comparison of an A/O process and biological aerated filters. Chemosphere 97, 130–134.

Gholami, A., Golestaneh, M., Andalib, Z., 2018. A new method for determination of cocamidopropyl betaine synthesized from coconut oil through spectral shift of Eriochrome Black T. Spectrochim. Acta Part A: Mol. and Biomol. Spectroscopy 192, 122–127.

Giahi, M., Habibi, S., Toutounchi, S., Khavei, M., 2012. Photocatalytic degradation of anionic surfactant using zinc oxide nanoparticles. Russ. J. Phys. Ch. 86, 689–693.

Haq, N., Siddiqui, N.A., Alam, P., Shakeel, F., Alanazi, F.K., Alsarra, I.A., 2018. Estimation of sodium lauryl sulphate concentration in marketed formulations by stability indicating 'green' planar chromatographic method. Chiang Mai J. Sci. 45 (3), 1531–1542.

Hassouna, Md.E.M., Abdelrahman, M.M., Mohamed, M.A., 2017. Simultaneous determination of azelastine hydrochloride and benzalkonium chloride by RP-HPLC method in their ophthalmic solution. J. Forensic Sci. Criminal Inves 1 (3). Available from: https://doi.org/10.19080/JFSCI.2017.01.555565.

Hong, J.-M., Xia, Y.-F., Zhang, Q., Chen, B.-Y., 2017. Oxidation of benzalkonium chloride in aqueous solution by $S_2O_8^{2-}/Fe^{2+}$ process: degradation pathway, and toxicity evaluation. J. Taiwan Inst. Chem. Eng. 78, 230–239.

Huang, F., Guo, X., Jia, L., Yang, R., 2014. A novel method for cetylpyridinium bromide determination in aqueous solution based on fluorescence quenching of dye. Anal. Methods 6 (5), 1435. Available from: https://doi.org/10.1039/c3ay41692e.

Huang, N., Wang, T., Wang, W.-L., Wu, Q.-Y., Li, A., Hu, H.-Y., 2017. UV/chlorine as an advanced oxidation process for the degradation of benzalkonium chloride: synergistic effect, transformation products and toxicity evaluation. Water Res. 114, 246–253.

Huntsman, B.E., Staples, C.A., Naylor, C.G., Williams, J.B., 2006. Treatability of nonylphenol ethoxylate surfactants in on-site wastewater disposal systems. Water Environ. Res. 78 (12), 2397–2404.

Ibrahim, Q., Ali, I.H., Kareem, S.H., 2016. Removal of cetyldimethylbenzylammonium chloride surfactant from aqueous solution by adsorption onto mesoporous silica. Int. J. Sci. Res. 5 (12), 1536–1542.

Idkowiak, J., Zgoła-Grześkowiak, A., Karbowska, B., Plackowski, R., Wyrwas, B., 2017. Determination of cationic surfactants in soil samples by the disulphine blue active substance (DBAS) procedure. J. Anal. Chem. 72 (7), 745–750.

Ismail, Z.Z., Tezel, U., Pavlostathis, S.G., 2010. Sorption of quaternary ammonium compounds to municipal sludge. Water Res. 44 (7), 2303–2313.

Jian-Hong, X., Lin, S., Li-Ming, D., Cai-Ping, C., Hao, W., Yin-Xia, C., 2016. Determination of cetylpyridinium chloride employing a sensitive fluorescent probe. Anal. Chem. Ind. J. 16 (14), 107–123.

Kabdaşli, I., Ecer, Ç., Olmez-Hanci, T., Tünay, O., 2015. A comparative study of $HO\bullet^-$ and $SO_4\bullet^-$-based AOPs for the degradation of non-ionic surfactant Brij 30. Water Sci Technol 72 (2), 194–202.

Karthikeyan, S., Ranjith, P., 2007. Degradation studies on anionic and non-ionic surfactants by ozonation. J. Indust. Poll. Cont. 23 (1), 37–42.

Khan, A.H., Kim, J., Sumarah, M., Macfie, S.M., Ray, M.B., 2017. Toxicity reduction and improved biodegradability of benzalkonium chlorides by ozone/hydrogen peroxide advanced oxidation process. Separ. Purific. Technol. 185, 72–82.

Kos, L., Michalska, K., Perkowski, J., 2014. Decomposition of non-ionic surfactant Tergitol TMN-10 by the Fenton process in the presence of iron oxide nanoparticles. Environ. Sci. Pollut. Res. 21 (21), 12223–12232.

Lee, M.-Y., Wang, W.-L., Xu, Z.-B., Ye, B., Wu, Q.-Y., Hu, H.-Y., 2018. The application of UV/PS oxidation for removal of a quaternary ammonium compound of dodecyl trimethyl ammonium chloride (DTAC): the kinetics and mechanism. Sci. Total Environ. 655, 1261–1269.

Levine, L.H., Garland, J.L., Johnson, J.V., 2005. Simultaneous quantification of polydispersed anionic, amphoteric and non-ionic surfactants in simulated wastewater samples using C18 high-performance liquid chromatography–quadrupole ion-trap mass spectrometry. J. Chromatogr. A 1062 (2), 217–225.

Li, N., Hao, X., Kang, B.H., Li, N.B., Luo, H.Q., 2015. Sensitive and selective turn-on fluorescence method for cetyltrimethylammonium bromide determination based on acridine orange-polystyrene sulfonate complex. Luminescence 31 (4), 1025–1030.

Li, S., Zhao, S., 2004. Spectrophotometric determination of cationic surfactants with benzothiaxolyldiazoaminoazobenzene. Analytica. Chimica. Acta. 501 (1), 99–102.

López, L.E., Fiol, P.S., Senn, A., Curutchet, G., Candal, R., Litter, M.I., 2012. TiO_2-photocatalytic treatment coupled with biological systems for the elimination of benzalkonium chloride in water. Separ. Purific. Technol. 91, 108–116.

Maduniccacic, D., Sakbosnar, M., Galovic, O., Sakac, N., Matesicpuac, R., 2008. Determination of cationic surfactants in pharmaceutical disinfectants using a new sensitive potentiometric sensor. Talanta 76 (2), 259–264.

Malakootian, M., Farzadeh, N.J., Dehdarirad, A., 2016. Efficiency investigation of photo-Fenton process in removal of sodium dodecyl sulphate from aqueous solutions. Desalin. Water Treat. 57, 24444–24449.

Malakootian, M., Yaghmaeian, K., Momenzadeh, R., 2015. Efficiencv of titanium dioxide photocatalytic activity in removing anionic surfactant of sodium dodecyl sulfate from waste water. Koomesh 16, 648–654.

Menzies, J.Z., McDonough, K., McAvoy, D., Federle, T.W., 2017. Biodegradation of nonionic and anionic surfactants in domestic wastewater under simulated sewer conditions. Biodegradation 28 (1), 1–14.

Merkova, M., Zalesak, M., Ringlova, E., Julinova, M., Ruzicka, J., 2018. Degradation of the surfactant cocamidopropyl betaine by two bacterial strains isolated from activated sludge. Int. Biodeterior. Biodegrad. 127, 236–240.

Mohammad, A., Mobin, R., 2015. Eco-favourable mobile phase in thin layer chromatographic analysis of surfactants: resolution of coexisting alkyl dimethylbenzyl ammonium chloride, cetyltrimethyl ammonium bromide and Triton X 100. Tenside Surfactants Deterg. 52 (5), 414–423.

Moradidoost, A., Giahi, M., Bagherinia, M.A., 2012. Photochemical studies on degradation of cetyl pyridinium chloride (cationic surfactant) in aqueous phase using different photocatalysts. J. Phys. Theor. Chem. IAU Iran 8 (4), 311–316.

Motia, S., Tudor, I.A., Popescu, L.M., Piticescu, R.M., Bouchikhi, B., Bari, N.E., 2018. Development of a novel electrochemical sensor based on electropolymerized molecularly imprinted polymer for selective detection of sodium lauryl sulfate in environmental waters and cosmetic products. J. Electroanal. Chem. 823, 553–562.

Nguyen, H.M., Phan, C.M., Sen, T., 2016. Degradation of sodium dodecyl sulfate by photoelectrochemical and electrochemical processes. Chem. Eng. J. 287, 633–639.

Nurdin, M., Ramadhan, L.O.A.N., Darmawati, D., Maulidiyah, M., Wibowo, D., 2018. Synthesis of Ni, N co-doped TiO_2 using microwave-assisted method for sodium lauryl sulfate degradation by photocatalyst. J. Coat. Technol. Rese. 15, 395–402.

Obeid, L., El Kolli, N., Dali, N., Talbot, D., Abramson, S., Welschbillig, M., et al., 2014. Adsorption of a cationic surfactant by a magsorbent based on magnetic alginate beads. J. Colloid Interface Sci. 432, 182–189.

Ono, E., Tokumura, M., Kawase, Y., 2012. Photo-Fenton degradation of non-ionic surfactant and its mixture with cationic or anionic surfactant. J. Environ. Sci. Health A Tox. Hazard Subst. Environ. Eng. 47 (8), 1087–1095.

Ourari, A., Flilissa, A., Boutahala, M., Ilikti, H., 2014. Removal of cetylpyridinium bromide by adsorption onto maghnite: application to paper deinking. J. Surfactant Deterg. 17 (4), 785–793.

Parham, H., Pourreza, N., Moradi, D., 2011. Development of a flotation-spectrophotometric method for determination of cetylpyridinium chloride in pharmaceutical products. Química Nova 34 (5), 884–887.

Petrović, M., Barceló, D., 2000. Determination of anionic and non-ionic surfactants, their degradation products, and endocrine-disrupting compounds in sewage sludge by liquid chromatography/mass spectrometry. Anal. Chem. 72 (19), 4560–4567.

Ramcharan, T., Bissessu, A., 2016. Analysis of linear alkylbenzene sulfonate in laundry wastewater by HPLC–UV and UV–Vis Spectrophotometry. J. Surfactant. Deterg. 19, 209–218.

Rao, P., He, M., 2006. Adsorption of anionic and non-ionic surfactant mixtures from synthetic detergents on soils. Chemosphere 63 (7), 1214–1221.

Rao, N.N., Dube, S., 1996. Photocatalytic degradation of mixed surfactants and some commercial soap/detergent products using suspended TiO_2 catalysts. J. Mol. Catal. A: Chem. 104 (3), L197–L199.

Ren, R., Liu, D.F., Li, K., Sun, J., Zhang, C., 2011. Adsorption of quaternary ammonium compounds onto activated sludge. J. Water Resour. Prot. 3, 105–113.

Sablayrolles, C., Montréjaud-Vignoles, M., Silvestre, J., Treilhou, M., 2009. Trace determination of linear alkylbenzene sulfonates: application in artificially polluted soil-carrots system. Int. J. Anal. Chem. 2009, 1–6.

Sakac, N., Jozanovic, M., Karnas, M., Sak-Bosnar, M., 2017. A new sensor for determination of anionic surfactants in detergent products with carbon nanotubes as solid contact. J. Surfactant Deterg. 20, 881–889.

Samadi, M.T., Dorraji, M.S.S., Atashi, Z., Rahmani, A.R., 2014. Photocatalytic removal of sodium dodecyl sulfate from aquatic solutions with prepared ZnO nanocrystals and UV irradiation. Avicenna. J. Environ. Health Eng. 5. Available from: https://doi.org/10.17795/ajehe-166.

Shrivas, K., Sahu, S., Ghorai, A., Shankar, R., 2015. Gold nanoparticles-based colorimetric determination of cationic surfactants in environmental water samples via both electrostatic and hydrophobic interactions. Microchimica Acta 183 (2), 827–836.

Singhal, B., Porwal, A., Sharma, A., Ameta, R., Ameta, S.C., 1997. Photocatalytic degradation of cetylpyridinium chloride over titanium dioxide powder. J. Photochem. Photobiol. A: Chem. 108, 85–88.

Singla, R., Grieser, F., Ashokkumar, M., 2009. Kinetics and mechanism for the sonochemical degradation of a non-ionic surfactant. J. Phys. Chem. A. 113 (12), 2865–2872.

Singla, R., Grieser, F., Ashokkumar, M., 2011. The mechanism of sonochemical degradation of a cationic surfactant in aqueous solution. Ultrasonics Sonochem. 18 (2), 484–488.

Steuer, C., Schütz, P., Bernasconi, L., Huber, A.R., 2016. Simultaneous determination of phosphatidylcholine-derived quaternary ammonium compounds by a LC–MS/MS method in human blood plasma, serum and urine samples. J. Chromatog. B 1008, 206–211.

Tada, H., Kubo, Y., Akazawa, M., Ito, S., 2000. SiOx monolayer overcoating effect on the TiO_2 photocatalytic oxidation of cetylpyridinium bromide. J. Colloid Interface Sci. 221 (2), 316–319.

Takayanagi, A., Kobayashi, M., Kawase, Y., 2017. Removal of anionic surfactant sodium dodecyl benzene sulfonate (SDBS) from wastewaters by zero-valent iron (ZVI): predominant removal mechanism for effective SDBS removal. Environ. Sci. Poll. Res. 24, 8087–8097.

Tsai, P.-C., Ding, W.-H., 2004. Determination of alkyltrimethylammonium surfactants in hair conditioners and fabric softeners by gas chromatography–mass spectrometry with electron-impact and chemical ionization. J. Chromatogr. A 1027 (1-2), 103–108.

Wahyuni, E.T., Roto, R., Sabrina, M., Anggraini, V., Leswana, N.F., Vionita, A.C., 2016. Photodegradation of detergent anionic surfactant in wastewater using $UV/TiO_2/H_2O_2$ and $UV/Fe^{2+}/H_2O_2$ processes. Am. J. Appl. Chem. 4, 174–180.

Wang, P.G., Zhou, W., 2016. Rapid determination of cocamidopropyl betaine impurities in cosmetic products by core-shell hydrophilic interaction liquid chromatography-tandem mass spectrometry. J. Chromatog. A 1461, 78–83.

Wang, Y., Zhi, J., Liu, Y., Zhang, J., 2011. Electrochemical detection of surfactant cetylpyridinium bromide using boron-doped diamond as electrode. Electrochem. Commun. 13 (1), 82–85.

Xu, Z., Zhang, X., Huang, N., Hu, H., 2017. Oxidation of benzalkonium chloride by gamma irradiation: kinetics and decrease in toxicity. J. Radioanal. Nucl. Chem. 312 (3), 631–637.

Yang, W., Lin, X., Wang, H., Yang, W., 2013. Ferrate(VI): a novel oxidant for degradation of cationic surfactant – cetylpyridinium bromide. Water Sci. Technol 67 (10), 2184–2189.

Yoshikawa, N., Kimura, T., Kawase, Y., 2008. Oxidative degradation of non-ionic surfactants with TiO_2 photocatalyst in a bubble column reactor. Canad. J. Chem. Eng. 81 (3-4), 719–724.

Zembrzuska, J., Budnik, I., Lukaszewski, Z., 2016. Monitoring of selected non-ionic surfactants in river water by liquid chromatography–tandem mass spectrometry. J. Environ. Manage. 169, 247–252.

CHAPTER

5 Risks and challenges of pesticides in aquatic environments

Afsane Chavoshani[1], Majid Hashemi[2,3], Mohammad Mehdi Amin[1,4] and Suresh C. Ameta[5]

[1]*Department of Environmental Health Engineering, School of Health, Isfahan University of Medical Sciences, Isfahan, Iran*
[2]*Environmental Health Engineering, School of Public Health, Kerman University of Medical Sciences, Kerman, Iran*
[3]*Environmental Health Engineering Research Center, Kerman University of Medical Sciences, Kerman, Iran*
[4]*Environment Research Center, Research Institute for Primordial Prevention of Non-Communicable Disease, Isfahan University of Medical Sciences, Isfahan, Iran*
[5]*Department of Chemistry, PAHER University, Udaipur, India*

5.1 Introduction

The term "pesticide" refers a vast range of compounds, including rodenticides, fungicides, insecticides, acaricides, molluscicides, nematocides, and herbicides. Pesticides are not chemical compounds of natural or synthetic but may be microorganisms (e.g., fungi or bacteria) or components there of (e.g., endotoxins from *Bacillus thuringiensis*), or even so-called "macroorganisms," for example, predatory wasps such as *Trichogramma evanescens*. Pesticides are used widespread to fight insects, nematodes, fungi, and other micro- and macroorganisms that affect human health and agricultural crops (Hamilton and Crossley, 2003).

Worldwide pesticide usage has increased dramatically during the last two decades, coinciding with changes in farming practices and the increasingly intensive agriculture. This widespread use of pesticides for agricultural and nonagricultural purposes has resulted in the presence of their residues in various environmental matrices. Pesticide contamination of surface waters has been well documented worldwide and constitute a major issue that gives rise to concerns at local, regional, national, and global scales. Pesticide residues enter the aquatic environment through direct run-off, leaching, careless disposal of empty containers, equipment washing, etc. (Parastar et al., 2018).

Because of their high toxicity, hydrophobicity, bioaccumulation, and biodegradation resistance, pesticides are well recognized as an important group of environmental contaminants. They have been watched in the list of priority pollutants

Micropollutants and Challenges. DOI: https://doi.org/10.1016/B978-0-12-818612-1.00005-2

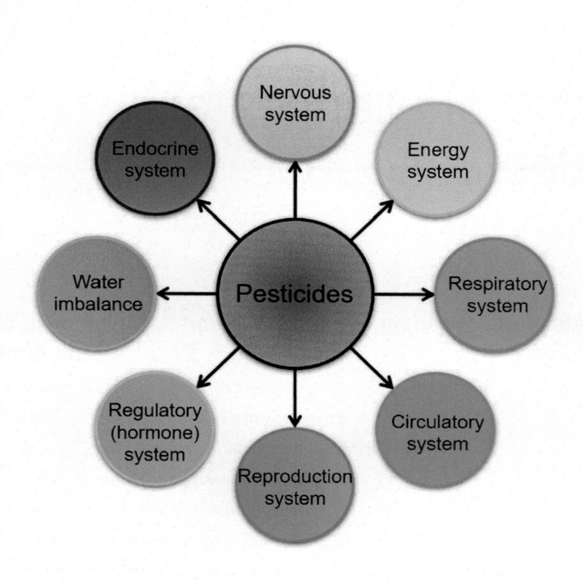

FIGURE 5.1

Human-health related adverse effects of pesticides on different living being systems (Bilal et al., 2019).

With permission.

reported by the US Environmental Protection Agency (EPA) and in the list of priority pollutants in Chinese waters (Lu et al., 2007). Pollution of aquatic systems by pesticide residues has been confirmed by sea food resources, aquaculture, and fisheries. Recently reports have showed the many side effects of pesticides on human and other living organisms (Pereira et al., 2015); therefore it is cleared that widespread exposure to pesticides results in serious health risks, even directly or indirectly lead to death.(Bilal et al., 2019). Due to the well documented toxicity and adverse effects of pesticides on different living being systems (Fig. 5.1), researchers have found their interest in pesticides removal from the different environments by physical, chemical, and biological methods (Maqboll et al., 2016, Mir-Tutusaus et al., 2018).

With regard to their high toxicity and environmental hazardous, occurrence and fate of pesticides in water and wastewater environments is very necessary for protection of public health. The EPA differentiated toxicity classes of pesticides are shown in Fig. 5.2 (Bilal et al., 2019).

Prediction of water quality impacts of pesticides is a necessary element for the development of approaches of pesticide pollution control. The aim of the present chapter is to compile all available data on pesticide risk in the aquatic environments to discuss type of pesticide, pesticide health effects, pesticide fate, detection and removal methods of pesticide, and their management and control in aquatic environments.

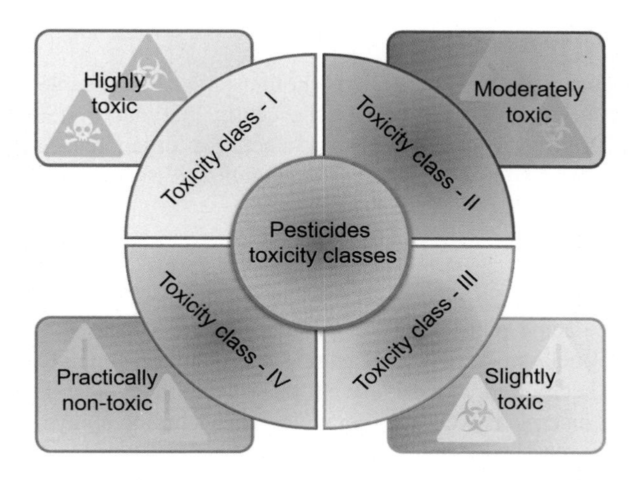

FIGURE 5.2

The EPA categorized toxicity classes of pesticides. The outer boxes represent overall toxicity ratings along with their representative labels (Bilal et al., 2019). *EPA*, Environmental protection agency.

With permission.

5.2 Classification based on chemical composition of pesticides

The most common and useful method of classifying pesticide is based on their chemical composition and nature of active ingredients. It is such kind of classification that gives the clue about the efficacy, physical and chemical properties of the respective pesticides. The information on chemical and physical characteristics of pesticides is very useful in determining the mode of application, precautions that need to be taken during application and the application rates. Based on chemical classification, pesticides are classified into four main groups namely: organochlorines, organophosphorous, carbamates, pyrethrin, and pyrethroids.

5.2.1 Organochlorines

Organochlorinated compounds have been used globally for many years as solvents, fumigants, and insecticides. In the 1930s, Swiss chemist and Nobel Prize recipient Paul Muller first discovered that the most well-known organochlorine agent dichlorodiphenyltrichloroethane (DDT) had significant insecticide properties (Genuis et al., 2016; Nollet and Rathore, 2016). Organochlorine pesticide groupings based on chemical structure are classified into (i) DDT and analogues including DDT,DDE, DDD, Methoxychlor, (ii) Hexachlorobenzene including Hexachlorobenzene, (iii) Hexachlorocyclohexane including α-HCH,β-HCH,σ-HCH, and Ɣ-HCH, (iv) Cyclodiene (Endosulfan I and endosulfan II, Heptachlor, Aldrin, Dieldrin, Endrin, (v) Chlordecone, Kelevan and Mirex, (vi) Toxaphene (Genuis et al., 2016).

5.2.2 Organophosphates

Organophosphates (OP) are inhibitors of cholinesterase that disable cholinesterase (cholinesterase an enzyme essential for the central nervous system to function). Application of OP insecticides increased when the organochlorine insecticides were banned in the 1970s. In following has mentioned to main organophosphates pesticides (Matthews, 2018).

5.2.2.1 Malathion

Malathion is far less toxic than parathion. According to the WHO classification of pesticides, the acute oral toxicity of parathion is 3–6 mg/kg (in class I), while malathion was unclassified with an acute toxicity to mammals of 1400 mg/kg. Malathion has been used extensively in public health against mosquitoes and other vectors of disease as well as on many crops, including sprays, with a protein hydrolysate, or a yeast bait for fruit flies (Matthews, 2018).

5.2.2.2 Temephos

Temephos is low toxic and has been used in rivers to control blackfly (*Simulium spp.*) larvae to reduce transmission of onchocerciasis.

5.2.2.3 Dimethoate

Dimethoate, as both an insecticide and acaricide, is readily absorbed and distributed through plant tissues and degrades quite quickly.

5.2.2.4 Phorate

Phorate (*O*,*O*-diethyl *S*-ethylthiomethyl phosphorodithioate) was marketed as a systemic insecticide and acaricide in 1954. It was used as a seed treatment as it gave up to 8 weeks' control of sucking pests such as aphids, thrips, and leaf hoppers (Matthews, 2018).

5.2.2.5 Dichlorvos

Dichlorvos (2,2-dichlorovinyl dimethyl phosphate, abbreviated as DDVP) due to its vapour action it became widely used against household and public health pests. Safety concerns have reduced its use, and it has been banned in Europe since 1998. However, in the USA, a low dose of naled (Dibrom), dimethyl 1,2-dibromo-2,2-dichloroethylphosphate, is applied in an aerial spray to control mosquitoes.

5.2.2.6 Trichlorfon

Trichlorfon (dimethyl 1-hydroxy-2,2,2-trichloro ethanephosphonate) is a non-systemic insecticide, rapidly hydrolyzed in plants. Dipterex was one of the trade names used.

5.2.2.7 Fenthion

Fenthion (*O*,*O*-Dimethyl *O*-[3-methyl-4-(methylsulfanyl)phenyl] phosphorothioate) is in some countries, but it is banned due to its impact on bird populations.

5.2.2.8 Phosphamidon

Phosphamidon ((E/Z)-[3-Chloro-4-(diethylamino)-4-oxobut-2-en-2-yl])dimethyl phosphate was marketed as Dimecron in 1956. It is a highly hazardous insecticide, and it is now included in the Rotterdam Convention and requires prior informed consent (PIC) before it can be exported to a country.

5.2.3 Carbamates

Carbamate insecticides are derivative of carbamic acids(NH_2COOH) and the first carbamate insecticide, carbazyl was introduced in 1956. They inhibit the AChE enzyme and cause over stimulation of nervous system. Carbaryl (1-naphthyl N-methylcarbamate), broad spectrum carbamate insecticide is extensively used worldwide for more than 120 different crops and ornamental plants. Because of very low mammalian toxicity together with short half-life in the environment carbaryl are the most popular insecticide and effectively acts against 160 harmful insects. Carbaryl is the second most widely detected insecticide in surface waters in the United States. It has been observed that carbamate insecticide carbaryl at lower concentrations ($<$10mg/L) exhibited the stimulatory effect on the growth and nitrogen fixation in cyanobacteria while higher concentrations ($>$20ppm) showed the inhibitory effects. Furthermore, it has been reported that carbaryl significantly increased the cyanobacterial populations in soil at 0.5kg/ha under non-flooded conditions and the concentration 1.0kg/ha was harmless while the higher concentrations exerted pronounced toxicity. In flooded soil, carbaryl application at 0.5kg/ha was found to be non-toxic, but all other applied concentrations (>0.5kg/ha) were significantly toxic to the native cyanobacterial populations. This was an expected result because carbaryl is quickly degraded with the half-life of about 4days in aerobic soil, while in anaerobic soil its degradation occurs slowly with an estimated half-life of 72days. The autotrophic growth response of the cyanobacterial species, that is, *Anabaena sp., A. cylindrica, Nostoc punctiforme, and Nostoc sp.* after carbaryl treatment was investigated in terms of chlorophyll content under laboratory conditions. Except *N. punctiforme*, the growths of all the cyanobacterial species were significantly inhibited by the insecticide carbaryl even at the concentration 5μg/mL. As 5μg/mLcarbaryl was stimulatory for the cyanobacteria under the field conditions, it seems that in the culture conditions (in vitro) carbaryl exhibits its toxic effects, even below the doses at which cyanobacteria tolerate in the field (Tiwari et al., 2019).

5.2.4 Pyrethrin, and pyrethroids

Pyrethrins are the insecticidal compounds obtained from the flowers of the plant *Tanacetum cinerariaefolium*, also called *Chrysanthemum cinerariaefolium* or *Pyrethrum cinerariaefolium*. Pyrethrum comes from extracts from the flowers that contain the active pyrethrin compounds. The use of pyrethrum in insecticide preparations dates back to Persia, about 400 BC. Pyrethroids are synthetic analogs of pyrethrins. Because of stability problems with the natural pyrethrins, these insec-ticides were replaced by the more stable organophosphate and organochlorine insecticides developed after World War II. As a result of the toxicity and environmental contamination associated with the organophosphate and organochlorine insecticides, interest in the use of pyrethrins and pyrethroids reemerged in the 1970s.Pyrethrin and pyrethroid insecticides are effective against a variety of insect pests on companion animals and live-stock, and are used on farms, in the home and garden and have many public health applications because of the safety associated with these compounds. The pyrethroids are considerably safer than the organochlorines. The neonicotinoids were developed to replace the pyrethrins.

There are six compounds that comprise the natural pyrethrins: pyrethrin I and II, jasmolin I and II, and cinerin Iand II. Synthetic pyrethroids have been developed because the natural pyrethrins tend to break down quickly when exposed to air, light and heat. The synthetic pyrethroids can be classified as first and second generation. First-generation pyrethroids are esters of chrysanthemicacid and an alcohol, having a furan ring and terminal sidechain moieties. Second-generation pyrethrins have 3-phenoxybenzyl alcohols derivatives in the alcohol moiety and have had some of the terminal side chain moieties replaced with a dichlorovinyl or dibromovinyl substitute and aromatic rings (Ensley, 2018).

5.2.5 Other pesticides

There are many more pesticides used in agricultural practice. Heavy metals have found vast use as pesticides. Elements like iron, lead, sulfur, arsenic, mercury, zinc, tin, etc. have been used in inorganic or organic metal form. Methyl mercuric chloride, sodium arsenate, calcium arsenate, and zinc phosphide are some of the compounds that fall under this category.

5.3 Overview of the fate of pesticides in the environment and water

When a pesticide is discharged into the environment, many parameters are important for change of pesticide fate in the different environments. Pesticide properties such as water solubility, pesticide persistence, tendency to adsorb to the soil, and soil features (clay, sand, and organic matter) are important in determining the fate of these compounds in the environment. Widely use of pesticides/chemical fertilizers in modern agriculture practices has resulted in pollution of different environmental matrices, including air, land, and water. Consequently the contaminated environmental matrices adversely affect human health and nontargeted animals in several ways (Bilal et al., 2019, Bret et al., 2011, Honeycutt, 2018) (Fig. 5.3).

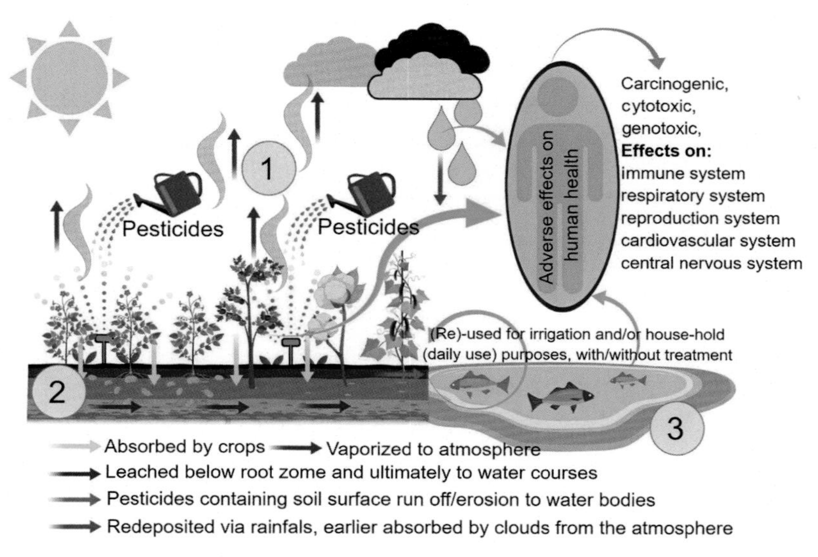

FIGURE 5.3

Pesticide drift and adverse consequences of excessive use of pesticides/chemical fertilizers in modern agriculture practices that can lead to contamination of different media including (1) air, (2) land, and (3) water (Liu et al., 2019).

From Liu, L., Bilal, M., Duan, X., Iqbal, H.M., 2019. Mitigation of environmental pollution by genetically engineered bacteria—current challenges and future perspectives Sci. Total Environ. 667, 444–454. With permission.

The environmental fate of pesticides is influenced by many processes defined their persistence and biodegradation. The interaction of pesticides with groundwater, surface water, and soils is very complex. Prediction of the occurrence, fate, and distribution of pesticides in bed sediment and aquatic environments needs considering the pesticide sources, transformation, and transportation processes (Gavrilescu, 2005).

The Júcar River in Eastern Spain is expected to suffer a decline in water quality and quantity as a consequence of the climate change. Table 5.1 shows pesticide concentration in water and fish tissues of this river (Belenguer et al., 2014).

5.4 Ecotoxicological risk assessment of pesticide

All pesticides are toxic to some degree. For some of them, only a very small amount is needed to cause acute or chronic effects; others are no more toxic than many materials in everyday use. To be prudent, inadvertent water contamination with any pesticide should be avoided. About 800 million pounds of pesticides are used annually in the United States. The major classes of pesticides used in agriculture are herbicides, insecticides, and fungicides. Herbicide use has increased most rapidly

Table 5.1 Pesticide concentrations in water (W, in ng/L) and fish (F, in ng/g) by study site in the Júcar River (from JUC-I to JUC-V) corresponding to the 23 pesticides detected (Belenguer et al., 2014).

	Sites JUC-I		JUC-II		JUC-III		JUC-IV		JUC-V	
Pesticide	**W**	**F**	**W**	**F**	**W**	**F**	**W**	**F**	**W**	**F**
Atrazine					7.97					
Atrazine-desethyl					8.65		10.61		4.68	
Atrazine desisopropyl						21.34–39.39		23.83		
Azinphos ethyl		2.52		2.36–46.63		86.17		65.64		
Buprofezin	14.07		13.06		11.68		13.27		12.82	Trace
Carbofuran		518.9		Trace						
Chlorfenvinphos			93.34		83.07		78.08		96.68	
Chlorpyriphos	6.84	Trace	16.99	Trace	32.14	24.42	2.23		36.23	7.13
Diazinon	11.94	0.92–3.53	0.44	1.04–2.31	8.59	0.37–2.36	6.31	1.33	8.87	0.87–5.83
Dichlofenthion	44.08		35.11		50.85		43.22		39.43	
Dimethoate		0.18		Trace		1.64	9.87	Trace		Trace
Ethion	0.09	Trace	2.45	Trace	7.07		12.9	13.76		0.48
Hexythiazox	17.71				25.52		48.94		50.66	
Imazalil			12.62		10.72				8.75	
Omethoate						78.82				4.32
Parathion ethyl					32.47		31.9		34.25	
Prochloraz	79.9		73.85		82.79		66.99		76.69	
Propazine				1.42						
Pyriproxyfen	99.59		89.95				82.92		88.43	Trace
Tolclofos methyl	28.64			12.63					27.57	

With permission.

during the last 15 years. Today herbicides account for about 85 percent of the pesticides used in agriculture, insecticides 13 percent, and fungicides the remaining 2 percent in the United States. Contamination of surface water is less serious than is the case for ground water. Most surface waters (except deep lakes) have a rapid turnover rate, which means that fresh water dilutes the concentration of the contaminant quickly. In addition, most surface waters contain free oxygen, which enhances the rate at which pesticides are broken down by microorganisms. Contamination of surface waters should not be treated casually, however. An extremely toxic pesticide can cause the death of fish and other aquatic organisms even at low concentrations. Pesticides in groundwater are an extremely serious problem. The turnover rate for groundwater may be as short as a few months, but more commonly years and decades are needed to replace the water in an underground aquifer. Oxygen is absent in ground water and the microorganisms that live in an oxygen free environment are much less effective in breaking down pesticide chemicals. Extremely slow dilution and breakdown means that the contaminant will be present (Aydinalp and Porca, 2004). Although the pesticides toxicity has been studied for numerous species of fish and crustacean, it is difficult to directly compare the available results because analyzing methods, study conditions, type and age of organisms are different. LC_{50} as a pesticide toxicity index (PTI) is available for the toxicity of organophosphate, carbamate, organochlorine, and pyrethroid in freshwater fish. Sucahyo et al. (2008) compared the LC_{50}s of the pesticides tested for a number of different organisms, including fish and crustaceans (Table 5.2). They did not found any data on the toxicity of 2,4-D-dimethylamine to other aquatic organisms. For diazinon and paraquat, *C. laevis* was more than 100 times more susceptible than fish. For the other pesticides, differences in sensitivity between the different species were relatively small with crustaceans generally being somewhat more sensitive than fish. Only exception was copper hydroxide, for which *C. laevis* seems less sensitive than fish, but only few toxicity data were available on this pesticide. For diazinon and lambda cyhalothrin, *C. laevis* had more or less the same sensitivity as *D. magna*. This might reflect similarities in physiology between the two crustaceans not shared by fish, even though *C. laevis* and *D. magna* belong to entirely different crustacean lineages. In case of endosulfan and paraquat, however, *C. laevis* was 50–75 times more sensitive, while for carbofuran it was a factor of 30 less sensitive than *D. magna*. This suggests that it is not possible to conclude that tropical species are more or less sensitive than the standard temperate species currently used for determining pesticide toxicity.

5.5 Routs of human exposure to pesticides

Exposure to a particular pesticide may occur through multiple exposure routes (Skin, oral, and respiration, eyes) depending on the type and use of the pesticide. Common sources of exposure to pesticides for the general population and occupational include residues in food and drinking water.

Table 5.2 Comparison of LC_{50}s for the toxicity of selected pesticides to freshwater species (Sucahyo et al., 2008).

Pesticide	Test organism	LC_{50} (μg a.i./L)			
		24 h	48 h	72 h	96 h
Carbofuran	*Lepomis macrochirus*	–	–	–	78–240
	Oncorhynchus mykiss	536–863	–	–	272–531
	Oreochromis niloticus	250			200
	Daphnia magna	92	30	–	–
	Caridina laevis	1700	–	–	950
Diazinon	*Danio rerio*	–	11,000	–	–
	Cyprinus carpio	3690	–	–	1530
	Oryzias latipes	6800	4400	–	–
	Oncorhynchus mykiss	1000–8100	–	–	230–700
	Lepomis macrochirus	270–480	–	–	120–220
	Tilapia mosssambica	1580–6390	–	2730–3080	–
	Silurus glanis	12,500	–	–	4140
	Daphnia magna		0.8–2.4		
	Ceriodaphnia cornuta		0.46		
	Caridina laevis	2.81–3.26			1.32–1.58
Endosulfan	*Oncorhynchus mykiss*	2.1–2.5			1.5–1.9
	Lepomis macrochirus	2.3–4.8			0.9–1.7
	Oreochromis niloticus	10.3			
	Channa punctatus	19.7			7.8
	Daphnia magna	54–82			52.9
	Caridina weberi	10.4			5.1
	Caridina leavis	2.38			1.02

(*Continued*)

Table 5.2 Comparison of LC_{50}s for the toxicity of selected pesticides to freshwater species (Sucahyo et al., 2008). *Continued*

Pesticide	Test organism	LC_{50} (μg a.i./L) 24 h	48 h	72 h	96 h
Lambda cyhalothrin	*Oncorhynchus mykiss*				0.24
	Lepomis macrochirus				0.21
	Danio rerio	8.26			1.94
	Gammarus pulex		0.031		0.024
	Daphnia galeata		0.40		
	Daphnia magna		0.39		
	Streptocephalus sudanicus	0.028	–	–	–
	Macrobrachium nippoensis	0.05			0.04
	Caridina laevis	0.87			0.33
$Cu(OH)_2$	*Oncorhynchus mykiss*	1420–2420			290–550
	Caridina laevis	7930			3390
Paraquat	*Cnesterodon decemmaculatus*	144,000–172,000			52,500–67,400
	Oreochromis niloticus				11,800
	Bryconamericus iheringii	40,470			20,210
	Daphnia magna				2680
	Cardina laevis	87.6			35.8

With permission.

5.5.1 Occupational exposure

Exposure occurs during the manufacture, transport, or use of pesticides, and relevant packaging and protective measures are necessary for all these situations. Major occupational exposure routes are by direct contact of material with the skin, eyes, and respiratory tract. All these routes may be involved with airborne pesticide resulting from spraying or dust generation and the skin and eyes are principal routes for exposure to nonvolatile liquids and solids that are not sprayed for application. Additionally the alimentary tract may be a route of exposure from the swallowing of contaminated saliva or coughed mucus.

5.5.2 General public exposure (none occupational exposure)

Although the general public appear to regard exposure to pesticides from residues in food and, perhaps, water of greatest concern, there are multiple other sources of exposure which can compound with those from residues. These include hand-to-mouth contact from pesticides used within buildings, veterinary medicines used against domestic pets (e.g., flea sprays), and contamination of food and working surfaces from the residential use of pesticides (e.g., control of insects). Thus although the oral route is probably the major route of exposure for the general public, the skin and eyes probably are also significant, and inhalation is the least. It should however be recognized that data on exposure by pathways other than food and drink is often very poor. An interesting intermediate between occupational and general population exposures is the "take-home pathway," in which workers exposed to pesticides at work may take them back into their homes and contaminate members of the family (Marrs and Ballantyne, 2004).

The results confirmed the existence of the take-home pathway of pesticide exposure, and accord with the fact that pesticides in soil and house dust were significantly higher in the homes of agricultural workers compared with nonagricultural reference homes. It is notable that many employers did not provide resources for hand washing.

5.6 Types of human exposure to pesticide

After exposure to pesticide, humans may be injured and poisoned. Therefore it is important to take appropriate decision for the treatment. Emergency treatments depend on type of exposure to contaminants. Type of human exposure to pesticide can be dermal (absorption through the skin or eyes), oral (ingestion by mouth), and respiratory (inhalation through the lungs) (Debnath and Khan, 2017).

- *Dermal*

  Dermal exposure to pesticide is an important route for pesticide applicators. Dermal absorption may occur as a result of a splash, spill, or spray drift when mixing, loading, disposing, and/or cleaning of pesticides. Absorption by human especially farmers may also result from exposure to large amounts of residue. Pesticide formulations vary broadly in physicochemical properties and in their capacity to be absorbed through the skin, which can be influenced by the amount and duration of exposure, the presence of other materials on the skin, temperature and humidity, and the use of personal protective equipment. In general solid forms of pesticides (e.g., powders, dusts, and granules) are not as readily absorbed through the skin and other body tissues as liquid formulations. However, the hazard from skin absorption increases when workers are handling (e.g., mixing) concentrated pesticides (e.g., one containing a high percentage of active ingredients). Certain areas of the body (such as the genital areas and ear canal) are more susceptible to pesticide absorption than other areas of the body. As such the rate at which dermal absorption proceeds differs for each part of the body (Fig. 5.4) (Kim et al., 2017).

- *Oral*

  Intake may result from swallowing of saliva contaminated with airborne material and eating food or drinking water contaminated at work. Oral exposure may also result from transfer from contaminated hands, for example, from eating or smoking.

- *Respiratory*

  This occurs principally from material present in the atmosphere resulting from spraying or drift of pesticide. The atmospheric concentration will be affected by rate of application, type of formulation application (aerosol, dust), and meteorological conditions, principally air movement (Marrs and Ballantyne, 2004).

5.7 Pesticides removal methods from aquatic environments

The majority of polar and semi polar pesticides will remain partitioned in the aqueous phase due to their relatively high water solubility; hence their removal by physical processes such as sedimentation and flocculation is not effective and has been reported to be less than 10%. For removal of pesticides, it was used from physical treatment processes such as membrane, reverse osmosis, ultrafiltration, microfiltration, nanofiltration, and adsorption processes.

5.7.1 Pesticides removal by membrane filtration

Many NF/RO membranes have been used for the removal of fungicides, herbicides, insecticides, and miscellaneous pesticides from aqueous solutions. Main effective factors on the pesticides removal by NF/RO membranes are membrane filtration characteristics and feed water composition.

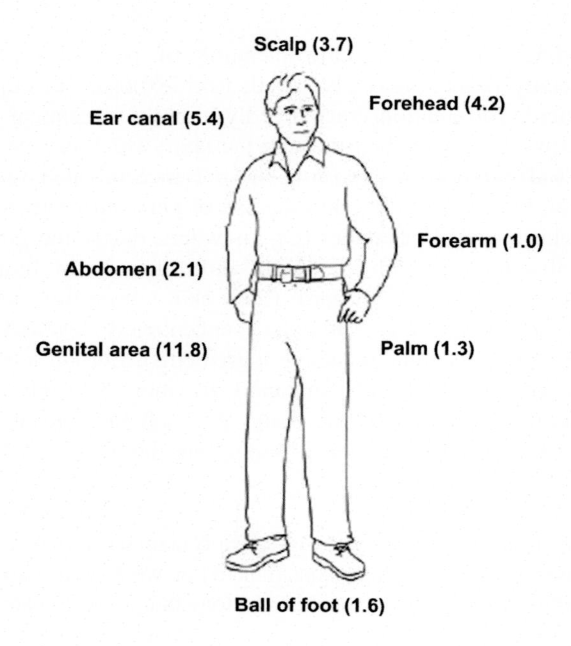

FIGURE 5.4

Relative dermal absorption of pesticides on the different parts of the body in compared to the forearm (with dermal absorption = 1).

Modified after Ministry of Agricultur (MOA), 2015. British Columbia, Canada. Available at: <http://www.agf.gov.bc.ca/pesticides/b_2.htm#2>.

5.7.1.1 Membrane filtration characteristics

The removal of pesticides from potable water by membrane processes is related to the type of membrane applied. Important factors to consider and to choose a suitable membrane are the molecular weight cut-off (MWCO), which is expressed in Dalton indicating the membrane porosity, the molecular weight of a hypothetical noncharged solute that is 90% rejected, the surface charge and the membrane composition, and the degree of ionic species rejection. The significance degree of each parameter on removal of pesticides is directly related to the particular solute characteristics (molecular size, molecular weight, hydrophobicity/hydrophilicity—logKow, and acid disassociation constant-p*K*a) which determine the strength of the contaminant rejection by the membrane.

5.7.1.2 Membrane molecular weight cut-off and other pore size parameters

The many of pesticides and their residues (based on the molecular weight) detected in the water body. Membranes with a MWCO of 200–400 Da are considered suitable for sufficient removal of such contaminants from water. It is reveal that the larger the pesticide molecule, the greater the sieving effect and pollutant retention. The retention of smaller pesticide molecules by relatively wide pore-size membranes can be affected not only by the sieving factors (i.e., relative pore sizes of pesticide and membrane) but also by physicochemical interactions occur between the organic materials and the membrane surfaces. For example in a study, two membranes examined for the pesticide removal which Desal DK membranes obtained the best retention results for all pesticides due to their smaller MWCO (150–300 Da) compared to NF200 (300 Da) membranes. MWCO provides only a rough estimate of the membrane capability to retain dissolved uncharged compounds. However, pore size of a membrane, usually refers to the smallest pore size in the membrane matrix, and the porosity, expressed as pore density, pore size distribution (PSD), or effective number of pores (N) in the membrane top layer (skin) have been considered as representative factors for predicting the rejection of different organic compounds such as or particles. For example, the rejection of uncharged pesticide is correlated with membrane porosity parameters (PSD and N) (Plakas and Karabelas, 2012).

- *Degree of membrane desalination*

  The separation efficiency of RO and NF membranes is determined by their salt rejection action, rather than by MWCO. The membrane desalination degree of is usually considered as the stabilized salt rejection of a 2000 mg/L sodium chloride or magnesium sulfate solution, and/or a 500 mg/L calcium chloride solution. The desalination degree is a suitable parameter in estimating the pesticides rejection, because the MWCO of a membrane is often unknown and manufacturer specific, whereas PSD and porosity determination require the performance of specific filtration experiments or the application of special analytical techniques, that is, atomic force microscopy, bubble point, gas adsorption/desorption, thermoporometry, etc. and the nonphenylic pesticides rejection is correlated with the desalination degree of membranes.

- *Membrane material*

  Membrane material is also a main parameter factor pesticide rejection and influences the membrane performance based on physicochemical interactions. Many studies showed that composite polyamide membranes have the higher rejection for pesticides compared to the cellulose acetate membranes. The lower efficiency of cellulose acetate membranes is related to the higher polarity of cellulose acetate membrane which is responsible for the low rejection (pesticides are highly polar). The sulfonated polyethersulfone (PES) membranes show lower rejection of pesticides compared to poly (vinylalcohol)/polyamide ones although their desalination capabilities are equal.

- *Membrane charge*

  The membranes with a negative charge tends to minimize the adsorption of negatively charged contaminants. The electrostatic repulsion of negatively charged pesticides at the membrane surface enhances rejection efficiency.

5.7.1.3 Effect of pesticides properties on retention

The selection of a suitable membrane should be occur on the pesticide properties, such as hydrophobicity/hydrophilicity (logKow), the polarity (dipole moment), the molecular weight and size, and the acid dissociation constant (p*K*a) (Košutić et al., 2005; Mehta et al., 2015).

- *Pesticide molecular weight and size*

  Size exclusion is the most main mechanism of pesticide retention. The Stokes diameter (ds), the molecular weight (MW), the diameter derived from the molar volume (dm), and the molecular length and width, are related with pesticide rejection. According to previous studies, the higher rejection of atrazine and cyanazine was related to molecular weight. Molecular length and width are also considered to be realistic measures of molecular size and important factor for estimating the rejection of different groups of organic compounds by NF/RO membranes. For instance the aromatic pesticides rejection is related to their molecular length rather than their molecular width.

  The rejection of aromatic (phenylic) and the nonphenylic pesticides is not correlated only with a molecular size parameter, but it is related the sorption capacity of these molecules on the membrane (Košutić et al., 2005; Mehta et al., 2015; Plakas and Karabelas, 2012).

- *Pesticide hydrophobicity/hydrophilicity*

  The hydrophobic interactions are mostly responsible for pesticide adsorption on the membrane surfaces, which is considered to be the first step of the rejection mechanism. A measure of solute hydrophobicity/hydrophilicity is the octanol/water partition coefficient (logKow or logP), while the hydrophobic nature of a membrane is commonly characterized by its contact angle value. logKow values of trace organic molecules vary between -3 and 7 as higher values are related to hydrophobic compounds. The presence of a phenyl in the aromatic pesticides increases its adsorption capacity, while alkyl groups can have negative effects on the interaction between a phenyl group and the membrane.

- *Pesticide polarity*

  One of the most important physicochemical criteria in NF and RO separation of organic micropollutants such as pesticides from aqueous solutions is the "Polar Effect" of the organic micropollutants. The separation of pesticides by NF/RO membranes is affected by charge repulsion, size exclusion, and polar interactions between solutes and the membrane surface. A important explanation for this behavior is electrostatic interactions because the dipole region with the opposite charge is closer to the membrane. The dipole is directed towards the pore and enters more easily into the membrane pore structure. The effect of polarity is same for positively and negatively charged membranes since the only change occurring is the direction of the dipole. The effect of solute polarity is important for membranes with an average pore size larger than the size of compounds to be retained (Košutić et al., 2005; Mehta et al., 2015; Plakas and Karabelas, 2012).

5.7.1.4 Effect of the feed water composition

Membrane filtration tests have shown that pesticide rejection is depended on the feed water composition. Specifically ionic strength, pH, and the presence of organic matter have an influence on pesticide rejection.

- *Influence of water pH*

  The pH effect in pesticide rejection is associated mainly with the changes take place in the membrane charge and surface. Due to the dissociation of functional groups, pH has a main effect on the membrane charge. Zeta potential becomes more negative when the pH is increased, and functional groups deprotonate. Moreover, pore enlargement or shrinkage can occur depending upon the electrostatic interactions between the dissociated functional groups of the membrane material. Pesticides are polar and show a lower rejection with increasing membrane charge because such molecules tend to preferentially orient themselves so that the dipole end with a charge opposite to that of the membrane charge is the closest to the membrane surface. Consequently this preferential orientation results in an increased attraction, an increased permeation, and thus a lower rejection. At lower pH, the same effect might occur with H^+ ions. For example, increasing the pH of solution enhanced atrazine and dimethoate rejection but degraded the permeate flux performance for NF200, NF270, and DK membranes.
- *Influence of solute concentration*

  Filtration examinations with atrazine and prometryn in various concentrations (10–700 μg/L) exhibited small variations in their rejection by NF membranes. The fluids filtration with lower feed concentrations led to a slight decrease of triazine retention could be related to the triazines amount adsorbed on the membranes. Specifically the smaller triazine concentration may be associated with a slightly smaller adsorption, in comparison to the results obtained with greater feed concentrations, something that was more pronounced in the case of the less tight NF membranes.

- *Influence of the ionic environment*

  The retention of pesticides can be moderately influenced by the presence of dissolved salts in the feed solutions due to the interactions occur between the ions and the membrane surfaces; at higher ionic concentrations, reduction takes place in the electrostatic forces inside the membrane which finally may cause a reduction of the actual size of the pores, leading to a reduced membrane permeability. Consequently a better rejection of pesticides accompanied by a reduced water flux could be observed. The presence of divalent cations (calcium) in the feed solution appeared to exercise little influence on pesticide rejection.

- *Effect of membrane fouling*
  In association with fouling, two points are important. First, if the membrane rejects solutes better than the deposited layer, hindered back-diffusion of solutes (by the fouling layer) would cause solute accumulation near the membrane surface. This cake-enhanced concentration polarization results in greater concentration gradient across the membrane and, hence, a decrease in solute retention. Second, if solutes are rejected better by the deposited layer than the membrane, the fouling layer controls solute retention which tends to improve. The membrane fouling may significantly affect the retention of low MW organic compounds depending on the concentration and characteristics of the foulants, the membrane properties, and the chemical composition of feed water.
- *Influence of the filtration system operating parameters*
  Rejection of pesticides is also found to be influenced by system operating parameters, such as the water permeation flux and the feed-stream velocity in the cross-flow mode of filtration. According to literature, rejection of pesticides is dependent on operating flux and recovery. In particular the greatest percent rejection occurred at high flux and low recovery, whereas the smallest percent rejection took place at low flux and high recovery, which is in accord with the solution–diffusion theory. According to the previous studies, the retention of dimethoate and atrazine was improved when the pressure was increased and consequently increased water flux.
  Košutić et al. (2005) have conducted a study about pesticides removal from drinking water by nanofiltration. The results have shown that the membrane material and the membrane's pore size distribution effect the uncharged organic molecules rejections. The pesticides' rejections are reasonably high, and besides the effect of size, the specific physicochemical phenomena should be considered in order to understand their rejection by the nanofiltration membranes.
  The results of Mehta et al. (2015) study have shown that rejection of two phenyl urea pesticides (diuron and isoproturon) dissolved in deionized by thin film composite polyamide reverse osmosis from field water was more than 95%. Membrane fouling and permeate flux decrease depends on type of water. The presence of mono-(Na^+) and bi-valent (Ca^{2+}) ions in feed water influences membrane efficiency. Depending upon the pesticide, membrane property influences permeate flux and water permeability (Košutić et al., 2005; Mehta et al., 2015; Plakas and Karabelas, 2012).
  Zhang et al. (2004) research shows that atrazine and simazine rejection by nanofiltration was the highest at pH 8 and decreased at higher and lower pH. In this study different water matrix has been considered. The rejection of pesticides was higher in river water and tap water than in distilled water, but the water flux was lower. This was mainly explained by ion adsorption inside the membrane pores.

5.7.2 Pesticides removal by photobased methods

As alternatives, photochemical methods are based on the utilization of light radiation as the source of energy. The source of light can be a solar light or an external UV light (like Xenon or mercury lamps). Usually absorption of radiant energy, that is, photons by reactant molecules can bring in both photophysical and photochemical changes. In the photochemical processes, the transformation of the reactant molecules is achieved mainly by two methods. If the reactant molecules absorb the radiant light and transforms, the process is called photolysis. In this process, the reactant molecules (or sensitizers) absorb the light energy and are transformed into other chemical forms through activation. This process thus involves the homolytic cleavage of activated molecules to form the degradation products. In fact the prolific dividends of photobased methods in the treatment of pesticides are diverse enough to include low-cost operations, easy handling, high efficiency, etc (Chavoshani et al., 2018). These methods not only provoked pesticide treatment but also facilitated complete mineralization.

- *Photolysis*

  Photolytic degradation of pesticides aims to induce the chemical change through the absorption of radiation in both direct and indirect routes. In the direct photolysis, a chromophore in the structure of pesticide absorbs the radiant energy to form an activated molecule of pesticide. Such excited molecules may then undergo processes like homolysis, heterolysis, or photoionization. Hence highly recalcitrant compounds that do not undergo any chemical change with other methods (or components) may be degraded by this process. Although this route is fairly simple in the treatment of pesticides, it is not easy to fulfill complete mineralization.

- *Photolysis combined with oxidants (H_2O_2/O_3)*

  The efficiency of photolysis, if combined with chemical oxidants like hydrogen peroxide or ozone, can be improved to enhance the rate of pesticide degradation further. For example in the synergetic process between UV/H_2O_2, the radiation can be used to cleave the O—O bond of hydrogen peroxide to yield the $^{\cdot}OH$ radicals. The formation of the $^{\cdot}OH$ radical is the fundamental principle of the photochemical advanced oxidation processes (AOP). Many studies have been carried out to learn the significance of oxidants (like ozone) addition relative to conventional photolysis.

- *PhotoFenton methods*

  In a typical photonFenton system, the irradiation of Fenton reaction systems with UV/visible light takes place. Upon absorption of the light, the photochemical reduction of Fe(III) to Fe(II) followed by reaction with hydrogen peroxide leads to the formation of a powerful oxidizing agent like the hydroxyl radical. Thus the resultant reactive oxidants (like hydroxyl radicals) will eventually promote the photochemical oxidation process to degrade biologically and chemically recalcitrant pesticide compounds in water bodies at circum-neutral pH. In a study photoFenton degradation has been evaluated against a mixture of commercial pesticides typically used to maintain greenhouse agriculture (10% oxamyl, 20% methomyl, 20% imidacloprid, 40% dimethoate, and 40% pyrimethanil).

In recent years, the application of solar photoFenton plus biological treatment has also been sought actively for the treatment of recalcitrant compounds. Martin et al. carried out the photoFenton approach coupled with biotreatment for four pesticides (laition, metasystox, sevnol, and ultra acid). It was found that the addition of this pretreatment (prior to the activated biosludge degradation) rapidly enhanced the efficiency and rate of degradation. Complete degradation of a pesticide mixture has been accomplished by a combination of a photoFenton pretreatment and an activated-sludge batch reactor (31% of mineralization). Therefore the combined process can be applied effectively by rapidly degrading pesticides in wastewater to aim for the complete removal of parent compounds.

- *Semiconductor based photocatalysis*

  The basis of photocatalysis is the photoexcitation of a semiconductor due to the absorption of electromagnetic radiation in either UV or visible spectrum. Upon absorption of light, electrons in the valence band of a semiconductor are excited to the conduction band, leaving a positive hole in the valance band. The empty hole on the valence band (+ charge) and electron on the conduction band (− charge) are able to induce reduction or oxidation of the pesticide or any targeted contaminant contained in aqueous solution. Based on this principle, a wide range of pesticides have been treated by photocatalytic degradation. Indeed there are many types of semiconductor materials (ZnO, TiO_2, and WO_3) to use for photocatalytic purposes. Among them titanium dioxide has been employed most extensively because of its favorable catalytic properties. In general titania exists in three forms (i.e., anatase, brookite, and rutile) wherein anatase has been employed most commonly in ambient conditions due to its stability. OCPs can also be treated by the photocatalysis method in a very short duration. For instance, photocatalytic degradation of dicofol was investigated on TiO_2 nanoparticles (TiO_2-NPs) under UV light irradiation. In addition to titania many other photocatalysts (ZnO, WO_3, etc.) have also been tested for their treatment potential for pesticides. In another work the ZnO-based degradation of diazinion in aqueous solution was studied under UV irradiation.

5.7.3 Adsorption for pesticides removal

5.7.3.1 Biosorption method using low-cost sorbent

Biosorption is a rapid sorption process include physicochemical and ion exchange interactions occurring at the cell surface between a sorbate and live, dead, or inactive biomass. It is important to note, when using live biomass, that biodegradation may also occur concurrently with sorption phenomena, and the total removal observed may consist of sorption and degradation contributions, and it is difficult to distinguish the contribution of each. Factors such as temperature, contact time, pH, sorbent dosage, and ionic strength, presence of competing agents, sorbent and sorbate types, and sorbent specific surface area affect biosorption efficiency at various extents (Abdi and Kazemi, 2015).

- *pH effects on pesticide's biosorption*

  The pH of the medium in which biosorption occurs has a pronounced effect on biosorption efficiency. Gupta observed that biosorption of dichlorodiphenyldrichloroethane (DDD) and dichlorodiphenyldichloroethylene (DDE) has not the same trend in the range of pH 2.0–9.0 by bagasse fly ash. This means that increasing the pH to a given value increases biosorption capacity, after which the reverse trend starts. Herein by increasing pH from 2 to 7, the biosorption efficiencies increased from 0.21 to 3.6 mg/g for DDD and from 0.11 to 3.42 mg/g for DDE.

  Jianlong et al. studied the behavior of adsorption of PCP at pH 6–8 onto activated-sludge biomass. They reported that at 258°C and 125 rpm, increasing the pH has negative effect on PCP biosorption, so that at pH 6, biosorption capacity dropped to at initial pH 8. They gave reason that at pH 6, the surface charge of the sorbent becomes more positive while PCP is negatively charged, and these lead to improvement in biosorption efficiency.

- *Effect of temperature*

  It observed that as the temperature raises removal efficiency of some pesticides decreases, which is explained by greater solubility of pesticides at higher temperature reducing the affinity of pesticides such as endosulfan sulfate for biosorption on the surface of the sorbents. Ghosh et al. changed the temperature in the range from 20°C to 408°C. They observed that the temperature variation has no effect on the removal efficiency of lindane by *R. oryzae* biomass.

- *Effect of initial sorbate concentration*

  The other parameter that can affect on the efficiency of pesticide removal is initial sorbate concentration. In many cases suppression of the all mass transfer resistances of the sorbate between the aqueous and solid phases can be provided as an important driving force by initial sorbate concentration parameter. Then adsorption process enhancement may occur at the higher initial concentration of pesticide. So it is essential to increase sorbate concentration relative to sorbent dosage to find out the concentration in which sorbent sorption sites become saturated. This enables us to find practical sorbent sorption capacity.

- *Effect of chemical treatment*

  Some studies deal with the treatment of biosorbents to modify structure and, thereby, improve the removal efficiency. Using chemical treatment, for example, 0.1 M HNO_3, is led to minerals washing out from the sorbent surface and increases the surface area, that increase adsorption potential of pesticide from aquatic environments by sorbent. Chemical treatment included submerging the particles sorbent in H_2O_2 and then reacting with 0.1M HNO_3 and finally soaking in methanol (Zolgharnein et al., 2011).

- *Effect of sorbent dosage*

  One of the important parameters is sorbent dosage because this determines the capacity of a sorbent for a given initial concentration of sorbate. Some researchers devoted a section to examine the effect of changing the ratio of sorbent initial amount to sorbate concentration on removal efficiency. They wanted to know if it is possible to use lower values of this ratio while keeping the removal efficiencies satisfactory. Increasing the sorbent mass in a solution having fixed concentration of solute will increase the sorbent efficiency usually defined as mg of sorbate/g sorbent. The higher sorption capacity means that the sorbent has been utilized more efficiently since most of the sorbent removal capacity is exhausted and that less of the waste is produced. In most cases removal efficiency rises with the increasing sorbate dosage. Memon et al. reported that rapidly increasing the sorption up to 90% by increasing the amount of sorbent from 0.1 to 0.4 g, and then it remains almost constant up to 1 g of sorbent dosage.

5.7.3.2 Pesticides adsorption on carbonaceous materials

The most commonly used adsorbent in adsorption processes is activated carbon. Generally carbonaceous adsorbents have a special application among the main adsorbents, as they are capable adsorbing diverse organic compounds. However, the operation cost in full-scale plants is high because carbon loss during the regeneration process (thermal desorption or combustion of toxicants). A diversity of activated carbon have been applied, for example, granular activated carbon (GAC), carbon cloth, powdered activated carbon (PAC), fibers, carbon cloth electrodes, black carbon from carbonaceous residues (WC), carbon black and commercial activated carbon (AC). The forms granular activated carbon (GAC) and powdered activated carbon (PAC) are the most used due to they are capable to the adsorption of the different pesticides. The adsorption process of powdered activated carbon has been used as an effective method, to remove residual pesticide from water during treatment process. At pesticide concentration in the range of 0.20–0.50 mg/L, 2–50 mg/L (200 ppt–50 ppm) PAC are required for the achievement of the upper permitted detection limit (0.1 mg/L or ppb). The pesticides adsorption onto PAC is dependent on their physicochemical characteristics. The use of residues of agricultural products for the production of cheap and effective activated carbon has been reported in the literature. Generally the activation method consists of two steps: a carbonization step under inert atmosphere, followed by an activation step under high temperature in the presence of an activated agent such as carbon dioxide, steam or air (Abdi and Kazemi, 2015).

5.7.3.3 Pesticide adsorption on biochar

Biochar generally has strong sorption capability for pesticides, due to its specific physicochemical properties which largely depend on its feedstock (such as, pinewood, wheat straw, rice husk, dairy manure, sugar beet tailing, and sewage sludge) and the pyrolysis conditions (such as temperature, heating rate, and residence time). The dominant properties affecting pesticide sorption–desorption by biochar have presented as follows:

- *Porosity and surface area*

  Porosity and surface area are two primary properties which affect the sorption capacity of biochar for organic pollutants including pesticides through pore filling and surface sorption, respectively. More porous structures and higher surface area will result in higher sorption capacities. Pore structure can be formed in biochar during the pyrolysis of biomass due to water loss in dehydration process, volatilization of organic matter, fracture, as well as collapse. Generally pore size plays a crucial role in pesticide sorption. Therefore biochar with small pore size cannot trap large pesticide molecules, regardless of their charges or polarity. Biochar porosity and surface area vary substantially with pyrolysis temperature. Apart from pyrolysis temperature, the composition of biochar feedstock is also an important factor affecting its properties.

- *pH*

  Higher pH of biochar can accelerate the hydrolysis of organophosphorus and carbamate pesticides through alkali catalysis mechanism. Akin to porosity and surface area, biochar pH is also influenced by pyrolysis temperature and feedstock. Increasing temperature generally results in higher ash content, which contributes to the higher pH of biochar.

- *Surface functional groups*

  Surface functional groups including carboxylic (–COOH), hydroxyl (–OH), lactonic, amide and amine groups are essential for the sorption capacity of biochar. Generally pyrolysis temperature and raw material are two crucial factors influencing the quantities of functional groups on biochar surface. However, unlike the increasing tendency of surface area, porosity, and pH, the higher temperature usually leads to decrease in the ratios of H/C, O/C, and N/C indicating a reduction in abundance of functional groups on biochar. The FTIR (Fourier Transform Infrared Spectroscopy) spectra, which have been extensively used to characterize the surface functional groups of biochar, showed a significant difference in composition in biochars produced at different temperatures.

- *Carbon content and aromatic structure*

  Both carbon content and aromatic structure are important factors affecting sorption capability of biochar for pesticides, while aromatic structure also plays an important role in the sustainability of biochar.

- *Mineralogical composition*
  The mineralogical composition which is responsible for the high cation exchange capacity of biochar is also considered to be very important for pesticide sorption. They can reduce the availability of pesticides through surface chelation and/or surface acidity mechanisms. Both pyrolysis temperature and raw material influence the concentrations of mineral components in biochar. Higher temperature enriches minerals in biochar. For example, biochar derived from higher pyrolysis temperature can provide more sites for pesticide sorption due to the higher surface area. While, at the same time, it may lead to lower pesticide sorption by reducing functional groups on the surface.
- *Modification of biochar to enhance pesticide sorption*
  Biochar sorption capacity is actually not as high as other types of sorbents, such as clays (e.g., stevensite, smectite), and activated carbon. Therefore modification of biochar, for example, increasing surface area, porosity, and/or functional groups, to enhance its sorption capacity has received increasing attention in recent years. Biochar modification methods such as loading with organic functional groups, nanoparticles, and activation with alkali have been well reported. Biochar modification also includes loading biochar with different minerals such as hematite, magnetite, zero valent Fe, hydrous Mn oxide, calcium oxide, and birnessite. The loading can be performed during the whole pyrolysis process of the raw material. Various exogenous functional groups have also been grafted to biochar. For example, amino and hydroxyl groups were added to biochars via polyethylenimine and chitosan modification. Furthermore, increasing the surface area of biochar by incorporating nanoparticles, such as ZnS nanocrystals, can enhance its sorption capacity for pesticides. Similarly activation with alkali solution can also increase the surface area of biochar, resulting in enhanced pesticide sorption capacity. Moreover, methanol modification can observably alter the functional groups and O-containing group properties of pristine biochar, which may play a dominant role in pesticide sorption compared to unmodified biochar (Ahmad et al., 2014).

5.8 Analysis of pesticides

5.8.1 Extraction procedures

The determination of pesticide residues in environmental samples, for example, water and soil, and agricultural products has been a major subject for many years because of their potential risk for human health, persistence and tendency to bioaccumulate. For water samples, analytical methods usually include extraction and enrichment steps for determining pesticide residues at very low level ($<$ sub-μg/L) (Ochiai et al., 2006). A suitable sample preparation procedure is an inevitable part of many chemical analyses. Typically the sample preparation step consists of extraction of target analytes from a sample matrix (Chau, 2018; Florindo et al., 2017; Tadeo, 2008). Traditional solvent extraction has been applied for many years to a wide range of compounds. The common extraction methods have been presented in the following:

5.8.1.1 Dispersive liquid–liquid microextraction method (theory and fundamentals)

In dispersive liquid–liquid microextraction (DLLME) a water-miscible and polar disperser solvent containing a water-immiscible extraction solvent is injected into the aqueous solution of analytes. A cloudy solution (a mixture of water, disperser solvent, and extraction solvent) is formed. Analytes experience enrichment in the dispersed fine droplets of the extraction solvent which is then separated by centrifugation. The advantages of this microextraction technique are simplicity of operation, low cost, rapidity of extraction, and high enrichment factor. The DLLME method was performed in a narrowbore tube containing aqueous sample. Acetonitrile and a mixture of *n*-hexanol and *n*-hexane (75:25, v/v) were used as disperser and extraction solvents, respectively, in the literature. The effect of several factors that influence performance of the method, including the chemical nature and volume of the disperser and extraction solvents, number of extraction, pH and salt addition, was investigated and optimized in the literatures (Farajzadeh et al., 2012; Farajzadeh and Khoshmaram, 2015; Farajzadeh et al., 2016a,b).

DLLME uses a ternary solvent system in which a small amount of a blend of extraction and disperser solvents is rapidly injected into an aqueous sample containing the analyte. After shaking the mixture a cloudy solution is obtained, and tiny fine droplets of the extraction solvent are formed. The surface area between water and the extraction solvent becomes infinitely large, so rapid, effective mass transfer occurs. The mixture is then centrifuged, and the sedimented phase is collected with a micro syringe for subsequent analysis (Ahmad et al., 2015; Farajzadeh et al., 2012; Farajzadeh and Khoshmaram, 2015; Farajzadeh et al., 2016a,b).

5.8.1.1.1 Analytical parameters affecting extraction efficiency of dispersive liquid–liquid microextraction

Development of strategies and techniques allowing the selection of representative samples continues to be the main focus of research in pretreatment and clean-up. In DLLME several factors influence the extraction efficiency, including type and volume of the extraction solvent, type and volume of the disperser solvent, pH of the sample, effect of salt, extraction time, centrifugation time, and sample volume. These parameters have to be optimized for high extraction efficiency. The most important task in the process is the selection of an appropriate extraction solvent, based on several requirements. It must be immiscible in water and should form tiny droplets in it. It must have higher affinity towards the analyte. Its volume should be carefully optimized, and its compatibility with the desired instrument is also considered. Optimizing the disperser-solvent type and volume is as important as optimizing the extraction solvent. The function of the disperser solvent is to empower the extracting solvent to partition itself uniformly in the aqueous sample, in order to achieve good extraction efficiency. The ratio of volume of extracting solvent to disperser solvent should be carefully considered. The volume of the sedimented phase is significantly influenced by disperser type and volume. Acetone, methanol, and acetonitrile are commonly used as disperser solvents (Watanabe et al., 2014). Optimal pH and salt concentration should be established. Extraction and centrifugation times should be carefully optimized. Extraction time is defined as the time interval between the injection of the mixture of disperser and extraction solvent and centrifugation. Centrifugation is important for phase separation and is time consuming compared to other parameters. It is usually 5–10 min. A long centrifugation time causes phase separation to dissolve (Ahmad et al., 2015; Farajzadeh et al., 2012; Farajzadeh and Khoshmaram, 2015; Farajzadeh et al., 2016a,b).

5.8.1.1.2 Classification in dispersive liquid–liquid microextraction

Several new advances occurred in time to overcome possible disadvantages and drawbacks of the process, hence leading to different modifications in DLLME, and, most often, for each modification a different acronym was assigned by the researcher. Sometimes there are more than two or three acronyms for the same DLLME method, so it is often difficult to differentiate them or it leads to complications. The four bases of classification are:

1. mixed mode extraction;
2. extraction based on assisting dispersion;
3. extraction based on use of ionic liquids (ILs); and,
4. extraction based on solvent density in its acronym, and any other type not fitting into the other three groups (Sajid and Alhooshani, 2018; Ahmad et al., 2015; Farajzadeh et al., 2012; Farajzadeh and Khoshmaram, 2015; Farajzadeh et al., 2016a,b).

5.8.1.1.3 Analytical applications of dispersive liquid–liquid microextraction in the analysis of pesticides

Classification of applications of DLLME continues to play a key role in searching for structure in data. Thus it allows meaningful generalization about large amounts of data to be performed by recognizing a few basic patterns among them. The techniques coupled with DLLME techniques can be summarized as follows.

5.8.1.1.4 Dispersive liquid–liquid microextraction combined with gas chromatography

It is evident from the large number of publications that DLLME is mainly used for the analysis of pesticides. Moreover, the most favorable analytical technique is gas chromatography (GC), since it has rapidly developed in a short time. The extraction solvent in DLLME should be less soluble or immiscible with water in order to achieve adequate phase separation, followed by direct injection into GC.

Pesticides are mostly analyzed in an aqueous matrix because they are directly introduced to water for agricultural purposes and with personal care products (PCPs), such as shampoos and sprays. Due to their continual supply, pesticides have become persistent in the environment, causing adverse effects to health. The World Health Organization (WHO) and various national governmental institutions have established residue limits and published guidelines and policies for quantification of pesticide residues in different kinds of waters, including environmental, drinking and irrigation waters. Alves et al. determined six organophosphorus pesticides (OPPs) in water utilizing DLLME prior to GC–mass spectrometry (GC–MS). In their method, chloroform was used as the extraction solvent and 2-propanol as disperser. The limit of detection (LOD) of the method was 1.5–9.1 ng/L. The method was validated for tap, well, and irrigation waters with relative recoveries 46.1%–129.4%.

A new method, ultrasound-vortex-assisted DLLME (US-VA-DLLME) for OPPs and triazines, resulted in excellent LODs of 0.007–0.07 ng/mL. Similarly OPPs were analyzed by combining supercritical fluid extraction (SFE) with DLLME in marine sediment samples. This method has been applied to real soil and marine samples with satisfactory results. Sousa et al. developed a methodology for the analysis of carbamates and carbofuran in water using DLLME–GC–MS. Chlorobenzene was used as extracting solvent and methanol as disperser solvent. A total sample volume of 10 mL satisfactorily obtained LODs of 0.04 μg/L for carbamates and 0.02 μg/L for organophosphorus pesticides. Similarly Chen et al. proposed a low density extraction solvent-based solvent terminated DLLME (ST-DLLME) combined with GC–tandem MS (GC-TMS) for determination of carbamate pesticides in water samples with good recoveries of 94.5%–104% at all spiked levels for all carbamates used. (Farajzadeh et al., 2012; Farajzadeh and Khoshmaram, 2015; Farajzadeh et al., 2016a,b).

5.8.1.1.5 Advantages of dispersive liquid–liquid microextraction

DLLME offers rapid, low cost, short time, reliable, simple operation, high preconcentration and recovery factors and environment friendliness. LODs in DLLME are much better and feasible compared to other liquid-phase microextraction techniques. In DLLME a syringe is employed for collection and injection of the extract, so problems are avoided, whereas, in SDME, the problem of drop dislodgment is common due to the use of a syringe as the drop holder during extraction. The performance of DLLME is much faster than cloud-point extraction (CPE) because, in many cases, CPE requires heating of the aqueous solutions for long periods to achieve the cloud-point temperature. DLLME has been successfully combined with extraction techniques (e.g., SPE, SFE, SBSE, and some nanotechniques such as D-μ-SPE). Combination of D-μ-SPE with DLLME leads to higher levels of selectivity and sensitivity and extends the analytical utility of

DLLME in complex matrices. The analytical applications of DLLME in analysis of pesticides are limited since the main disadvantage of DLLME is the consumption of relatively large volumes (i.e., mL) of disperser solvents, which usually decrease the partition coefficient of analytes into the extractant solvent. This problem is avoided by using ultrasonic energy or cationic surfactant to disperse the extraction solvent instead of disperser solvent. Recently some interesting modifications were introduced to avoid the problems and to expand the analytical utility of DLLME, and there is an increasing number of applications of DLLME techniques. (Farajzadeh et al., 2012; Farajzadeh and Khoshmaram, 2015; Farajzadeh et al., 2016a,b).

5.8.1.2 Solid-phase extraction

Solid-phase extraction (SPE) method has received wide acceptance for extraction of pollutants including pesticide compounds. It is used on adsorption of analytes onto solid sorbents, thus selecting a suitable sorbent is very important and depends on the interaction between the sorbent and the analytes itself. Considered as a convenient pesticide extraction method, numerous researches have been conducted based on SPE approach. Recently, Rashidi Nodeh, Aini, Afzal, and Sanagi have applied SPE for OPs pesticides analysis from different water samples. A graphene-based tetraethoxysilane methyltrimethoxysilane sol–gel hybrid magnetic (Fe_3O_4 @G-TEOS-MTMOS) nanocomposite was synthesized and applied in magnetic SPE (MSPE) as adsorbent. GC-μECD was used to determine and quantify the analytes. High percentages of recovery (83%–105%) were obtained, and the obtained LOD was below the maximum residue limit (MRL) for the spiked OPs from water samples. SPE method was used as clean up procedure, and the elution ratio of acetone: toluene (3:1) was applied for better response. In combination with GC-ECD, the determination and quantification were carried out, and LOD was found lower than MRL set by the European Union (EU) in the range of 0.02–4.5 ng/g (Żwir-Ferenc and Biziuk, 2006; Beltran et al., 1998; Beltran et al., 2000).

5.8.1.3 Solid-phase microextraction

Solid-phase microextraction (SPME) is a rapid and solvent-free sample preconcentration method. This method was introduced by Arthur and Pawliszyn (1990) and had overcome lots of limitations faced by conventional techniques. As its outstanding features includes solventless or no solvent and simple and easiness in operating, SPME has received great applications in analytical separation. A newly synthesized hierarchical pores graphene framework (HPGF) has applied and coated on the SPME fiber as adsorbent. The proposed method is coupled with GC analysis and LODs obtained in the range of 0.08–0.80 ng/L and with a wider linearity of 10–30,000 ng/L. Li et al. applied a C18 composite SPME fiber with a new method for the analysis of OCs pesticides in water sample. They used coupling method with GC–MS and managed in getting good sensitivity for the determination of six OCs pesticides in water samples. The LODs for all pesticides

Table 5.3 Summarization of extraction methods for pesticides (Samsidar et al., 2018; Zhao and Lee, 2001).

Extraction method	Class of pesticides	Matrix
Liquid-liquid extraction (LLE) • *Advantages* ◦ Simple and reliable ◦ Adaptable to the various sample types and analytes ◦ Compatible with majority of analytical instruments. • *Drawbacks* ◦ Requires large volume of hazardous solvent ◦ Time consuming technique	Multiclass pesticides	Well water
Solid-phase extraction (SPE) • *Advantages* ◦ Less time consuming than LLE ◦ Has effective purification and preconcentration technique • *Drawbacks* ◦ Requires pretreatment and further needs toxic ◦ Organic solvent	Multiclass pesticides OCs Carbamate	Tap water River water Lake water sea Drinking and environmental water Lake water
Quick, easy, cheap, rugged, effective and safe(QuEChERS) • *Advantages* ◦ Wide scope of analytes (including high polarity, acidic, and basic pesticides) ◦ Requires low volume of solvents and glassware ◦ Simple instrumentation ◦ Flexible and effective • *Drawbacks* ◦ Low enrichment factors	OPs	Tap water
Solid-phase microextraction (SPME) • *Advantages* ◦ Solvent-free ◦ Simple and easy to use ◦ Fast ◦ Portable • *Drawbacks* ◦ Quite fragile of fiber ◦ Limited lifetime ◦ Difficulty in sample carry over	OCs OPs	River water Surface water and tap water River water and agricultural waste Water

(*Continued*)

Table 5.3 Summarization of extraction methods for pesticides (Samsidar et al., 2018; Zhao and Lee, 2001). *Continued*

Extraction method	Class of pesticides	Matrix

Dispersive liquid–liquid microextraction (DLLME)
• *Advantages*
  • Simplicity
  • Minimal volume of toxic solvents
  • High speed extraction
  • Inexpensive
• *Drawbacks*
  • Low efficiency of extraction
	OPs	Tap water
		Rain water
		River water

Single Drop Microextraction (SDME)
• *Advantages*
  • Quick and cheap
  • Easy to operate
  • Environment friendly as it requires little organic
  • Solvents
  • Renewability of extraction phase
• *Drawbacks*
  • Less stability of the suspending drop
  • Quite long extraction time
	Fungicides (Chlorothalonil, kresoxim-methyl, famoxadone)	Lake water
		River water
		Waste-water treatment plant

With permission.

tested (hexacholorbezene, transchlordane, cischlordane, *o,p*-DDT, *p,p*-DDT, and mirex) achieved in the ranges of 0.059–0.151 ng/L· (Henriksen et al., 2001; Psillakis and Kalogerakis, 2003; Samsidar et al., 2018; Aguilar et al., 1998; Lu et al., 2007; Yao et al., 2001). The common extraction methods and analysis instruments for pesticides have been shown in Tables 5.3 and 5.4.

5.9 Pesticide management and control

Recognition of pesticide abuse and environmental and public health impacts the European countries to adopt a variety of measures that include the following:

- reduction in use of pesticides (by up to 50% in some countries),
- bans on certain active ingredients,
- revised pesticide registration criteria,
- training and licensing of individuals that apply pesticides,

Table 5.4 Detection techniques of pesticide residues: Analytical and advanced techniques (Samsidar et al., 2018).

Detection method	Pesticides	Limit of detections (LODs)
GC-ECD	OCs	1.7 ng/L
GC-μECD	OPs	1.4–23.7 pg/mL
GC-ECD	Multiclass pesticides	0.02–4.5 ng/g
GC-ECD	Pyrethroids	0.1–0.5 ng/g
GC-ECD	OCs	0.52–3.21 ng/kg
GC-ECD	Multiclass pesticides	0.002–0.667 ng/mL
GC-ECD	Multiclass pesticides	0.4–37 μg/kg
GC-ECD	OCs and Pyrethrin	0.04–1 μg/L
GC-ECD	Ocs and Triazines	0.003–0.04 mg/L
GC-ECD	Multiclass pesticides	LOQ: 0.03–10.6 ng/g
GC-FID	OPs	0.82–2.72 ng/mL
GC-FID	Multiclass pesticides	0.34–5 μg/L
GC-FID	Pyrethroids	0.02–0.16 mg/kg
GC-FPD	OPs	0.043–0.085 μg/L
GC-FPD	OPs	0.02–0.61 mg/kg
GC-FPD	OPs	1.5–5.6 μg/L
GC-NPD	Multiclass pesticides	0.05–0.43 ng/mL
GC-NPD	OPs	0.02–0.1 μg/L
GC-NPD	Triazole	0.05–0.21 ng/g
GC-NPD	OPs	0.29 and 3.20 μg/kg
GC-NPD	OPs	0.1–2.0 ng/g
GC-NPD	OPs	1.85–7.32 μg/L
GC–MS	OCs	0.12–26.28 ng/L
GC–MS	Multiclass pesticides	0.5–1.0 ng/kg
GC–MS	OPs	1.2–5.1 ng/L
GC–MS	OPs	0.59–1.57 μg/kg
GC–MS	OCs	0.059–0.151 ng/L
GC–MS	OPs	0.012 and 0.2 ng/g
GC–MS	OCs	1.4–15 ng/L
GC–MS	OPs	0.4–1.7 μg/kg
GC–MS	Multiclass pesticides	0.026–33.515 ng/L
GC–ICD/MS	OPs	0.005–0.020 μg/L
GC–QqQ/MS	Organochlorines	0.39–0.70 ng/L
GC–MS/MS	Multiclass pesticides	0.005–0.025 mg/kg
GC–MS/MS	Multiclass pesticides	0.03–2.17 μg/kg
GC–EI/MS/MS	Ocs and OPs	Not stated
LC–MS/MS	Multiclass pesticides	Not stated
LC–MS/MS	Multiclass pesticides	Not stated
LC–MS/MS	Multiclass pesticide	0.003–0.35 mg/L

(Continued)

Table 5.4 Detection techniques of pesticide residues: Analytical and advanced techniques (Samsidar et al., 2018). *Continued*

Detection method	Pesticides	Limit of detections (LODs)
LC–MS/MS	OPs	0.016 and 50.85 μg/kg
HPLC	OPs	0.01–1.0 μg/L
HPLC	Carbamate	0.58–2.06 ng/g
UPLC–MS/MS	Carbamates	0.5–6.9 ng/L
HPLC–DAD	Acaricides	0.16–0.57 μg/L
HPLC–DAD	Pyrazole	0.3–1.5 μg/L
HPLC–DAD	Neonicotinoid (Thiamethoxam)	0.02 g/kg

With permission.

- reduction of dose and improved scheduling of pesticide application to more effectively meet crop needs and to reduce preventative spraying,
- testing and approval of spraying apparatus,
- limitations on aerial spraying,
- environmental tax on pesticides, and
- promote the use of mechanical and biological alternatives to pesticides.

5.10 Conclusion

Pesticides are emerging micropollutants in the environment and influence on human health. In the recent decade, the production and consumption of pesticides have been increased. The pesticide removal methods are very different also. Membrane filtration, adsorption, and advanced oxidation processes are the common methods for the removal of pesticides. The pesticides are existing to the aquatic environment in trace amounts. Detection of pesticides is based on advanced analysis by chromatography methods. Thus detection and removal of pesticides are necessary because pesticides make adverse impacts on human health. As well as improving pesticide application according to the existing regulations could contribute to the reduction of the pesticide side effects on human health and the aquatic environments.

References

Abdi, O., Kazemi, M., 2015. A review study of biosorption of heavy metals and comparison between different biosorbents. J. Mater. Environ. Sci. 6 (5), 1386–1399.

Aguilar, C., Penalver, S., Pocurull, E., Borrull, F., Marcé, R.M., 1998. Solid-phase microextraction and gas chromatography with mass spectrometric detection for the determination of pesticides in aqueous samples. J. Chromatogr. A 795 (1), 105–115.

Ahmad, M., Rajapaksha, A.U., Lim, J.E., Zhang, M., Bolan, N., Mohan, D., et al., 2014. Biochar as a sorbent for contaminant management in soil and water: a review. Chemosphere 99, 19–33.

Ahmad, W., Al-Sibaai, A.A., Bashammakh, A.S., Alwael, H., El-Shahawi, M.S., 2015. Recent advances in dispersive liquid-liquid microextraction for pesticide analysis. TrAC. Trends Anal. Chem. 72, 181–192.

Arthur, C.L., Pawliszyn, J., 1990. Solid phase microextraction with thermal desorption using fused silica optical fibers. Anal. Chem. 62, 2145–2148.

Aydinalp, C., Porca, M.M., 2004. The effects of pesticides in water resources. J. Cent. Eur. Agric 5, 5–12.

Belenguer, V., Martinez-Capel, F., Masiá, A., Picó, Y., 2014. Patterns of presence and concentration of pesticides in fish and waters of the Júcar River (Eastern Spain). J. Hazard. Mater. 265, 271–279.

Beltran, J., Lopez, F.J., Cepria, O., Hernandez, F., 1998. Solid-phase microextraction for quantitative analysis of organophosphorus pesticides in environmental water samples. J. Chromatogr. A 808 (1-2), 257–263.

Beltran, J., López, F.J., Hernández, F., 2000. Solid-phase microextraction in pesticide residue analysis. J. Chromatogr. A 885((1-2), 389–404.

Bilal, M., Iqbal, H.M., Barceló, D., 2019. Persistence of pesticides-based contaminants in the environment and their effective degradation using laccase-assisted biocatalytic systems. Sci. Total. Environ. 133896.

Bret, B., Potter, T., Gan, J., 2011. In: Goh, K.S. (Ed.), Pesticide Mitigation Strategies for Surface Water Quality. American Chemical Society.

Chau, A.S., 2018. Analysis of Pesticides in Water: Volume I: Significance, Principles, Techniques, and Chemistry of Pesticides. CRC press.

Chavoshani, A., Amin, M.M., Asgari, G., Seidmohammadi, A., Hashemi, M., 2018. Microwave/hydrogen peroxide processes. Advanced Oxidation Processes for Waste Water Treatment. Academic Press, pp. 215–255.

D'Andrea, M.F., Letourneau, G., Rousseau, A.N., Brodeur, J.C., 2020. Sensitivity analysis of the Pesticide in Water Calculator model for applications in the Pampa region of Argentina. Sci. Total. Environ. 698, 134232.

Debnath, M., Khan, M.S., 2017. Health concerns of pesticides. Pesticide Residue in Foods. Springer.

Ensley, S.M., 2018. Pyrethrins and Pyrethroids. In: Gupta, R.C. (ed.). Veterinary Toxicology: Basic and Clinical Principles. 515–520.

Farajzadeh, M.A., Khoshmaram, L., 2015. A rapid and sensitive method for the analysis of pyrethroid pesticides using the combination of liquid–liquid extraction and dispersive liquid–liquid microextraction. Clean–Soil Air Water 43 (1), 51–58.

Farajzadeh, M.A., Djozan, D., Khorram, P., 2012. Development of a new dispersive liquid–liquid microextraction method in a narrow-bore tube for preconcentration of triazole pesticides from aqueous samples. Anal. Chim. Acta 713, 70–78.

Farajzadeh, M.A., Asghari, A., Feriduni, B., 2016a. An efficient, rapid and microwave-accelerated dispersive liquid–liquid microextraction method for extraction and preconcentration of some organophosphorus pesticide residues from aqueous samples. J. Food Compos. Anal. 48, 73–80.

Farajzadeh, M.A., Mogaddam, M.R., Aghdam, S.R., Nouri, N., Bamorrowat, M., 2016b. Application of elevated temperature-dispersive liquid-liquid microextraction for determination of organophosphorus pesticides residues in aqueous samples followed by gas chromatography-flame ionization detection. Food Chem. 212, 198–204.

Florindo, C., Branco, L.C., Marrucho, I.M., 2017. Development of hydrophobic deep eutectic solvents for extraction of pesticides from aqueous environments. Fluid Phase Equilibria 448, 135–142.

Gavrilescu, M., 2005. Fate of pesticides in the environment and its bioremediation. Eng. Life Sci. 5, 497–526.

Genuis SJ, Lane K, Birkholz D. 2016. Human elimination of organochlorine pesticides: blood, urine, and sweat study. Biomed Res Int. https://doi:10.1155/2016/1624643.

Hamilton, D., Crossley, S., 2003. Pesticide Residues in Food and Drinking Water. John Wiley & Sons.

Henriksen, T., Svensmark, B., Lindhardt, B., Juhler, R.K., 2001. Analysis of acidic pesticides using in situ derivatization with alkylchloroformate and solid-phase microextraction (SPME) for GC–MS. Chemosphere 44 (7), 1531–1539.

Honeycutt, R., 2018. Mechanisms of Pesticide Movement Into Ground Water: 0. CRC Press.

Kim, K.-H., Kabir, E., Jahan, S.A., 2017. Exposure to pesticides and the associated human health effects. Sci. Total. Environ. 575, 525–535.

Košutić, K., Furač, L., Sipos, L., Kunst, B., 2005. Removal of arsenic and pesticides from drinking water by nanofiltration membranes. Sep. Purif. Technol. 42 (2), 137–144.

Liu, L., Bilal, M., Duan, X., Iqbal, H.M., 2019. Mitigation of environmental pollution by genetically engineered bacteria—Current challenges and future perspectives. Sci. Total Environ 667, 444–454.

Lu, J., Liu, J., Wei, Y., Jiang, K., Fan, S., Liu, J., et al., 2007. Preparation of single-walled carbon nanotube fiber coating for solid-phase microextraction of organochlorine pesticides in lake water and wastewater. J. Sep. Sci. 30, 2138–2143.

Matthews, G.A., 2018. A History of Pesticides. CABI.

Maqboll, Z., Hussain, S., Imran, M., Mahmood, F., Shahzad, T., Ahmed, Z., Azeem, F., Muzammil, S., 2016. Perspectives of using fungi as bioresource for bioremediation of pesticides in the environment: a critical review. Environ. Sci. Poll. Res 23, 16904–16925.

Marrs, T.C., Ballantyne, B. (Eds.), 2004. Pesticide Toxicology and International Regulation. John Wiley & Sons, Chichester.

Mehta, R., Brahmbhatt, H., Saha, N.K., Bhattacharya, A., 2015. Removal of substituted phenyl urea pesticides by reverse osmosis membranes: laboratory scale study for field water application. Desalination 358, 69–75.

Ministry of Agricultur (MOA), 2015. British Columbia, Canada. Available at: <http://www.agf.gov.bc.ca/pesticides/b_2.htm#2>.

Mir-Tutusaus, J.A., Baccar, R., Caminal, G., Sarra, M., 2018. Can white-rot fungi be a real wastewater treatment alternative for organic micropollutants removal? A review. Water Res 138, 137–151.

Nollet, L.M., Rathore, H.S., 2016. Handbook of Pesticides: Methods of Pesticide Residues Analysis. CRC press.

Ochiai, N., Sasamoto, K., Kanda, H., Nakamura, S., 2006. Fast screening of pesticide multiresidues in aqueous samples by dual stir bar sorptive extraction-thermal desorption-low thermal mass gas chromatography–mass spectrometry. J. Chromatogr. A 1130 (1), 83–90.

Parastar, S., Ebrahimpour, K., Hashemi, M., Maracy, M.R., Ebrahimi, A., Poursafa, P., et al., 2018. Association of urinary concentrations of four chlorophenol pesticides with cardiometabolic risk factors and obesity in children and adolescents. Environ. Sci. Pollut. Res. 25 (5), 4516–4523.

Pereira, L.C., de Souza, A.O., Bernardes, M.F.F., Pazin, M., Tasso, M.J., Pereira, P.H., et al., 2015. A perspective on the potential risks of emerging contaminants to human and environmental health. Environ. Sci. Pollut. Res. 22 (18), 13800–13823.

Plakas, K.V., Karabelas, A.J., 2012. Removal of pesticides from water by NF and RO membranes—a review. Desalination 287, 255–265.

Psillakis, E., Kalogerakis, N., 2003. Developments in liquid-phase microextraction. TrAC Trends in Anal. Chem. 22 (9), 565–574.

Sajid, M., Alhooshani, K., 2018. Dispersive liquid-liquid microextraction based binary extraction techniques prior to chromatographic analysis: a review. TrAC. Trends Anal. Chem. 108, 167–182.

Samsidar, A., Siddiquee, S., Shaarani, S.M., 2018. A review of extraction, analytical and advanced methods for determination of pesticides in environment and foodstuffs. Trends Food Sci. Technol. 71, 188–201.

Sucahyo, D., Van Straalen, N.M., Krave, A., Van Gestel, C.A., 2008. Acute toxicity of pesticides to the tropical freshwater shrimp Caridina laevis. Ecotoxicol. Environ. Saf. 69, 421–427.

Tadeo, J.L., 2008. Analysis of Pesticides in Food and Environmental Samples. CRC Press.

Tiwari, B., Kharwar, S., Tiwari, D.N., 2019. Pesticides and rice agriculture. In: A.K. Mishra, D.N. Tiwari, A.N. Rai (eds.) Cyanobacteria, 303–325.

Watanabe, E., Kobara, Y., Baba, K., Eun, H., 2014. Aqueous acetonitrile extraction for pesticide residue analysis in agricultural products with HPLC – DAD. Food Chem. 154, 7–12.

Yao, Z.W., Jiang, G.B., Liu, J.M., Cheng, W., 2001. Application of solid-phase microextraction for the determination of organophosphorous pesticides in aqueous samples by gas chromatography with flame photometric detector. Talanta 55 (4), 807–814.

Zhang, Y., Van der Bruggen, B., Chen, G.X., Braeken, L., Vandecasteele, C., 2004. Removal of pesticides by nanofiltration: effect of the water matrix. Sep. Purif. Technol. 38 (2), 163–172.

Zhao, L., Lee, H.K., 2001. Application of static liquid-phase microextraction to the analysis of organochlorine pesticides in water. J. Chromatogr. A 919 (2), 381–388.

Zolgharnein, J., Shahmoradi, A., Ghasemi, J., 2011. Pesticides removal using conventional and low-cost adsorbents: a review. Clean–Soil, Air, Water 39, 1105–1119.

Żwir-Ferenc, A., Biziuk, M., 2006. Solid phase extraction technique – trends, opportunities and applications. Pol. J. Environ. Stud. 15 (5).

CHAPTER

Other trace elements (heavy metals) and chemicals in aquatic environments

Bahare Dehdashti[1,2], Mohammad Mehdi Amin[1,2] and Afsane Chavoshani[1]

[1]*Department of Environmental Health Engineering, School of Health, Isfahan University of Medical Sciences, Isfahan, Iran*

[2]*Environment Research Center, Research Institute for Primordial Prevention of Non-communicable Disease, Isfahan University of Medical Sciences, Isfahan, Iran*

6.1 Introduction

According to recent estimates, 1.2 billion people in the worldwide are deprived of basic living conditions such as healthy drinking water. Industry development, energy projects, mining industry, and environmental events are the causes of toxic pollutants in different forms and concentrations in the environment. Approximately two million tonnes of industrial, sewage, and agricultural wastes are discharged daily into the aquatic ecosystem, causing many health problems. Microscopic, biological, inorganic, and organic pollutants are forms water pollutants. So, heavy metal ions that are naturally or anthropogenic (Abdi et al., 2018; Adam et al., 2019) are released into the environment through human activities and subsequently accumulate biologically by exposure and are not easily degradable. As a result, they are a major threat to human health, especially through the food chain. In recent years, extensive studies have been conducted on metals and semi-metals in various environments all over the world. That, attention to aquatic environments has increased due to increased pollutants from human activities such as agriculture and industry. The development of urbanization and the consequent increase in agriculture near aquatic ecosystems in developed countries have had adverse effects on the environment. For example, different fish species are widely exposed to these pollutants, which accumulate in their tissues through food chain (Abdi et al., 2018). Contaminants containing metals and trace elements (TEs) such as mercury, lead, cadmium, arsenic, chromium, and selenium have harmful effects on wildlife and are a major concern for many species. The adverse effects of this group of toxic pollutants such as mercury, arsenic, and selenium are related to their bioaccumulation and biomagnification properties, which increase their concentration through the food chain. As a result, monitoring and evaluation of these toxic elements in the environment is necessary to predict impacts, risk assessment and management (Alam et al., 2019). On

Micropollutants and Challenges. DOI: https://doi.org/10.1016/B978-0-12-818612-1.00006-4

Table 6.1 Summary of sources, health issues, and MCL standard for heavy metal in drinking water (Avigliano et al., 2019).

Heavy metals	Anthropogenic sources	Health issues	Typical concentration of heavy metals in industrial wastewater	MCL by WHO (WHO, 2011)
Arsenic, As (III), As(V)	Fertilizer and pesticide industries	Carcinogenic	0.5–100.0 mg/L (Bessbousse et al., 2008)	0.01 mg/L
Chromium, Cr(III), Cr(VI)	Plating and leather-tanning industries	Carcinogenic	5.0–220.0 mg/L (Bilal and Iqbal, 2019)	0.05 mg/L
Selenium, Se(IV), Se (VI)	Agricultural irrigation, mining	Gastrointestinal disturbances, bronchitis	0.1–100.0 mg/L (Bolisetty et al., 2019)	0.04 mg/L
Copper, Cu (II)	Electroplating industries	Anemia, kidney damage	8.5–21.0 (Engwa et al., 2019)	2.00 mg/L
Cadmium, Cd(II)	Plating and battery industries	Carcinogenic	1.0–4.0 mg/L (Esposito et al., 2018)	0.003 mg/L
Lead, Pb(II)	Mining and batteries	Carcinogenic	25.0–80.0 mg/L (Esposito et al., 2018)	0.01 mg/L
Mercury, Hg(II)	Battery industries	Chromosome breakage	1.0–3.0 mg/L (Esposito et al., 2018)	0.001 mg/L
Nickel, Ni(II)	Electroplating industries	Eczematous reaction	0.1–165.0 (Engwa et al., 2019)	0.02 mg/L
Zinc, Zn(II)	Battery and electroplating industries	Skin inflammation	19.0–72.0 (Engwa et al., 2019)	3.00 mg/L

the other hand, arsenic contamination has also been observed all over the world, with approximately 140 million people exposed to arsenic in 50 countries at regular levels above the WHO standard of 10 μg/L (Adam et al., 2019). Table 6.1 summarizes the MCL standards of heavy metal that set by WHO, anthropogenic sources and health issues of each heavy metal ions on human being (Avigliano et al., 2019) and Table 6.2 provides the chemical properties of common heavy metals found in the environment (Badry et al., 2019).

6.2 Trace elements: definition

According to the International Union of Pure and Applied Chemistry, TEs are any element having an average concentration of less than about 100 parts per million

Table 6.2 Chemical properties of heavy metals.

Heavy metal	Molecular weight ($g\ mol^{-1}$)	Oxidation state (S)[a]	van der Waals radius (10^{-12} m)	Electronegativity (Pauling scale)	Log Kow
Arsenic	74.9	−3, + **3**, + **5**	119	2.18	NA
Cadmium	112.4	+2	158	1.69	3.86 ± 0.36[b]
Chromium	52.0	0, +2, + **3**, + **6**	200	1.66	NA
Cobalt	58.9	−1, 0, + **2**, +3	200	1.88	NA
Copper	63.5	+1, + **2**	140	1.90	NA
Lead	207.2	+**2**, +4	202	2.33	4.02 ± 0.28[b]
Manganese	54.9	−1, 0, + **2**, +3, +4, +6, +7	205	1.55	3.98 ± 0.25[b]
Mercury	200.6	+1, + **2**	155	2.00	0.62[c]
Nickel	58.7	0, + **2**, +3	163	1.91	NA
Zinc	65.4	+2	139	1.65	NA

NA, Not available.

[a]*Bold values represent the most common oxidation state(s) for the heavy metal.*

[b]*Values determined experimentally by (Gifford et al., 2017).*

[c]*Values provided by Michigan Department of Environmental Quality.*

atoms (ppma) or less than 100 μg/g (Guo et al., 2019). So that the concentration of this group of rare pollutants in the water matrix is expressed in the pico moles per liter and lower. As a result, it can be difficult to analyze and track (Henze and Comeau, 2008).

Such a precise definition does not exist in earth sciences because the concentration of an element in a given phase can be so low that it is considered a TE, whereas the same element can constitute a main part of another phase (e.g., Fe and Al). Previously, scientists used the generic term "heavy metals" when referring to TEs. Today this appellation is discussed. Effectively, some metals are not particularly "heavy" (e.g., Al and Ni). In addition, some elements are not metals (e.g., As and Se). For these reasons, the majority of researchers prefer today the name "metallic trace elements" (if it is indeed metals) to the appellation "heavy metals", or the formula "trace elements" when they are not metals (e.g., As, Se, and B). TEs as any element, metallic or not, other than the few major constitutive ones (i.e., C, H, N, S, O, P, Si, Cl, K, Na, Ca, and Mg) forming the bulk of living and mineral (except Fe and Al) matter, whose concentrations are mostly below but sometimes above 100 ppma according to the matrix analyzed. TEs can either be essential or nonessential. TE essential elements recognized by the World Health Organization (WHO) are I, Zn, Se, Fe, Cu, Cr, and Mo, the latter playing an important role in biological systems. Others TEs may/could also be essential, such as Mn, Co, As, Ni, or V. For these elements among others, the essentiality is a characteristic which evolves according to our knowledge and to the sensitivity of the authors who have a propensity more or less strong to classify an element among the essential or not. Nonessential TEs such as Hg, Pb, or Cd play no physiological role, and are oten toxic even in very small quantity. For these nonessential TEs, only a threshold of toxicity exists, while essential TEs can be either deficient in too small quantities, either toxic when they are absorbed in high concentrations (Guo et al., 2019).

6.3 Environmental sources of trace elements

Natural and anthropogenic factors are the two main pathways for TEs to enter the aquatic environment. In spite of people's opinions, natural resources on a global scale are the main source of this group of pollutants in to the environment. But a serious threat to human health has been expressed through human resources. Impact of natural resources is on large areas with low concentrations and specific types of pollutants that it varies based on different characteristics such as area topography, soil type, flora, fauna, and human population. Among the natural factors, one of the most important way that TEs can entry in the aquatic environment is the activity of volcanoes. However, the impact of phenomena in the specific area and uninhabited is also significant. Precipitation has washed away soils and rocks that contain large amounts of TEs. Mercury and selenium compounds, which are volatile elements, can reach water by deposition. Biological activities can also affect the rate of species with vaporizable by increasing the mobility and rapid transfer of TEs. Other natural causes include forest fires, which contribute to the movement of TEs by wind, along with dust and ash particles. The flood also helps to transport these elements by erosion and weathering (Henze and Comeau, 2008).

Despite the limitations of natural resources that create TEs in aquatic environments, the anthropogenic range of these factors are highly diverse. The variety of combustion processes in the creation of TEs is probably the largest source of man-made materials. Combustion processes are associated with the production of dust and ash containing TEs. Waste incineration and biomass, coal and lignin combustion, or even cement production are major sources of pollution. Industries are widely the second largest source of anthropogenic TEs that can be found in non-volatile smelting, battery manufacturing, military industries, pharmaceuticals, or even bubble lamps as distributors of TEs in the environment. In this respect, waste that is inseparable produced by a variety of products is one of the biggest problems, for example nuclear power plants, military equipment, or measuring devices that produce radioactive TEs. Since wastewater treatment processes are based on biological treatment, so they are other anthropogenic sources for producing TEs. Although chemical and mechanical methods are also used in purification processes, they are insufficient for the stabilization of TEs, especially heavy metals. Also not all treatment plants contribute to the release of TEs into surface water. Large sewage sludge treatment plants, which generally collect large amounts of sewage, have this challenge. A large wastewater sludge treatment plant is 100–1000 times more likely than small sewage sludge treatment plants to produce wastewater. Human-propagating TEs are a major threat to aquatic environments. The use of new technologies without waste generation significantly reduces the release of this group of pollutants into the aquatic environment as well as limiting the use of renewable resources to generate human-contaminating energy with TEs (Henze and Comeau, 2008).

6.4 Production and uses

The world refinery and mine production of most TEs except a few (e.g., As, Cd, Pb, and Sn) have substantially increased these last decades (e.g., Fe, Al, and Mo) and particularly since the beginning of years 2000. World demand for minerals is affected by three general factors: (1) uses for mineral commodities; (2) the level of population that will consume these mineral commodities; and (3) the standard of living that will determine just how much each person consumes. Today, with the integration of India, the People's Republic of China and other populous developing and emerging countries (e.g., Brazil and Russia) into the world economy, more than 50% of the world's population (instead of the previous 20%) account for the largest part of raw materials consumption. The increasing demand for mineral raw materials further concerns numerous "emerging elements". These elements can be "truly emerging" because they have just gained entry to the environment (new commercial uses and industrial releases) or have nowadays become contaminants "of emerging concern" while they were not in the past (new advances in analytical chemistry, new knowledge on their toxicity, new environmental compartments explored). The use of these emerging chemicals are multiple and diverse. For example, V is regarded as one of the hardest of all metals. This ubiquitous TE is employed in a wide range of alloys for numerous commercial applications extending from train rails, tool steels, catalysts, to aerospace. Sb greatly increases the hardness and the mechanical strength of Pb, and is found in batteries, antifriction alloys, type metal, small arms and tracer bullets, and cable sheathing. It further has many uses as a flame retardant (in textiles, papers, plastics, and adhesives), as a paint pigment, ceramic opacifier, catalyst, mordant and glass decolorizer, and as an oxidation catalyst. Bi is largely consumed in low melting alloys and metallurgical additives, including electronic and thermoelectric applications. The remainder is used for catalysts, pearlescent pigments in cosmetics, pharmaceuticals, and industrial chemicals (Guo et al., 2019). Moreover e-waste refers to all types of electrical or electronic equipment (EEE) nearing the end of their useful life and is regarded as a critical environmental issue as well as a threat to human health. Approximately, 41.8 and 65.4 million tonnes of e-waste were generated globally in 2014 and in 2017, respectively. It was estimated that only 15% of global e-waste were fully recycled. China processed about 70% of the e-waste produced worldwide and struggled with e-waste burden due to previous illegal e-waste importation, domestic production and consumption of EEE, and unregulated informal handling and recycling. E-waste contains two major types of substances: metals (60%) and plastics (30%) (He et al., 2017).

6.5 Distributions of trace elements

6.5.1 Water and wastewater

Coastal ecosystems are heavily polluted by human activities. These pollutants reach the aquatic environment and accumulate in the sediment. Among the chemical pollutants, metals represent a notable hazard since they are not biodegradable and have the capability to bioaccumulate, resulting in toxic effects in both the short and long term (Hong et al., 2019). Trace metals such as lead (Pb), zinc (Zn), copper (Cu), chromium (Cr), and arsenic (As) are crucial chemical pollutants of aquatic ecosystems (sediments, water, and organisms) and this pollution from toxic metals has proven to be an increasing global problem due to the toxic metals, persistence, inherent toxicity, bioavailability for accumulation in the food chain, and potential threat to ecological systems and human health. Generally pollutants of both geogenic and anthropogenic origin could enter the fluvial systems by surface runoff or by direct atmospheric deposition (Jawed et al., 2020).

To describe the distribution of elements several definitions need explanation (Henze and Comeau, 2008):

- *Bio-Limiting elements*: elements which are those necessary to sustain life and which may exist low concentrations.
- *Conservative elements*: elements whose distribution is relatively homogenous and which maintain a constant ratio to one another, varying only as a result of the water mass mixing.
- *Recycled elements*: sometimes called nutrient type elements or biological elements. As the name suggests the distribution of these elements is controlled by biological cycling.
- *Scavenged elements*: these elements typically have a depth profile that shows some decrease with depth. This is a result of adsorption of the ions or ionic complexes onto particle surfaces, such as clay minerals, organic matter, bacteria, fecal pellets, and so on.
- *Nutrient distribution*: distribution of elements which are considered limiting nutrients. These elements are used up for photosynthesis almost to zero at low depth. Their concentration rises with increasing depth to a maximum of around 800 m. Below this, the concentration of these nutrients begins to fall and stabilizes around 2000 m depth.

Accumulated elements are characterized by (Henze and Comeau, 2008):

- Weak interactions with particles,
- Long residence time (much greater than the mixing time of the ocean),
- Uniform distribution,
- Geochemically controlled content,
- High concentrations in the oceans relative to their crustal abundance,
- Relatively constant concentration: salinity ratio.

Recycled elements are characterized by (Henze and Comeau, 2008):

- Intermediate concentrations and residence times (a few thousand to one hundred thousand years),
- Systematic variations in distribution in space and time,
- Active take up during formation of biological particles by organisms,
- Significantly involved with the internal cycles of biologically derived particulate material,
- Lowest concentrations in surface waters (due to assimilation by plankton and/ or adsorbed by biogenic particles),
- The most obvious elements are those which perform essential biological functions,
- The content is biologically controlled.

Scavenged elements are characterized by (Henze and Comeau, 2008):

- Very strong particle interactions,

- Low concentrations relative to crustal abundance,
- Maximal concentrations near major sources (rivers, atmospheric dust, bottom sediments, and hydrothermal vents),
- Short residence times (100–1000 years),
- Most effectively removed from seawater during formation of particulate matter,
- Heterogeneous distribution,
- Biologically controlled content.

In aquatic systems, major factors influencing the dynamics and extent of contaminant bioaccumulation in organisms include (Henze and Comeau, 2008):

- Element properties: solubility and speciation form,
- Sediment parameters: TOC content, particle size distribution, clay type, clay content, and pH,
- Water characteristics: pH, DOC, temperature, dissolved oxygen, and alkalinity,
- Biological characteristics: organism behavior, modes and rates of feeding, and source of water,
- Organism parameters: age and size.

Organisms (depending on their position in the trophic chain) can metabolize TEs in one of two ways (Henze and Comeau, 2008):

- Directly by absorption through the gills and skin (water pathway),
- Indirectly via the food chain (sediment pathway).

Organisms appear to have evolved a number of mechanisms to limit the uptake of contaminants (Henze and Comeau, 2008):

- Altering the chemical speciation of elements in the environment to reduce their bioavailability,
- Complexing elements at the surface of the organism,
- Decreasing the permeability of the epithelial surfaces by introducing extracellular barriers,
- Reducing transport into the cell across the lipid bilayer,
- Undertaking behavioral avoidance activity.

In the study in 2019, Three sediment cores taken from the Yangtze River Estuary (YRE) and the East China Sea (ECS) in 2016 were analyzed for trace metals (Pb, Zn, Cu, Cr, and As), major elements (Al, Fe, and Mn), sediment composition, chemical properties (Eh and pH), and natural/artificial radionuclides ($^{210}Pb/^{137}Cs$) to decipher the high resolution historical variation in anthropogenic metals over the past 60 years. The results showed that anthropogenic Pb was primarily derived from atmospheric deposition, while anthropogenic Zn, Cu, and As were carried out by fluvial discharge. In the YRE, the recent decrease in sedimentary metals could be largely a result of intensified erosion, which was triggered by the construction of dams in particular the Three Gorges Dam (TGD) in 2003 and soil conservation projects. In comparison, sedimentary records in the deposition-dominated environment of the ECS showed that anthropogenic Pb and Cr began to gradually increase following 1970, but a remarkable increase in anthropogenic Zn and Cu occurred in 1990, probably reflect the different origins and dispersion pathways of these elements. Anthropogenic Pb and Cr exhibited a sharp decrease near early 2000s due to increasing investment in treating pollution. Being sensitive to the substantial reduction in riverine particulate flux of Zn and Cu from the Yangtze River, the concentrations of anthropogenic Zn and Cu in the cores decreased by 5%–27% in 2005–2016. Furthermore, a prominent increase trend of as record after 1990, probably due to the excessive pesticides used in agricultural activity (Jawed et al., 2020).

During 2019, in Spanish Mediterranean coast, the presence of eight TEs (Cr, Cd, Ni, Pb, Cu, Hg, Zn, and As) was determined. In this area, the presence of the contaminants is due to both natural and anthropogenic sources. The results obtained allowed, first, to establish nearness reference values of the area under study, second, to use several pollution indices (contamination factor, enrichment factor, geo accumulation factor, Nemerow pollution index, and modified pollution index) to determine contamination levels in the area, and finally to select the best index to apply in this coastal zone. The best indices to use in this region according to Fig. 6.1 are EF and MPI since both take into consideration the natural contributions of the elements studied. The results revealed that according to the index used only two studied zones are classified as heavily and severely polluted. The remaining zones (between 25% and 29%) were classified as moderately or moderately to heavily polluted and most of the zones (63%–100%) were classified as unpolluted/low polluted and unpolluted/slightly polluted. The outcomes obtained with this work indicate that in general, the Valencian coast does not present significant levels of pollution due to the studied TEs (Hong et al., 2019).

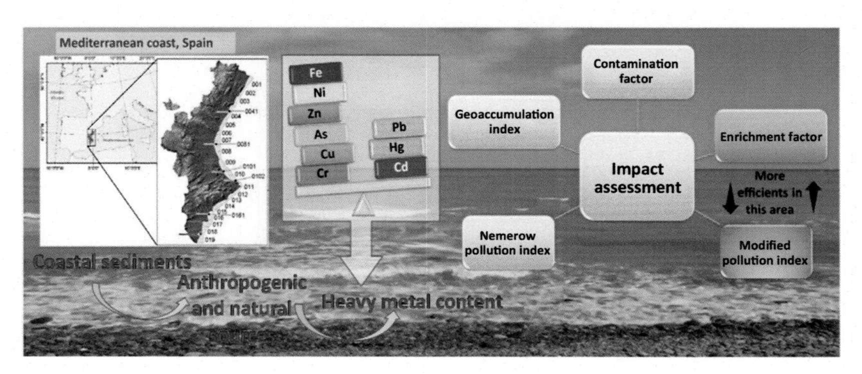

FIGURE 6.1

Mediterranean Coast, Spain (Hong et al., 2019).

Laundry lint is generated during the drying of clothes and linen in a clothes dryer and accumulates on a screen or filter. Lint is composed of fibers of clothing–linen as well as extraneous material that is associated with laundry, like dust, pollen, skin, hair, food, microorganisms and soil, and, potentially, residues of laundry detergent (Fig. 6.2).

Using dryer lint as a novel and noninvasive proxy of household contamination and nondietary exposure since samples are readily collected and material on clothing may reflect a composite of individual chemical sources within and external to the household. To this end, the authors determined the concentration of Pb in dryer lint samples collected by participants from three different groups: a control, comprising university members of staff with no known occupational or recreational exposure to the metal; an urban composite, consisting of samples taken from communal dryers in a city center; and an occupational, made up of radiator shop employees whose managers were aware of potentially high Pb exposure. Arithmetic mean values of Pb in lint were about 20 μg/g for the former groups and about 60 μg/g for secondary radiator shop employees (i.e., clerical or managerial staff) and 350 μg/g for primary shop employees (i.e., those directly involved with soldering). Some participants or family members of the latter category exhibited Pb blood levels above recommended action thresholds. It was analyzed samples of domestic dryer lint and detected various brominated flame retardants. Their presence was attributed to household dust (and associated flame retardants)

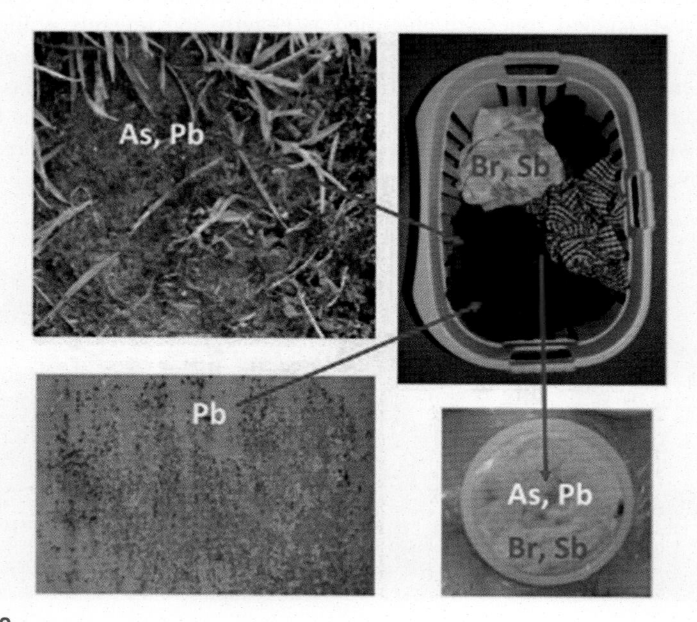

FIGURE 6.2

Trace elements in laundry dryer lint.

being picked up by clothing, although contributions from the dryers themselves could not be ruled out.

More recently, a spectrum of chemicals in domestic laundry dryer lint has been identified that included pesticides, therapeutic and illicit drugs, fragrances, plasticizers and compound metabolites. Significantly, the concentrations of some compounds, including one commonly employed as an insect repellent, were considerably higher in lint than in household dust. Given that the composition of dryer lint reflects the signature of chemicals shed in solid form (and mainly as fibrous microplastics) from the washing of clothes and linen, analysis of lint can also be used to gain semi-quantitative information on chemicals discharged to the environment from this process. Thus, while there has been considerable recent interest in laundry as a source of microplastics to water treatment plants and aquatic systems, analysis of lint may afford a simple means of chemically characterizing these plastics (Jiang et al., 2017).

6.5.2 Sea foods

Over the past few decades, dramatic increase of pollutants, which have strongly affected the health of aquatic ecosystems. This has been reflected in the rapid growth of a wide range of fish and shellfish diseases due to exposure to contaminants, such as inorganic compounds and xenobiotics. Aquatic organisms can accumulate polluting substances from the surrounding aquatic environment in their tissue; the extent of this accumulation is strictly related to the geographic location, species, animal size, feeding patterns, solubility, and lipophilicity of the chemicals, as well as their persistence in the environment. Many elements, such as selenium, iron, zinc, and copper, which are present in fish and shellfish, are essential for humans at low concentrations. However, some of them can over-accumulate from the aquatic environment and can pose a food safety concern. On the other hand, nonessential elements, such as cadmium, lead, mercury, and metalloids—such as arsenic—are toxic even at lower levels of exposure. Moreover, the synergist action of essential and nonessential elements can occur, with negative consequences for human health. TEs find different routes into the aquatic environment; bio concentration can occur directly from water and through the uptake of suspended particles, and increasing levels of metals may be found in predatory species as a result of biomagnification. The presence of chemical compounds can be determined using widespread cosmopolitan bio indicators. An organism is considered a suitable bio indicator if it is able to accumulate high levels of pollutants; it is sessile or constrained to a particular location in order to reflect local pollution; it is relevant in the food chain, abundant and widespread. Moreover, good bio indicators should be easy to sample and identify. It is cleared that plants, invertebrates, and fish are the main pollution bio indicators. Among invertebrates, molluscs play a primary role in assessing water contamination levels due to their ability to concentrate contaminants. Bivalve molluscs have developed some subcellular systems for the accumulation, regulation and immobilization of excessive amounts of essential and nonessential metal elements. Exposure of these invertebrates to pollutants results in the capture of these elements through mechanisms related to their ability to filter the water. Therefore, bivalve molluscs can bioaccumulate substantial amounts of toxic metals into their organs without any apparent negative effects. Wetlands and coastal lagoons are very productive ecosystems but are also valuable and sensitive sites due to their unique ecological conditions. Thus, in moderate or heavily polluted areas that do not have sufficient exchange with the seas, metal concentrations found in seafood may exceed safe limits (Joseph et al., 2019).

One of the elements of highest concern is Hg, which is considered as one of the top ten chemicals or groups of chemicals of major public health concern. It is listed as a priority substance in the field of water policy within the European Union. In many countries of the world the use of mercury has been restricted, but in other parts of the world, huge amounts of the toxic metal are still emitted into the air and into wastewater streams. Hg in fish poses a threat not only to the health of an aquatic ecosystem itself, but also to predatory organisms occupying higher trophic levels in the food web, including fish as well as fish-eating mammals and birds. High Hg levels in fish also endanger the health of humans in many parts of the world where fish is often an important source of protein for the local population. Fish is the main source of foodborne Hg to humans. Therefore, maximum allowed Hg levels have been set for fish consumed by humans in many parts of the world. In the European Union, the water framework directive sets an environmental quality standard level for mercury of 0.020 mg/kgww in fish as an indicator organism. The Canadian Tissue Residue Guidelines for the Protection of Wildlife Consumers of Aquatic Biota recommend a maximum level of 0.033 mg/kgww MeHg in the diet of fish-consuming birds to ensure no negative effect. One TE that has a positive effect regarding mercury levels in food and its uptake into the body is selenium. Back in 1972, Ganther et al. (1972) found a positive effect of Se on quails fed a MeHg diet. Since then, numerous studies have been conducted on the assumed positive effect of Se on Hg toxicity, showing that it moderates the uptake of MeHg and counteracts its toxicity in many different animal species, including fish and humans.

The exact mechanisms of this interaction remain unknown, but most proposed mechanisms imply the formation of Hg\Se compounds. In general, a molar Se:Hg ratio of 1 or higher is considered to be sufficient to lower the toxic effects of MeHg to both fish and fish consumers. Silver After a decrease in the discharge of ionic Ag during the last two decades due to the rise of digital photography and the resulting reduced use of Ag halides in photographic films, the use of nanoparticles of this metal in consumer products is currently booming in many countries. This boom of sanitary and cosmetic products containing Ag nanoparticles also includes South Africa. Silver nanoparticles are increasingly present in wastewater streams, mostly ending up in aquatic environments. Ionic silver is highly toxic to aquatic life, including fish, but the concentrations in the aquatic environment were low, so that little attention has been paid to this element. Until now, very

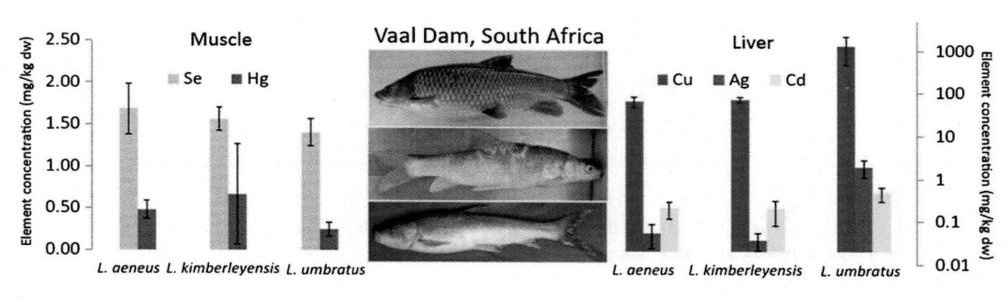

FIGURE 6.3

Trace elements in three cyprinid fish species from the Vaal Dam, South Africa.

little is known about the consequences and the toxicology of silver nanoparticles to the aquatic environment. There are two main advantages of measuring TE levels in fish: First, fish is an important food source for humans in many parts of the world, and monitoring their TE levels is therefore important to ensure food safety. Second, fish provide an integrated insight into possible pollution of their environment over longer periods and are therefore often-used bio indicators for environmental pollution.

The study in 2019 in South Africa the levels of Cr, Cu, Zn, Se, Ag, Cd, Hg, and Pb were determined in muscle and liver samples from 30 specimens of fish belonging to the species Labeobarbus aeneus, Labeobarbus kimberleyensis, and Labeo umbratus from the Vaal Dam. This is the first comprehensive report on Hg levels in fish from this lake. Mean concentrations ranging from 0.247 to 0.481 mg/kg DW in muscle and from 0.170 to 0.363 mg/kg dw in liver clearly show a contamination with this element (Fig. 6.3). Levels of Cu were remarkably high in the liver of L. Umbratus, calling for further investigation on this species. Cadmium levels were above the maximum allowances for fish consumption in the liver of all three species (means between 0.190 and 0.460 mg/kg dw), but below the LOD in all muscle and intestine samples. This is also the first report of Ag in fish from South Africa. Levels were below the LOD in muscle, but well detectable in liver (Konieczka et al., 2018).

6.5.3 Personal care products

Cosmetics have been used as a part of routine body care by all classes of people throughout the world. They are classified as any item intended to be rubbed, poured, sprinkled or sprayed on, or introduced into or otherwise applied to the human body or any part of the body for cleansing, beautifying, promoting attractiveness, or altering the appearance, and include any item intended for use as a component of cosmetics. Cosmetics are mixtures of some surfactants, oils, and other ingredients and are required to be effective, long-lasting, stable, and safe to human use. With many new products released into the market every season, it is hard to keep track of the safety of every product and some products may carry carcinogenic contaminants. There are concerns regarding the presence of harmful chemicals, including heavy metals, in these products. Many cosmetic products contain heavy metals such as lead, cadmium, chromium, arsenic, mercury, cobalt, and nickel as ingredients or impurities. Recent research has reported that these metals can easily cause many types of skin problems. The use of some heavy metals in cosmetic has been controversial due to the biological accumulation of those metals and their toxicity in human body. In most countries, it is legally prohibited to use lead, arsenic, and mercury in skin cosmetic products. It is also reported that these metals can cause allergic contact dermatitis or other skin problems. Since the issue of heavy metals as deliberate cosmetic ingredients has been addressed, attention is turned to the presence of these substances as impurities. The metals of primary toxicological concern in cosmetics are lead, cadmium, arsenic, chromium, mercury, and antimony. Dermal exposure is expected to be the most significant cosmetics are applied to the skin (Lau et al., 2018).

Oral exposure can occur for cosmetics used in and around the mouth as well as from hand to mouth contact after exposure to cosmetics containing heavy metal impurities. However, inhalation exposure is typically considered to be negligible. At higher concentrations, heavy metals have been shown to have negative effects. Lead, which may be an impurity, is a proven neurotoxin linked to learning, language, and behavioral problems. It has also been linked to miscarriage, reduced fertility in men and women, hormonal changes, menstrual irregularities, and delays in puberty onset in girls. Pregnant women and young children are also vulnerable because lead crosses the placenta and may enter fetal brain. Cadmium found in body and hair creams are absorbed into the body through dermal contact and stored in the kidney and the liver, although it can be found in almost all adult tissues. It is considered to be "carcinogenic to humans" and its compounds, categorized as known human carcinogens by the United States Department of Health and Human Services (Lau et al., 2018).

Ingestion of high levels of cadmium can lead to severe stomach irritation, vomiting, and diarrhea, while exposure to lower level for a long time can lead to kidney damage, bone deformity, and the ability of bones to break easily. Chromium is also corrosive and causes allergy to the skin. Ant route for cosmetic products since the majority of adverse effects of the chromium on the skin may include ulcerations, dermatitis, and allergic skin reactions. Mercury is linked to nervous system toxicity, as well as reproductive, immune, and respiratory toxicity. It is also found in thiomersal, which is a mercury based (Lau et al., 2018).

Preservative and used as direct ingredients or impurities. But the high toxicity of this metal means that the presence of mercury in any cosmetic is a concern. Other heavy metals show a similar tendency to be toxic. A recent assessment by WHO reported that mercury in skin lightening creams and soaps that are commonly used in Asian and Central African nations is potentially dangerous as they have serious side effects and can be fatal (Lau et al., 2018).

Fig. 6.4 illustrates a schematic representation of the Food and Drug Administration (FDA) categorized classification of cosmetics. After absorption into the body via dermal penetration (or other sources), these cosmetic products can act as endocrine disruptors, carcinogens, mutagens, neuro-, and reproductive toxins (Lingamdinne et al., 2016).

Lip cosmetics have become one of the most commonly used cosmetics worldwide among modern women to moisturize their lips or appearing more attractive and charming. Nevertheless, some chemical additives such as a dye, shine, natural mineral mica, and other colorants are incorporated in formulating these products to increase the brightness of lip cosmetics achieving healthier quality and boosted effects. Addition of these compounds result in the introduction of Trace metals (TMs) into lip cosmetics, including antimony (Sb), arsenic (As), cadmium (Cd), chromium (Cr), cobalt (Co), copper (Cu), nickel (Ni), manganese (Mn), and lead (Pb). Several adverse effects of above coated TMs available as active ingredients in cosmetics are displayed in Fig. 6.5. Following ingestion, the release of significant amounts of TMs from the lip cosmetics to the digestive tract damages various vital organs once reaching into systemic circulation. It is worth mentioning that the release of TMs by using lip cosmetics is a more active exposure amid female consumers in contrast to unintentional contact to TMs in dust and metal jewelry (Lingamdinne et al., 2016).

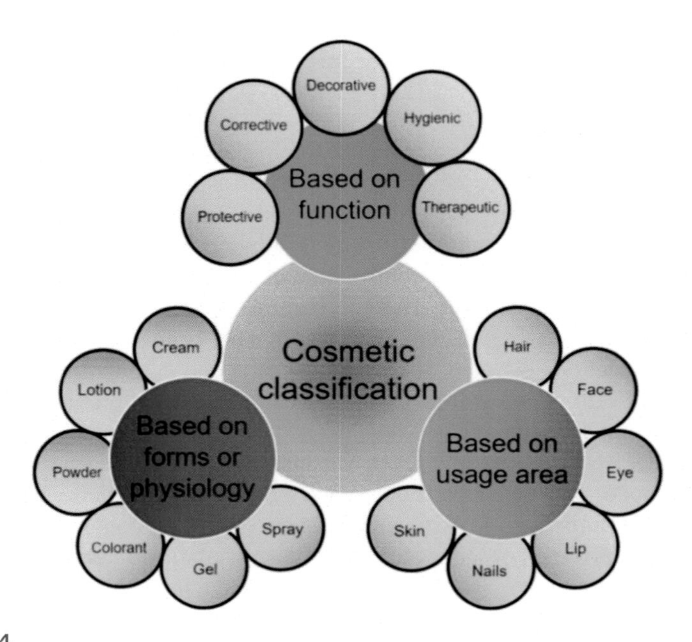

FIGURE 6.4

A schematic representation of FDA categorized classification of cosmetics. *FDA*, Food and Drug Administration.

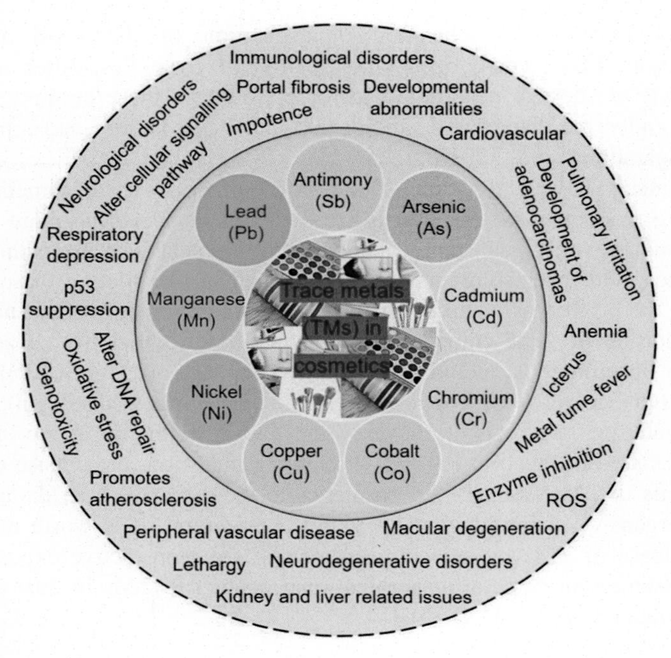

FIGURE 6.5

Several adverse effects of TMs in cosmetics. *TMs,* Trace metals.

Therefore, long term use of lip cosmetics results in high amounts of TMs reaching the systemic circulation and consequently damaging various vital organs depending on the chemical nature of the TMs. Among the TMs mentioned above, Cd, Cr, and Ni are classified as Group 1 human carcinogens by the International Agency for Research on Cancer (IARC). Several reports demonstrated that prolonged exposure to As, Cu, Ni, Pb, Cd, and Cr have been related with a greater risk of many infectious diseases, such as neurologic and cardiovascular disorders. Arsenic used in skin cream and make-up powder causes skin problems, lung cancer, circulatory and peripheral neuropathy, and increased risk of gastrointestinal and urinary tract cancers. Its low level was also detected in the eyeshadows, eye pencils, shampoo, hair gel/conditioner, and showering products. However, as exhibits a high affinity for the skin, and therefore long term contact of skin to this metal results in adverse consequences including variability of skin eruptions and skin cancer. Inhalation of Sb metal, primarily found in lipsticks, face powder, skin creams, eye pencils, and eyeshadows, could cause respiratory disorders (bronchitis, emphysema, altered pulmonary functions, and pneumoconiosis) and adverse gastrointestinal effects, that is, vomiting, diarrhea, ulcers, and abdominal pain (Lingamdinne et al., 2016).

Sb level in the range of 0.008−0.39 μg/g was examined in lipsticks but in another report, it was detected low concentrations of Sb including 0.179−0.46, 0.0134−0.265 and 0.157−0.76 μg/g for makeup powder, skin cream, and eyeshadow, respectively. Owing to its coloring property, Cd is used in different cosmetic products, that is, lipsticks, skin, and hair creams. Found Cd concentrations ranging from 4.9 to 10.6 μg/g in lip cosmetics. Likewise, low contents of Cd have also been identified in shampoos, soaps, hair gels, body shower, and skin creams. The chronic exposure to Cd may cause skin tumors and kidney damage. It can also cause bone brittleness and therefore increase in the risk of bone fractures following chronic exposure to low levels of Cd. Various sources of acquaintance to chromium exist; including lipstick, eye pencil, eyeliner, eyeshadow, and makeup powder. Reports have investigated the lip cosmetic samples harboring Cr levels ranging from the LOD −9.72 μg/g. However, the concentration of Cr was detected to be very low ($\leq$0.4 μg/g) in shampoos, hair conditioners/gels/creams, soaps, and shower emulsions. Similarly, skin creams and skin emulsions also revealed the minute amount of Cr ($<$0.5 μg/g). Cr exists in two valence states, and both oxidation states, that is, Cr^{3+} and Cr^{6+} can act as potential haptens causing contact dermatitis allergies. Distinguished Cr^{6+} from Cr, where the Cr^{6+} level in colored cosmetics ranged from not-detection to 39.2 μg/g. Notably, Cr^{6+} is more poisonous than Cr^{3+}, nevertheless, the Cr^{6+} contents as cosmetics constituent are not restricted by regulation authorities. Therefore, it is of great concern to regulate Cr as a cosmetics contaminant due to its harmful effect on human's health (Lingamdinne et al., 2016).

The main sources of mercury include numerous cosmetic products such as lipsticks, shampoos, hair conditioners/gels, nail polishes, eyeliners/shadows. In a study, Murphy et al. (2009) found varying Hg levels (between 0.01 and 12,590 μg/g) when analyzed 19 skin creams in Cambodia. McKelvey et al. (2010) analyzed high concentrations of Hg (between 204 and 4700 μg/g) in soaps, cream, and germicidal ointments. Importantly, the risks of Hg contents should be considered during cosmetics application, because it is toxic to human health and has the pronounced capability to cause neurological, nephro-, and gastrointestinal disorders. Cobalt and nickel metals commonly found in lipstick, eyeshadow, face paint, and hair cream associated with contact dermatitis. Cobalt and its salts are widely used as coloring agents in makeup and lightbrown hair dyes. Generally, substantial levels of Co were detected in face paints, eyeshadows, henna dyes, and hair creams. The highest levels of Co, between 122.78 and 253.33 μg/g were present in the eyeshadows of Chinese products. A Co concentration ranging from 0.055 to 0.105, 1.05−2.42, and 3.21−5.64 μg/g has been recorded for lipstick, eyeshadow, and powdered makeup, respectively. Some studies reported very low or even undetectable Co levels in body creams, soaps, lips products, and nail polish. Inorganic lead is found in numerous cosmetics such as hair dyes (lead acetate) and lipsticks, eyeshadow, eye pencil, hair cream. Pb content ranging from 0.25 to 81.50 μg/g was determined for eyeshadows in Italian, Chinese, and USA samples. Among 400 lipsticks samples analyzed, 13 lipsticks consisted of a maximum Pb level of 7.19 μg/g according to the FDA report. A high concentration of Pb was recorded in shampoos (0.98–1.59 μg/g) and soaps (3.8–4.63 μg/g). Oral ingestion or inhalation is the major route reported for Pb entry into the body. Although the skin hardly absorbs it, significant attention is required for monitoring absorption through the skin. Ingestion of large quantities of the lead (Pb) may interfere with heme and calcium channels synthesis, which are important for nerve transmission. In addition, reports have suggested that exposure to Pb perturbs the fetus and central nervous system (CNS) in children. In short, exposure to heavy metals might exhibit a threat for humans, because they are extensively employed in a variety of everyday cosmetic products, and thus can accumulate and concentrate on the human body (Lingamdinne et al., 2016).

6.5.4 Human samples

Heavy metals lead to serious environmental pollution and pose a great threat to human health. In addition, heavy metals are not only a class of nonbiodegradable pollutants, but they can also be stored in body tissues or organs via the food chain, causing adverse effects on the body. Moreover, distribution and excretion rate of heavy metals are different in diverse organs of the body. In blood, the half-life of some heavy metals are short. For instance, both blood lead (Pb) and cadmium (Cd) have a half- life of 35 days. However, the half-life of some heavy metals in other tissues or organs of the body is extremely long. For example, the half-life for Pb in bone reaches 30 years, a half-life for Cd in kidney ranges from 15 to 30 years and they can also be released into blood over time. Exposure to heavy metals can cause toxicity to a variety of tissues, organs, and systems such as circulatory, respiratory, endocrine, immune, nervous, urinary, and reproductive systems (He et al., 2017).

Metal ions can be combined on a variety of molecules in the body, such as proteins, fats, and polysaccharides in body tissues and organs. In addition, metal ions are also involved in diverse physiological and pathological reactions. Some of the heavy metals themselves are carcinogens such as cadmium, chromium, nickel, and arsenic. In addition, heavy metals can cause toxicity to humans through both direct and indirect mechanisms of promoting oxidative stress and inflammation. Exposure to heavy metals can cause adverse effects on human health. A recent study from Taiwan region of China demonstrates that heavy metals are associated with child developmental delays and health status evaluated using the pediatric quality of life (PedsQL) inventory for health related quality of life (HRQOL). Another study from South Korea shows that both extreme level of maternal blood Mn level was associated with lower birth weight outcome in a nonlinear fashion (He et al., 2017).

Early childhood has been recognized as a critical period for cognitive, emotional, and physical growth and development of children. In addition, children are more susceptible and vulnerable to environmental heavy metal exposure when compared to adults, and need widespread concern and protection from the general public, governments, and scientific communities around the world. One of the previous studies demonstrated that blood Pb was negatively correlated with height and weight in Guiyu children. However, there was no significant association between blood Pb and height and weight based on the linear regression analysis. Association between more heavy metals and additional growth and development indicators in children is still unclear and needs to be confirmed (He et al., 2017).

Moreover TEs play a central role in the growth and development of organisms, the proper functioning of carbohydrates, and lipid metabolism. However, many studies found that insufficient or excessive TEs could initiate or exacerbate adverse health effects on lipid metabolism, blood glucose concentrations, and cardiovascular function. Heavy metals, commonly used in or produced from industrial activities, can get accumulated in the human body and influence glucose homeostasis, hyperglycemia, carotid atherosclerosis (Maher et al., 2014) and led to metabolic syndrome. Metabolic syndrome (MetS) is a global health problem with an increasing prevalence. However, effects of TEs and heavy metals on MetS and the mechanism underlying this effect are poorly understood. Significantly higher blood concentrations of lead (Pb), cadmium (Cd), copper (Cu), and selenium (Se) were observed in the MetS group (Maher et al., 2014).

6.5.4.1 Route of exposure of heavy metals in humans

Humans may directly get in contact with heavy metals by consuming contaminated food stuffs, sea animals, and drinking of water, through inhalation of polluted air as dust fumes, or through occupational exposure at workplace. The contamination chain of heavy metals almost usually follows this cyclic order: from industry, to the atmosphere, soil, water and foods then human. These heavy metals can be taken up through several routes. Some heavy metals such as lead, cadmium, manganese, arsenic can enter the body through the gastrointestinal route; that is, through the mouth when eating food, fruits, vegetables or drinking water or other beverages. Others can enter the body by inhalation while others such as lead can be absorbed through the skin. Most heavy metals are distributed in the body through blood to tissues. Lead is carried by red blood cells to the liver and kidney and subsequently redistributed to the teeth, bone and hair mostly as phosphate salt. Cadmium initially binds to blood cells and albumin, and subsequently binds to metallothionein in kidney and liver tissue. Following its distribution from blood to the lungs, manganese vapor diffuses across the lung membrane to the CNS. Organic salts of manganese which are lipid soluble are distributed in the intestine for fecal elimination while inorganic manganese salts which are water soluble are distributed in plasma and kidney for renal elimination. Arsenic is distributed in blood and accumulates in heart, lung, liver, kidney, muscle and neural tissues and also in the skin, nails and hair (Manjuladevi and Sri, 2017).

6.5.4.2 Mechanism of heavy metal toxicity

Some heavy metals such as iron and manganese are essential for certain biochemical and physiological activities in the body, elevated level in the body can have delirious health effects. Most of the other heavy metals are generally toxic to the body at very low level. The main mechanism of heavy metal toxicity include the generation of free radicals to cause oxidative stress, damage of biological molecules such as enzymes, proteins, lipids, and nucleic acids, damage of DNA which is key to carcinogenesis as well as neurotoxicity. Some of the heavy metal toxicity could be acute while others could be chronic after long term exposure which may lead to the damage of several organs in the body such as the brain, lungs, liver, and kidney causing diseases in the body (Manjuladevi and Sri, 2017).

6.5.5 Review of removal processes from aquatic environments

There are various stages and technologies for water purification. Primary, secondary, and tertiary treatment processes constitute the stages of urban and industrial water treatment to remove organic, inorganic, and biological pollutants. Industrial wastewater pollutants and streams containing heavy metals with suspended solids, oxidized metals, and minerals require pretreatment prior to tertiary treatment to avoid membrane fouling that increases high operating costs. Chemical precipitation processes, such as lime deposition, are common methods for the removal of heavy metals in pretreatment processes (Adam et al., 2019).

6.5.5.1 Primary treatment

In primary treatment, the most important primary technology for the removal of heavy metals is coagulation and flocculation, chemical filters, and microfiltration (Adam et al., 2019).

6.5.5.1.1 Microfiltration

Among the membrane filtration methods, microfiltration has the largest cavity size range. In addition to the heavy metals that soluble solids, it also removes bacteria, algae and smaller microorganisms from the virus. Microfilters are also used in tertiary purification such as adsorption, ion exchange and membrane filtration (Adam et al., 2019).

6.5.5.1.2 Coagulation and flocculation

Coagulation and flocculation Adding coagulants or flocculants is one of the important methods in wastewater treatment. Its method is to stabilize suspended particles. Ferric or aluminum salts are used as coagulants and flocculants. Its advantages include removing contaminants and improving sludge sedimentation (Adam et al., 2019).

6.5.5.2 Secondary treatment

The secondary treatment consists of aerobic and anaerobic groups. Based on the performance of microorganisms in converting pollutants into simpler and safer materials. In aerobic conditions, microorganisms convert pollutants to CO_2 and biological mass using oxygen. Continuous presence of oxygen in this process is essential. In anaerobic conditions, in the absence of oxygen, pollutants convert biogas, such as methane, which is used as an energy source. Secondary purification is used to remove organic contaminants. Microorganisms have also been effective in removing toxic and nontoxic metals from industrial wastewater (Adam et al., 2019).

6.5.5.3 Tertiary treatment

6.5.5.3.1 Chemical oxidation

One of the most reliable technologies with low chemical oxidation equipment is chemical oxidation. This process is based on the oxidation of agents in wastewater based on electron transport. Oxidation has been effective in removing organic and inorganic compounds. Chlorine, chlorine dioxide, permanganate, oxygen, and ozone are common oxidants in purification processes. The toxic by-products of chemical oxidation can be eliminated by advanced oxidation and adsorption methods on activated carbon (Adam et al., 2019).

6.5.5.3.2 Electrochemical precipitation

Another method of treatment is electrochemical processes known as simple and environmentally friendly methods. This method has been very effective in removing the chemical oxygen demand, although it has not been as effective in removing heavy metal ions as in other methods (Adam et al., 2019).

6.5.5.3.3 Crystallization and distillation

Thermal and crystallization processes are also considered as tertiary treatment methods. In this case, by using the energy of the water containing the pollutant, it reaches the boiling temperature and after that by separating the pollutant and then cooling the pure water is extracted. Crystallization increases the concentration of contaminants. This technique has been effective in removing ammonia, nitrogen, and phosphorus from wastewater (Adam et al., 2019).

6.5.5.3.4 Photo catalysis

In photocatalytic technology, light sources by photon generation and the effect of adsorption on the surface of catalysts are produced free radical such as hydroxyl radicals. Photons or ultraviolet light and the use of titanium dioxide (TiO_2) catalysts photocatalytic reduction and oxidation are formed (Adam et al., 2019). In a study, Zhang et al. used titanium dioxide photocatalytic particles to remove arsenic (Martínez-Guijarro et al., 2019). The basis for the use of TiO_2 using UV light to removing arsenic was that arsenite was first converted to arsenate and then removed by the adsorption process (Martínez-Guijarro et al., 2019). Under the influence of UV light, TiO_2 receives the required energy and, as a result, the electron is released from the valence band and transferred to the conduction band, that leads to creating hole or a positive radical that contacts the surface of the TiO_2 catalyst or water converted to hydroxyl radical. One of the effective factors in the removal of contaminants is hydroxyl radical or •OH. In the process of metal removal by photocatalysts, the pH of the solution, the initial metal ion concentration, the light intensity and the mass of the photocatalyst are very important (Adam et al., 2019).

6.5.5.4 New technologies for tertiary treatment

Deposition using chemical materials, use of activated carbon as an adsorbent, use of ion exchange on resin and membrane processes are techniques available in water treatment design (Nasir et al., 2019).

6.5.5.4.1 Adsorption

Removal of toxic metals using adsorption, has been effectively applied in recent decades. Adsorption methods can usually be classified as either chemisorption or physisorption. Simple physical adsorption depends on materials having a suitable porous structure to achieve removal. However, this is not generally sufficient to achieve optimal removal and therefore, chemical adsorption has received increasing attention for wastewater treatment. Heavy metal ions or complexes can adsorb to the surface of functional adsorbents, via electrostatic attraction or chemical bonding forces, achieving effective separation. Meanwhile, ion exchange and inner/outer-sphere complexation mechanisms are often involved in the chemisorption of heavy metals to date. A range of adsorbents have been investigated for use as adsorbents, requiring characteristics such as a large specific surface area, appropriate pore and surface structure, easy to manufacture and regenerate, as well as good mechanical properties. Commonly used adsorbents include carbon materials, silica gel, molecular sieves, natural clay, or other emerging biomaterials (Nasir et al., 2018).

6.5.5.4.2 Carbon-based nanosorbents

Nanotechnology, the conscious control of matter to restrict its size less than 100 nm, has led to developing new materials, which have certain relative unique properties that cannot be found in bulk-sized materials. Usefulness of nanomaterials has been reported by various researchers due to their large surface area and surface free energy, small size, active atomicity, and reactivity. The high surface area to volume ratio of nanomaterials immensely improvises its sorption efficiency. Further, surface modification is a powerful tool which improvises the properties of a material through imparting the desired functional groups on its surface without compromising with the bulk properties. Surface modification of nanomaterials by forming various functional groups, that is, amine (−NH2), carboxylic (−COOH), thiol (−SH), and methyl on their surfaces have been explored and resulted in enhanced performances. These features led to the vast applications of engineering nanomaterials (Plessl et al., 2019).

Adsorption using carbonaceous materials as adsorbents (carbon adsorbents) is perhaps one of the most cost-effective methods for removal of heavy metals from aqueous solutions. Carbonaceous materials including activated carbon (AC), biochar, carbon nanotubes (CNTs), and graphene oxide (GO) have been widely studied for adsorption of various environmental contaminants. AC is the most widely used carbon adsorbent for water and wastewater treatment. A wide range of AC adsorbents can be prepared to suit for various environmental applications including the removal of heavy metals from aqueous solutions. Due to the high cost associated with production of coal-based AC, biochar has recently emerged as a low-cost alternative of AC with comparable or superior performance for heavy metal adsorption. A variety of woody biomass including agricultural wastes or by-products such as peanut hull and dairy manure can be used to develop biochar. Its multi functionalities including carbon sequestration, soil fertility improvement, and environmental remediation are also well recognized. With the advent of nanotechnology, graphene-based materials including GO or reduced GO and CNTs have been studied intensively for their potential applications in removal of heavy metals including GO or reduced GO and CNTs have been studied intensively for their potential applications in removal of heavy metals. Graphene is a single sheet of 2D hexagonal network of carbon and GO is the oxidized form of graphene synthesized through chemical or thermal reduction processes. GO has high specific surface area and rich surface functional groups, ideal for adsorption of heavy metals. CNTs are cylindrical carbon tubes rolled from one or multiple sheets of graphene. They feature well-defined cylindrical hollow structure, large surface area, hydrophobic wall, and easily modified surfaces. Both single walled and multi-walled CNTs have been tested as adsorbent materials. With excellent electrical, thermal, and mechanical properties, CNTs and GO possess great potential in adsorption technology for removal of heavy metals (Richir and Gobert, 2016).

While every carbon adsorbent has its unique structure characteristics and functionality, one common feature shared by all carbon adsorbents is that they all contain rich active surface functional groups, which are crucial for the surface chemistry of carbon materials and the adsorption of heavy metals. It is generally believed that the chemical/physical interactions between heavy metals and functional groups of adsorbents contribute significantly to the adsorption of heavy metals. Functional groups are typically bonded with heteroatoms on carbon surfaces, commonly oxygen, nitrogen, sulfur, phosphorus, and halogens. Therefore, functional groups are conventionally classified according to heteroatoms bonded to the surface of carbon that is, oxygen-containing functional groups, nitrogen-containing functional groups, and sulfur containing functional groups. The functionality and quantity of each type of functional groups can be enhanced by chemically and/or physically modifying the surface of carbon materials to introduce the desired heteroatoms on carbon surfaces. The mechanisms by which heavy metals are adsorbed onto carbon adsorbents may involve physical adsorption, electrostatic interaction, ion exchange, surface complexation, and precipitation (Fig. 6.6) (Yang et al., 2019; Richer and Gobert, 2016).

The surface chemistry of carbons is determined, to a large extent, by the number and the nature of surface functional groups on carbon surface. Several studies have provided evidences of the enhanced heavy metal adsorption capacity with

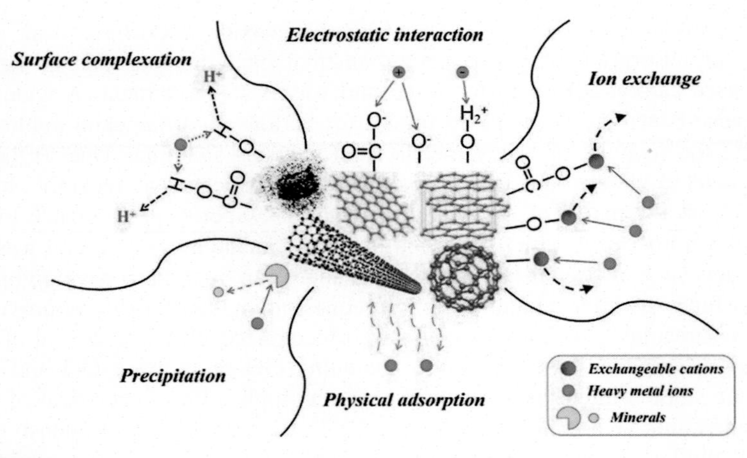

FIGURE 6.6

Schematics of adsorption mechanisms of heavy metals onto carbon adsorbents.

modifications of carbon adsorbents with functional groups the method of modification and the type of raw carbon material from which the final adsorbent is prepared influence the nature of these functional groups. Selection of modification methods depends on the intended application and the chemical characteristics of target contaminants. Generally oxidation, nitrogenation, and sulfuration are the common modification techniques that are employed to introduce oxygen, nitrogen, and sulfur heteroatoms on carbon surface for generation of functional groups. Fig. 6.7 is a graphic exhibition of these modification techniques, depicting the chemical agents used in each technique and the desired functional groups generated (Yang et al., 2019, Richer and Gobert, 2016).

6.5.5.4.3 Zeolite based nanosorbents

Zeolites are attractive adsorbent because of their unique characteristics: ion exchange capacity, affinity for heavy metal cations, and proven chemical and mechanical stability. Zeolite are microporous alumina silicate frameworks with three-dimensional networks of corner sharing (TO_4) tetrahedra, in which T usually represents silicon or aluminum. The framework composed of purely [SiO_4] units is neutral. When Al with a charge of 3 + is an isomorphous substitute for Si with a charge of 4 +, the framework becomes negatively charged compensated by extra-framework cations, leading to its cation exchange capacity. Currently, more than 220 distinct zeolite framework types are known (Sajjad et al., 2019).

6.5.5.4.4 Metal-based nanosorbents

The sorption process has to be an effective and efficient technique for treatment of heavy metals from water with easy handling and availability of a wide range of adsorbents. But, some difficulties are faced in the filtration and regeneration of

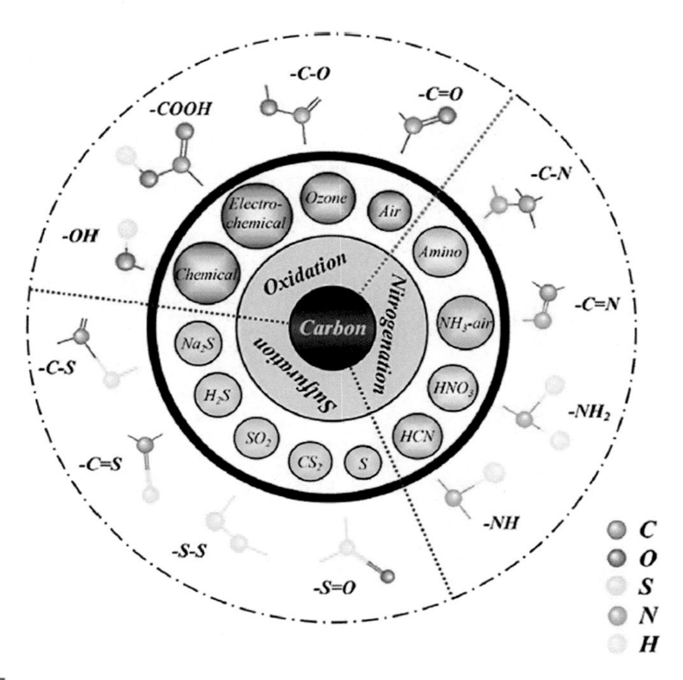

FIGURE 6.7

Modification techniques to functionalize carbon adsorbents with various functional groups.

adsorbent for their subsequent applications and to overcome these difficulties, novel sorbents designs are necessary. In the recent years, magnetic materials have been widely used as an adsorbent for treatment of toxic contaminates. Moreover, magnetite metal oxides or various ferrites are frequently used in the hydrometallurgical extraction process for recovery of precious metals and heavy metals, and regulating digestion of high silica bauxite. Among the available magnetic sorbents, inorganic spinel ferrites and their composite materials were widely used in water treatment due to its unique advantages of magnetic and chemical stability (Sakultantimetha et al., 2009). hybrid sorbents that could simultaneously remove or reduce organic and inorganic contaminants have been fabricated by combining Nanosized zero valent iron and polymers, or incorporating nanoscale metal (hydr) oxides in biochar, chitosan, or activated carbon (Sun et al., 2019).

6.5.5.4.5 Membrane processing

Membrane separation processing includes several advanced and diverse technologies showing great potential for removing various types of pollutants with high efficiency. When used for heavy metal removal, membrane processing can provide a reliable solution, which generally does not require chemical additives or

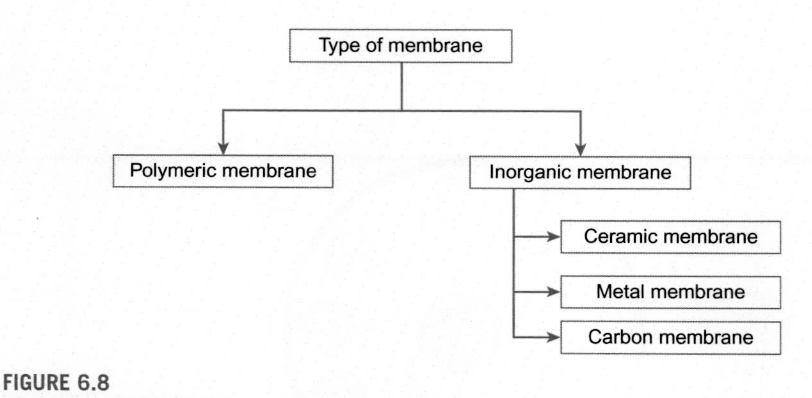

FIGURE 6.8

Type of membrane separation.

thermal inputs, does not involve phase changes, is environmentally friendly, and relatively simple in fabrication, operation, scale up and control (Adam et al., 2019).

Nowadays, the membrane from various polymeric materials has been widely commercialized worldwide. According to some sources in term of geometry, hollow fiber membrane has received much attention due to its highest surface area (surface area to volume up to 9000 m^2/m^3). This kind of membrane can be produced via phase inversion process. There are two types of membranes; polymeric membrane and inorganic membrane, as illustrated in Fig. 6.8. In term of pore sizes, there are four types of membrane which are microfiltration (0.05–1.0 μm), ultrafiltration (0.005–0.5 μm), nanofiltration (0.0005–0.01 μm), and reverse osmosis (0.0001–0.001 μm). Accordingly, heavy metals ions were tiny and sometimes they are soluble, such as arsenic, in which it is necessary to reverse the osmosis membrane's size to treat the water. Additionally, the water treated through the RO process may not contained precious minerals such as calcium and magnesium in which concerned by a human being through drinking water (Turner, 2019).

6.5.5.4.6 Ultrafiltration

Ultrafiltration (UF) is one of the most prevailing processes for water and wastewater treatment. Over the past decade, this technology has experienced rapid development that has spanned a wide range of applications in various fields. Some of the most promising applications range from drinking water production, wastewater treatment, juice concentration, dairy production, and chemical recovery. UF membrane has been feasibly used as an alternative to the existing conventional water treatment technologies as it has proven to have greater removal efficiency to physically separate particulate and colloidal matter from wastewater. Additionally, it permits the clarification and disinfection of wastewater in a single step (Wang et al., 2015).

UF membranes have pore sizes ranging from 10 to 100 nm and typically are used to remove viruses, bacteria, color pigments, and some natural organic colloids. UF has the further advantage to be an energy saver technology compared to reverse osmosis (RO), where the considerable higher transmembrane pressure requires substantially larger energy consumption (Adam et al., 2019). In recent years, the potential use of UF membranes has been demonstrated in the separation of heavy metal ions from water sources. However, as the pore size of microporous membranes is significantly larger compared to the molecular size of ions, such membranes are notable to remove them based on a sieving effect. A new type of adsorptive UF membrane has thus been developed to address the issue, while solving the main technical challenge faced by the typical adsorption process, that is, difficulty in separating adsorbent from treated water after the adsorption process (Wang et al., 2015). However, since UF membranes' pore size is larger than metal ions' hydrated radius, metal ions generally pass through the membranes and the heavy metal ion rejection is not satisfactory. To obtain high metal separation efficiency, the addition of surfactants to wastewater becomes essential, realizing the so called "micellar-enhanced ultrafiltration (MEUF)", which may allow reaching heavy metal rejection of 99% and beyond. MEUF has several operating stages, including addition of surfactants with concentrations higher than their critical micelle concentration (CMC) to wastewater, micelle formation by surfactant monomer aggregation, binding of metal ions and micelles via electrostatic interactions and eventually their separation by a typical UF technique. Lin et al. for example, applied the MEUF system comprising cellulose UF membranes and sodium dodecyl sulfate (SDS) as a surfactant to achieve rejections above 90% for Cu^{2+}, Co^{2+}, Ni^{2+} (Adam et al., 2019).

Despite the increasing number of UF applications and advancements made in this field, this process is still largely limited by the trouble causing fouling phenomenon. Upon long-term operation, different foulants can deposit on the membrane surface via various fouling mechanisms, such as biofouling and organic fouling, etc. The existence of fouling on membrane surfaces has always been associated with a decline in quality and flux of permeate, an increase in operational costs mainly due to higher energy consumption, additional maintenance and cleaning chemical costs, and reduced membrane lifespan. As such, effective and efficient countermeasures have been actively taken to control and minimize the impacts of fouling. Several straightforward strategies have been developed to mitigate fouling prior to its occurrence through pretreatment of the feed streams, design of antifouling membrane, as well as optimization of operational conditions. As various studies have confirmed that the intrinsic membrane surface property is one of the main factors contributing to fouling, the development of antifouling membranes has been widely attempted (Wang et al., 2015).

6.5.5.4.7 Nanofiltration

Nanofiltration (NF) membranes are designed to remove pollutants smaller than 10 nm and are best suited for water softening applications (Adam et al., 2019).

With a nominal pore size of 2 nm, NF falls between RO and UF in its separation characterization. NF needs lower operational pressures than RO, so it has a significantly lower electrical energy consumption. NF has two separation mechanisms; separation of uncharged solutes due to size effects (sieving) and separation of charged species such as ions because of electrical repulsion (Donnan and Dielectric effects). Generally some advantages of the NF membrane technology in the water and wastewater treatment are: elimination of divalent ions such as heavy metals without the need of secondary chemicals, capability of treating waters containing more than one kind of heavy metals, continuous water treatment and having automated process. There are some examples of using the NF membrane technology for the elimination of heavy metal ions such as lead and nickel from wastewaters in the other researches. For example, Jakobs and Baumgarten, discussed the use of NF for lead elimination in the picture tube industry waste treatment process. Murthy and Chaudhari, studied the effect of operating conditions on heavy metal removal of a NF membrane from single and mixed salt solutions. Wahab Mohammad and his coworkers, have used NF for treatment of electroplating rinse water (Yang et al., 2019).

6.5.5.4.8 Reverse osmosis

RO membrane technology is widely used in seawater desalination, drinking water production, brackish water treatment and wastewater treatment. RO is currently the most energy efficient technology for desalination, with energy cost about 1.8 kWh/m, which is much lower than that of other technologies. Also, RO membrane has the advantages of high water permeability and salt rejection, fulfillment of the most rigorous rules for public health, environmental protection and separation process (Zeng et al., 2019). In the case of heavy metal removal, RO is mostly used to treat large volumes of brackish groundwater and can be used to eliminate metallic ions directly without the need of secondary chemicals (Yang et al., 2019). A part from the notable exception of As(III), RO is very efficient in removing most ions, and it is widely used in aerospace, food, oil and gas, galvanic, dairy, pulp and paper industries, and power plants. In RO, ions are separated from water through a semipermeable membrane by applying a hydrostatic pressure against the osmotic pressure. Hence the threshold of hydrostatic pressure is settled by the osmotic pressure, although the hydrophilic nature of RO membranes definitely facilitates water transport through the membrane. However, high losses of processed water, high-energy consumption and large initial investments are the major drawbacks of RO. Furthermore, precisely because RO leads to a manifold concentration of the pollutant solution, efficient disposal of residual water volumes is also an important issue. Finally, specifically to drinking water, since RO removes ions in a nonselective way, the mineral composition of water oligo elements always needs posttreatment and readjustment (Adam et al., 2019).

6.5.5.4.9 Nanohybrid membranes

Nanohybrid membranes are a new generation of membranes, which are made by the incorporation of inorganic particles with the aim of improving the properties of the polymeric membrane precursors. Blending some polymers like polyvinyl alcohol with adsorptive polymers was used to make membranes with the aim of raising their separation properties to the removal of metal ions and dyes (Zhang and Itoh, 2006). There are two types of Nanohybrid membranes: mixed matrix membranes (MMMs) and Nanocomposite membranes (NCMs). MMMs are formed by the incorporation of high separation performance porous inorganic molecular sieves, such as zeolites, and carbon-based molecular sieves, in polymer matrixes to combine the advantages of polymeric membranes with the superior separation performances of inorganic molecular sieves (Adam et al., 2019). Utilizing nanoparticles in the polymeric membrane structures have found remarkable attention to improve adsorptive removal of pollutants. Some nanoparticles as adsorbent extensively have been employed in the preparation of membranes; such as Fe_3O_4, ZnO, and TiO_2 nanoparticles (Zhang and Itoh, 2006).

6.5.5.4.10 Electro dialysis

For over 60 years, ED is an established technology in treating industrial wastewater, brackish water, municipal wastewater and used in drug and food industries, chemical processes, table salt production, electronics, biotechnology, heavy metals removal, and acids and bases production via its capability to remove ionic and nonionic components under the effect of electric current. It has been driven by the development of IEM with enhanced electrochemical and physicochemical characteristics. The main advantages of ED are higher water recovery rates compared to RO, easy operation, long membrane lifetime, operation at high temperature, and unlike RO, it does not require extensive pretreatment or posttreatment. Also, it can accomplish a selective separation of monovalent ions (e.g., NO^{3-}, CL^-, NH_4^+, K^+, Na^+) against multivalent ions (e.g., PO_4^{3-}, SO_4^{2-}, Mg^{2+}, Ca^{2+}) for the production of irrigation water, with the use of monovalent permselective IEM (Zhu et al., 2019).

6.5.5.5 Ion exchange resins

Ion exchange is a well established method commonly applied to drinking water treatment for hardness removal, but it is also increasingly being studied for the removal of heavy metal ions (Adam et al., 2019). Ion exchange is a reversible chemical process that allows heavy metals that have similarly charged with ion exchange resin to attach each other (Turner, 2019). In this water purification technology, porous ion exchange microbeads trap the specific ions by releasing the presaturated nontoxic ions. Because the technology relies on exchange surfaces, as in adsorption, an important requirement is to have nanoporous resins with accessible high specific surface area. Anion and cation resins are most common exchangers which have the capacity to exchange negatively charged (arsenate,

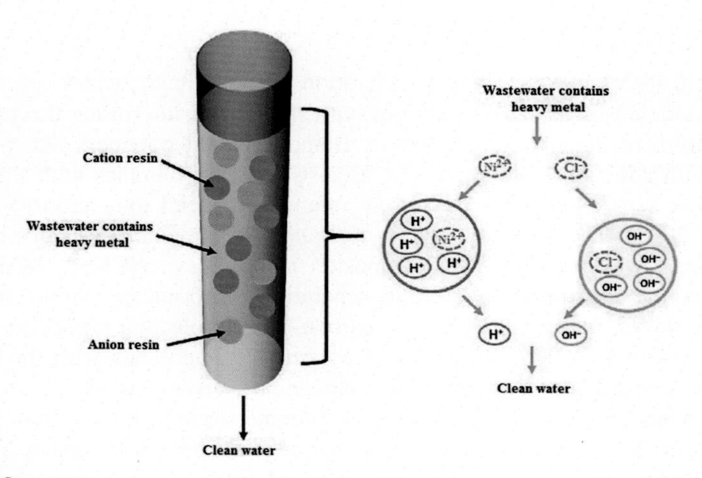

FIGURE 6.9

Mechanism of an ion-exchange process for heavy metal removal.

selenate, chromate, and uranium) and positively charged ions (barium, strontium, radium, calcium, and magnesium), respectively. These resins are categorized depending on their available functional groups as either strong or weak exchangers and may be acidic or basic in nature. Most common ion exchange resins available in the market are sodium silicates, zeolites, polystyrene sulfonic acid, and acrylic and methacrylic resins (Adam et al., 2019). Mechanism of an ion-exchange process for heavy metal removal in Fig. 6.9.

6.6 Conclusion

Heavy metals have varying concentrations in different environments due to different conditions, and many factors influence their distribution in the environment that are considered a serious threat to human health and environmental damage. This study reviews the presence of TEs and heavy metals in water and wastewater, seafood, human samples and personal care products, followed by methods and technologies available in water and wastewater treatment plans to eliminate this. The group of pollutants was expressed because of their harmful effects on human health and the environment.

References

Abdi, G., Alizadeh, A., Zinadini, S., Moradi, G., 2018. Removal of dye and heavy metal ion using a novel synthetic polyethersulfone nanofiltration membrane modified by magnetic graphene oxide/metformin hybrid. J. Membr. Sci 552, 326–335.

Adam, M.R., Hubadillah, S.K., Esham, M.I.M., Othman, M.H.D., Rahman, M.A., Ismail, A.F., et al., 2019. Adsorptive membranes for heavy metals removal from water. Membrane Separation Principles and Applications. Elsevier, pp. 361–400.

Alam, M., Akhter, M., Mazumder, B., Ferdous, A., Hossain, M., Dafader, N., et al., 2019. Assessment of some heavy metals in selected cosmetics commonly used in Bangladesh and human health risk. J. Anal. Sci. Technol. 10 (1), 2.

Avigliano, E., Monferran, M.V., Sanchez, S., Wunderlin, D.A., Gastaminza, J., Volpedo, A.V., 2019. Distribution and bioaccumulation of 12 trace elements in water, sediment and tissues of the main fishery from different environments of the La Plata basin (South America): risk assessment for human consumption. Chemosphere 236, 124394.

Badry, A., Palma, L., Beja, P., Ciesielski, T.M., Dias, A., Lierhagen, S., et al., 2019. Using an apex predator for large-scale monitoring of trace element contamination: associations with environmental, anthropogenic and dietary proxies. Sci. Total Env 676, 746–755.

Bessbousse, H., Rhlalou, T., Verchère, J.-F., Lebrun, L., 2008. Removal of heavy metal ions from aqueous solutions by filtration with a novel complexing membrane containing poly (ethyleneimine) in a poly (vinyl alcohol) matrix. J. Membr. Sci 307 (2), 249–259.

Bilal, M. &, Iqbal, H.M., 2019. An insight into toxicity and human-health-related adverse consequences of cosmeceuticals—a review. Sci. Total Environ. 670, 555–568.

Bolisetty, S., Peydayesh, M., Mezzenga, R., 2019. Sustainable technologies for water purification from heavy metals: review and analysis. Chem. Soc. Rev. 48 (2), 463–487.

Engwa, G.A., Ferdinand, P.U., Nwalo, F.N. &, Unachukwu, M.N., 2019. Mechanism and health effects of heavy metal toxicity in humans. Poisoning in the Modern World: New Tricks an Old Dog? IntechOpen.

Esposito, G., Meloni, D., Abete, M.C., Colombero, G., Mantia, M., Pastorino, P., et al., 2018. The bivalve Ruditapes decussatus: a biomonitor of trace elements pollution in Sardinian coastal lagoons (Italy). Environ. Pollut. 242, 1720–1728.

Gifford, M., Hristovski, K., Westerhoff, P., 2017. Ranking traditional and nano-enabled sorbents for simultaneous removal of arsenic and chromium from simulated groundwater. Sci. Total Environ. 601, 1008–1014.

Guo, X., Yang, Q., Zhang, W., Chen, Y., Ren, J., Gao, A., 2019. Associations of blood levels of trace elements and heavy metals with metabolic syndrome in Chinese male adults with microRNA as mediators involved. Environ. Pollut. 248, 66–73.

Henze, M., Comeau, Y., 2008. Wastewater characterization. Biol. Wastewater Treatment: Princ. Model Des. 33–52.

He, Y., Tang, Y.P., Ma, D., Chung, T.-S., 2017. UiO-66 incorporated thin-film nanocomposite membranes for efficient selenium and arsenic removal. J. Membr. Sci. 541, 262–270.

Hong, M., Yu, L., Wang, Y., Zhang, J., Chen, Z., Dong, L., et al., 2019. Heavy metal adsorption with zeolites: the role of hierarchical pore architecture. Chem. Eng. J. 359, 363–372.

Jawed, A., Saxena, V. &, Pandey, L.M., 2020. Engineered nanomaterials and their surface functionalization for the removal of heavy metals: a review. J. Water Process Eng. 33, 101009.

Jiang, S., Li, Y., Ladewig, B.P., 2017. A review of reverse osmosis membrane fouling and control strategies. Sci. Total Environ. 595, 567–583.

Joseph, L., Jun, B.-M., Flora, J.R., Park, C.M., Yoon, Y., 2019. Removal of heavy metals from water sources in the developing world using low-cost materials: a review. Chemosphere 229, 142–159.

Konieczka, P., Cieślik, B. &, Namieśnik, J., 2018. Trace elements in aquatic environments. Recent Adv. Trace Elem. 143.

Lau, W.-J., Emadzadeh, D., Shahrin, S., Goh, P.S., Ismail, A.F., 2018. Ultrafiltration Membranes Incorporated With Carbon-Based Nanomaterials for Antifouling Improvement and Heavy Metal Removal. Carbon-Based Polymer Nanocomposites for Environmental and Energy Applications. Elsevier, pp. 217–232.

Lingamdinne, L.P., Koduru, J.R., Choi, Y.-L., Chang, Y.-Y. &, Yang, J.-K., 2016. Studies on removal of Pb (II) and Cr (III) using graphene oxide based inverse spinel nickel ferrite nano-composite as sorbent. Hydrometallurgy 165, 64–72.

Maher, A., Sadeghi, M., Moheb, A., 2014. Heavy metal elimination from drinking water using nanofiltration membrane technology and process optimization using response surface methodology. Desalination 352, 166–173.

Manjuladevi, M. &, Sri, M.O., 2017. Heavy metals removal from industrial wastewater by nano adsorbent prepared from cucumis melopeel activated carbon. J. Nanomed. Res. 5 (1), 00102.

Martínez-Guijarro, R., Paches, M., Romero, I., Aguado, D., 2019. Enrichment and contamination level of trace metals in the Mediterranean marine sediments of Spain. Sci. Total Environ. 693, 133566.

Nasir, A.M., Goh, P.S., Abdullah, M.S., Cheer, N.B. &, Ismail, A.F., 2019. Adsorptive nanocomposite membranes for heavy metal remediation: recent progresses and challenges. Chemosphere 232, 96–112.

Nasir, A., Goh, P., Ismail, A., 2018. Novel synergistic hydrous iron-nickel-manganese (HINM) trimetal oxide for hazardous arsenite removal. Chemosphere 200, 504–512.

Plessl, C., Gilbert, B.M., Sigmund, M.F., Theiner, S., Avenant-Oldewage, A., Keppler, B. K., et al., 2019. Mercury, silver, selenium and other trace elements in three cyprinid fish species from the Vaal Dam, South Africa, including implications for fish consumers. Sci. Total Environ. 659, 1158–1167.

Richir, J., Gobert, S., 2016. Trace elements in marine environments: occurrence, threats and monitoring with special focus on the Costal Mediterranean. J. Environ. Anal. Toxicol. 6 (1), 349.

Sajjad, A.-A., Yunus, M.Y.B.M., Azoddein, A.A.M., Hassell, D.G., Dakhil, I.H. &, Hasan, H.A., 2019. Electrodialysis desalination for water and wastewater: a review. Chem. Eng. J. 380, 122231.

Sakultantimetha, A., Bangkedphol, S., Lauhachinda, N., Homchan, U., Songsasen, A., 2009. Environmental fate and transportation of cadmium, lead and manganese in a river environment using the EPISUITE Program. Kasetsart J. (Nat. Sci.) 43 (3), 620–627.

Sun, X., Fan, D., Liu, M., Liao, H. &, Tian, Y., 2019. Persistent impact of human activities on trace metals in the Yangtze River Estuary and the East China Sea: evidence from sedimentary records of the last 60 years. Sci. Total Environ. 654, 878–889.

Turner, A., 2019. Trace elements in laundry dryer lint: a proxy for household contamination and discharges to waste water. Sci. Total Environ. 665, 568–573.

Wang, Y., Liu, D., Lu, J. &, Huang, J., 2015. Enhanced adsorption of hexavalent chromium from aqueous solutions on facilely synthesized mesoporous iron–zirconium bimetal oxide. Colloids Surf A: Physicochem. Eng. Asp. 481, 133–142.

Yang, X., Wan, Y., Zheng, Y., He, F., Yu, Z., Huang, J., et al., 2019. Surface functional groups of carbon-based adsorbents and their roles in the removal of heavy metals from aqueous solutions: a critical review. Chem. Eng. J. 366, 608–621.

Zeng, X., Xu, X., Qin, Q., Ye, K., Wu, W., Huo, X., 2019. Heavy metal exposure has adverse effects on the growth and development of preschool children. Environ. Geochem. Health 41 (1), 309–321.

Zhang, F.-S. &, Itoh, H., 2006. Photocatalytic oxidation and removal of arsenite from water using slag-iron oxide-TiO_2 adsorbent. Chemosphere 65 (1), 125–131.

Zhu, Y., Fan, W., Zhou, T. &, Li, X., 2019. Removal of chelated heavy metals from aqueous solution: a review of current methods and mechanisms. Sci. Total Environ. 678, 253–266.

CHAPTER

Conclusions and future research

7

Afsane Chavoshani[1], Majid Hashemi[2,3], Mohammad Mehdi Amin[1,4] and Suresh C. Ameta[5]

[1]*Department of Environmental Health Engineering, School of Health, Isfahan University of Medical Sciences, Isfahan, Iran*

[2]*Environmental Health Engineering, School of Public Health, Kerman University of Medical Sciences, Kerman, Iran*

[3]*Environmental Health Engineering Research Center, Kerman University of Medical Sciences, Kerman, Iran*

[4]*Environment Research Center, Research Institute for Primordial Prevention of Non-Communicable Disease, Isfahan University of Medical Sciences, Isfahan, Iran*

[5]*Department of Chemistry, PAHER University, Udaipur, India*

7.1 General markers

This final chapter presents an overview on the occurrence and fate of emerging micropollutants (EMPs) in the aquatic environments, effects on humans and other organisms, analytical methodologies, and removal technologies. EMPs are defined as synthetic or natural compounds released from point and nonpoint resources and end up to the aquatic environments at low concentration (Geissen et al., 2015). In the recent years emerging contaminants, including pharmaceuticals, personal care products (PCPs), pesticides, surfactants, and other trace elements, such as heavy metals and microplastics have taken many attentions from the scientific and monitoring communities and the global media (Wanda et al., 2018). Most of the EMPs are persistent, bioaccumulative, toxic, and endocrine disruptor. Due to their lipophilic bioactive and water soluble features, they occur in the aquatic environments at trace concentrations (nanogram–microgram per liter and nanogram–microgram per gram) (Luo et al., 2014). However, likely due to the high number and variety of chemicals included in this group or because of the analytical method limitations (chemical and toxicological analyses), there are still many research gaps that are certainly essential to be considered to fully understand their fate (Cruz and Barceló, 2015). Considering the emerging threats posed by EMPs, we believe that this book will be a useful and effective tool to encourage further research on the fate, occurrence, threats, and remove of EMPs in the aquatic environments.

Micropollutants and Challenges. DOI: https://doi.org/10.1016/B978-0-12-818612-1.00007-6

7.2 Occurrence of emerging micropollutants in wastewater treatment plants

In this book it has been made a picture of the current EMPs distribution and occurrence in the aquatic environments. More than 80 compounds and several metabolites have been found in the aquatic environment which indicates that not all contaminants are removed during water treatment (Heberer, 2002). According to the varieties in EMPs resources, contamination of water media with these chemicals is not a surprise. It is cleared that EMPs discharge in water bodies is influenced by disposal of municipal, industrial, and agricultural wastes, excretion of pharmaceuticals, and accidental spills. In addition, future developments in society can result in the emission of new substances to the environment (Besha et al., 2017).

Insufficient removal of EMPs during conventional wastewater treatment processes and lack of precautions and monitoring actions for EMPs have been caused that many of these compounds act as a threat for human and wildlife in the aquatic environments. The discharge of wastewater treatment plant (WWTP) effluent into surface water has been considered as a main reason of the EMPs occurrence in surface water resources. In comparison to surface water, ground water was found to be less contaminated with EMPs (Wanda et al., 2017a). Hence, the presence of EMPs in groundwater mainly results from landfill leachate, groundwater–surface water interaction, infiltration of contaminated water from agricultural land or seepage of septic tanks, and sewer systems. The concentration of EMPs in drinking water is dependent on water resources and seasons, for example, water samples in winter showing higher concentrations in comparison with water samples in summer. Furthermore water treatment processes play a significant role in removal of EMPs from drinking water. Monitoring of EMPs in water treatment plants (WTPs) has shown the presence of EMPs in WTPs at concentration between part per billion and part per trillion (Luo et al., 2014).

It has been found that pharmaceuticals as an EMPs are used to ensure the protection of public and animal's health against unhealthy situation. Commercial and domestic usages and release of these chemicals into wastewater have contributed to their occurrence in aquatic environment (Ebele et al., 2017). Wastewater from medical facilities is usually discharged without pretreatment directly into the sewage system. Although the major part of wastewater from medical facilities is only a small fraction of the total volume of wastewater entering the WWTPs, it is being received many attentions by both scientists and the public (Cidlinová et al., 2018). They can reach to the aquatic environments by different routes such as human excretion, disposal of unused and expired drugs, agricultural, and livestock practices. Due to their environmental side effects in aquatic and terrestrial environments, in flora, biota, and human health, the occurrence of pharmaceuticals in the environment has received increased scientific attention. Wastewater resources (hospital, aquaculture, industrial, and domestic) are considered as one of the most important point sources of pharmaceuticals to release in the aquatic environment

(Mandaric et al., 2019). Because of nonbiodegradable feature and pseudopersistence, pharmaceutical removal is ineffective during the conventional wastewater treatments. Biodegradation and persistence of pharmaceuticals are influenced by molecular structures such as chloro-, nitro-, and fluorofunctions attached to aromatic rings (Patel et al., 2019).

PCPs refer to other products than pharmaceuticals consumed or used by an individual for personal health, hygiene, or cosmetic reasons. Triclocarban, triclosan, and parabens are usually used in personal hygiene products and cosmetics as preservatives or antimicrobials. These compounds are also known as endocrine disruption compounds. The widespread of the presence of PCPs in receiving water resources is a growing concern because of its impact on environmental and human health. The major sources of PCPs discharge into the aquatic environments are sewage treatment plants, WWTPs, and landfill leakage. PCPs are detected widely in wastewater, surface water, and groundwater. The excreted metabolites can become secondary pollutants and be further modified in receiving water bodies (Zhang et al., 2015).

Today, surfactants are classified in the EMPs group and found in groundwater, surface water, municipal wastewater, drinking water, and food sources. They are found in soap, detergent, dispersion, emulsifier, wetting agent, foaming agent, corrosion inhibitor, antistatic agent, bactericide, etc. As surfactants are hazardous for people and environment, detection/determination and removal/degradation of surfactants/detergent from wastewater are urgently needed (Ivanković and Hrenović, 2010). The degree of damage by surfactants to aquatic plants depends on its concentration in environment. If the concentration of surfactants is too high in the water, it will affect the growth of algae and other microorganisms in water. It resulted in a decrease in primary productivity of water bodies. Thus the food chain of aquatic organisms in water bodies is undermined. Surfactants can cause acute poisoning, which can lead to increase in permeability of membrane, resulting into gradually disintegration of material exosome and cell structure (Ivanković and Hrenović, 2010).

The persistence of heavy metals in wastewater is due to their nonbiodegradable and toxicity nature. Heavy metals at trace concentration can be released into the aquatic environments by both natural and anthropogenic processes. Natural emissions of heavy metals occur under different processes such as volcanic eruptions, sea-salt sprays, forest fires, rock weathering, biogenic sources, and windborne soil particles. The most common heavy metals released from natural emissions are lead (Pb), nickel (Ni), chromium (Cr), cadmium (Cd), arsenic (As), mercury (Hg), zinc (Zn), and copper (Cu). Some main anthropogenic sources of the heavy metal pollution in the aquatic environments are smelting which releases arsenic; automobile exhaust releases lead, copper, and zinc; insecticides release arsenic; and burning of fossil fuels releases nickel, vanadium, mercury, selenium, and tin. Although chemical and mechanical methods are also used in purification processes, they are insufficient for the stabilization of trace elements, especially heavy metals (Masindi and Muedi, 2018).

7.3 Ecotoxicity and risk assessment

In the past few years the ecotoxicological assessment of EMPs has received a great attention. Currently many researches are being conducted to investigate the effects of EMPs at low concentrations in groundwater, surface waters, and treated wastewaters. Despite their occurrence at low concentrations in aquatic environments, the EMPs are associated with several ecotoxicological effects, such as chemosensitization, endocrine disruption, disruption of the production of platelets, red and white blood cells, short- and long-term toxicity, antibiotic resistance of microorganisms, and insomnia among others. Severe ecotoxicological effects of EMPs have been reported to be as a result of the interactions of EMPs with human and/or aquatic life usually after long-term exposure to the EMPs (Wanda et al., 2017b).

Several researches were involved in the environmental risks evaluation posed by pharmaceuticals in hospital effluent for both the environment and the human health. The pharmaceutical risks depend not only on their concentration in effluents but also on their ecotoxicity. Therefore the hazard quotient (HQ) can be calculated for pharmaceuticals using their highest measured concentration in effluents from medical facilities, divided by the predicted no-effect concentration. According to the classification used in many studies, the risk is classified as high ($HQ \geq 1$), medium ($1 > HQ > 0.1$), and low ($HQ \leq 0.1$). Acute toxic effects of low concentrations of pharmaceuticals in the environment are not likely, but due to the lack of information, it is not possible to exclude adverse effects resulting from the long-term impact of low doses of pharmaceuticals (Cidlinová et al., 2018).

The widespread PCPs occurrence in the aquatic environments is becoming of a great concern. Up to now most of the (eco)toxicological tests are conducted using acute toxicity assays. However, as it was demonstrated by other emerging contaminants, pharmaceuticals, for instance, acute toxicity cannot necessarily use instead of chronic toxicity effects. It is common knowledge that certain compounds may indicate adverse effects after chronic exposure. Therefore chronic exposure assessment should be conducted as an essential part of the toxicity assessment of PCPs (Carmona et al., 2014).

Based on the standardized protocols, some toxicity assessments, such as EC_{50}, are usually obtained from the experimental test. However, due to cost and time consumption, it is impractical to identify all of the toxicity effects of PCPs using EC_{50} test. In this situation, the development of a predictive model provides a good occasion to fill date gaps related to environmental risk assessment of PCPs (Carmona et al., 2014).

Surfactants have some diverse effects on water bodies. Because of foam, sorption on solid particles preventing the sedimentation, reduction of river self-purification, surfactants led to decreasing of air, water oxygen transfer, and water quality. Many studies have assessed the ecotoxicity of anionic and nonionic surfactants, and therefore future research should be focused on the toxic effects of cationic and amphoteric surfactants whose ecotoxicological features are unknown. It is practically impossible to perform bioassays test for toxicity on all aquatic living organisms. Therefore microorganisms, algae, crustaceans (benthic and planktonic), and fish are used for bioassay test. The physicochemical properties of the surfactants (the size of aliphatic chain, type of surfactant, absorption capacity, and concentration) have a great influence of the toxicity level. Also the surfactant toxicity is influenced by biotic factors such as age of organisms, species type, species sensitivity, and adaptation at very low concentrations of surfactants (Gheorghe et al., 2013).

Although some heavy metals, called essential heavy metals, play important roles in biological systems, their toxicity for living organisms depends on dose and duration of exposure. It is a well-known fact in toxicology that "excess of everything is bad." Nonessential heavy metals (Cd, Pb, and Hg) and metalloids (As, etc.) may be toxic even at quite low concentrations. Essential heavy metals are required in trace quantities in the body but become toxic beyond certain limits or threshold concentrations. Heavy metals have been reported to be carcinogenic, mutagenic, and teratogenic. Oxidative stress caused by heavy metals in organisms leads to the development of various diseases and abnormal conditions (Ali et al., 2019).

7.4 Chemical analysis

Todays, water analysis of EMPs is not a complex task; however, the preparation and analysis of solid samples are still a challenge. Among solid samples, sewage sludge and biosolids display the most difficulty and are, thus, rarely considered. Most studies on aquatic biota have focused on fish and some on bivalves. Another problem is the lack of suitable reference for validation of methods which delays the new protocols development. Besides, there are not always isotopic-labeled compounds for use as Surrogates Internal Standards for all the target compounds, and their availability is commercially expensive (Carmona et al., 2014).

Detection of EMPs conduct by different extraction methods such as DLLME, solid-phase extraction, solid-phase microextraction, liquid–liquid extraction (LLE), and single-drop microextraction (Carasek et al., 2019). These methods can be combined with chromatography for analysis. DLLME offers rapid, low-cost, short time, reliable, simple operation, high preconcentration and recovery factors, and environment friendliness. LODs in DLLME are much better and feasible compared to other liquid-phase microextraction techniques. LLE has many disadvantages such as requires large volume of hazardous solvent and long time for extraction (Samsidar et al., 2018). Numerous techniques have been developed over the past decades for determination of pesticide residues in environment samples. Due to the wide variety of formulated EMPs, the analysis of these chemicals may be carried out using traditional analytical methods with different types of detectors including gas chromatography, high-performance liquid chromatography, and advanced methods including electrochemical sensors, optical sensor, and immunosensors (Samsidar et al., 2018).

7.5 Removal methods

EMPs removal efficiencies in conventional wastewater treatment processes are incomplete. It is commonly accepted that the major source of EMPs to the environment is WWTPs' effluents. The improvement in EMPs removal methods by the application of suitable technologies is necessary to avoid hard environmental problems. Todays, a number of new technologies have recently applied to remove EMPs from wastewater effluents (Carmona et al., 2014).

Both WWTPs and WTPs are designed to remove organic and inorganic suspended materials, flocculated matter, and pathogens from wastewater or drinking water sources; however, neither drinking nor wastewater treatment processes are specifically designed to remove PCPs from water. Owing to their high chemical stability in a reaction chemical and low biodegradability, most PCPs are impossible to be completely eliminated by conventional treatment processes. In recent years, several technologies were developed to provide more efficiency removal, including advanced oxidation processes (AOPs) and membrane filtration and activated carbon (Liang et al., 2014; Nakada et al., 2017).

Among the water treatment methods used up to now, AOPs show more potential for removing a wide range of priority and EMPs. For example, ozonation process has excellent removal efficiency for a wide range of pharmaceuticals and its efficiency increases by ozone combined with other oxidants such as H_2O_2. Similarly other AOPs processes (H_2O_2/UV, Fe^{2+}/H_2O_2, Fe^{2+}/ H_2O_2/UV, O_3/ H_2O_2, and O_3/UV) run high removal efficiency, but high energy, cost inputs, and formation of oxidative by-products limit their applications. However, high energy consumption and formation of oxidative by-products limit its applications.

Also among membrane processes only nanofiltration (NF)/reverse osmosis has been found to be efficient for the removal of pharmaceuticals and other dissolved pollutants from wastewater while microfiltration (MF) and ultrafiltration (UF) processes can only be used as pretreatment process. Although advance membrane filtrations such as NF, MF, and UF are able to remove EMPs with high efficiency, their application is limited by membrane fouling, high operational cost, and high energy demands. Further research is also required to evaluate the economics costs of EMPs analysis in the aquatic environments to better estimate the advantages of tertiary treatments application during EMPs removal in the aquatic environments.

References

Ali, H., Khan, E., Ilahi, I., 2019. Environmental chemistry and ecotoxicology of hazardous heavy metals: environmental persistence, toxicity, and bioaccumulation. J. Chem. 2019.

Besha, A.T., Gebreyohannes, A.Y., Tufa, R.A., Bekele, D.N., Curcio, E., Giorno, L., 2017. Removal of emerging micropollutants by activated sludge process and membrane bioreactors and the effects of micropollutants on membrane fouling: a review. J. Environ. Chem. Eng. 5, 2395–2414.

Carasek, E., Bernardi, G., Do Carmo, S.N., Vieira, C., 2019. Alternative green extraction phases applied to microextraction techniques for organic compound determination. Separations 6, 35.

Carmona, E., Andreu, V., Picó, Y., 2014. Occurrence of acidic pharmaceuticals and personal care products in Turia River Basin: from waste to drinking water. Sci. Total. Environ. 484, 53–63.

Cidlinová, A., Wittlingerová, Z., Zimová, M., Chrobáková, T., Petruželková, A., 2018. Ecotoxicity of wastewater from medical facilities: a review. Sci. Agr. Biochem. 49, 26–31.

Cruz, S., Barceló, D., 2015. Personal Care Products in the Aquatic Environment. Springer.

Ebele, A.J., Abdallah, M.A.-E., Harrad, S., 2017. Pharmaceuticals and personal care products (PPCPs) in the freshwater aquatic environment. Emerg. Contam. 3, 1–16.

Geissen, V., Mol, H., Klumpp, E., Umlauf, G., Nadal, M., Van Der Ploeg, M., et al., 2015. Emerging pollutants in the environment: a challenge for water resource management. Int. Soil Water Conserv. Res. 3, 57–65.

Gheorghe, S., Lucaciu, I., Paun, I., Stoica, C., Stanescu, E., 2013. Ecotoxicological behavior of some cationic and amphoteric surfactants (biodegradation, toxicity and risk assessment). Biodegr. Life Sci. 83–114.

Heberer, T., 2002. Occurrence, fate, and removal of pharmaceutical residues in the aquatic environment: a review of recent research data. Toxicol. Lett. 131, 5–17.

Ivanković, T., Hrenović, J., 2010. Surfactants in the environment. Arch. Ind. Hyg. Toxicol. 61, 95–110.

Liang, R., Hu, A., Hatat-Fraile, M., Zhou, N., 2014. Fundamentals on adsorption, membrane filtration, and advanced oxidation processes for water treatment. Nanotechnol. Water Treat. Purif. Springer.

Luo, Y., Guo, W., Ngo, H.H., Nghiem, L.D., Hai, F.I., Zhang, J., et al., 2014. A review on the occurrence of micropollutants in the aquatic environment and their fate and removal during wastewater treatment. Sci. Total. Environ. 473, 619–641.

Mandaric, L., Kalogianni, E., Skoulikidis, N., Petrovic, M., Sabater, S., 2019. Contamination patterns and attenuation of pharmaceuticals in a temporary Mediterranean river. Sci. Total. Environ. 647, 561–569.

Masindi, V., Muedi, K.L., 2018. Environmental contamination by heavy metals. Heavy Met. 19, 2019.

Nakada, N., Hanamoto, S., Jurgens, M.D., Johnson, A.C., Bowes, M.J., Tanaka, H., 2017. Assessing the population equivalent and performance of wastewater treatment through the ratios of pharmaceuticals and personal care products present in a river basin: application to the River Thames basin, UK. Sci. Total. Environ. 575, 1100–1108.

Patel, M., Kumar, R., Kishor, K., Mlsna, T., Pittman JR, C.U., Mohan, D., 2019. Pharmaceuticals of emerging concern in aquatic systems: chemistry, occurrence, effects, and removal methods. Chem. Rev. 119, 3510–3673.

Samsidar, A., Siddiquee, S., Shaarani, S.M., 2018. A review of extraction, analytical and advanced methods for determination of pesticides in environment and foodstuffs. Trends Food Sci. Technol. 71, 188–201.

Wanda, E., Nyoni, H., Mamba, B., Msagati, T., 2017a. Occurrence of emerging micropollutants in water systems in Gauteng, Mpumalanga, and North West Provinces, South Africa. Int. J. Environ. Res. Public Health 14, 79.

Wanda, E.M., Nyoni, H., Mamba, B.B., Msagati, T.A., 2017b. Occurrence of emerging micropollutants in water systems in Gauteng, Mpumalanga, and North West Provinces, South Africa. Int. J. Environ. Res. Public Health 14, 79.

Wanda, E.M., Nyoni, H., Mamba, B.B., Msagati, T.A., 2018. Application of silica and germanium dioxide nanoparticles/polyethersulfone blend membranes for removal of emerging micropollutants from water. Phys. Chem. Earth Parts A/B/C 108, 28–47.

Zhang, N.-S., Liu, Y.-S., Van Den Brink, P.J., Price, O.R., Ying, G.-G., 2015. Ecological risks of home and personal care products in the riverine environment of a rural region in South China without domestic wastewater treatment facilities. Ecotoxicol. Environ. Saf. 122, 417–425.

Glossary

A

Acridine orange (AO) An organic compound. It is used as a nucleic acid-selective fluorescent cationic dye useful for cell cycle determination.

Activated sludge process (ASP) A sewage treatment process in which air or oxygen is blown into raw, unsettled sewage to smash the solids and develop a biological 'soup' which digests the organic content and pollutants in the sewage.

Acute toxicity Acute toxicity describes the adverse effects of a substance that result either from a single exposure or from multiple exposures in a short period of time (usually less than 24 h).

Adenocarcinoma Malignant tumor formed from glandular epithelial tissue or formed in a glandular pattern.

Adolescence (n)/adolescent (adj) Stage of human development beginning with puberty and ending with adulthood.

Adsorbate Molecular species of gas, dissolved substance, or liquid that adheres to or is adsorbed in an extremely thin surface layer of a solid substance.

Adsorbent Condensed phase at the surface of which adsorption may occur.

Adsorption factor Amount of substance adsorbed at the interface of a condensed and a liquid or gaseous phase divided by the total amount of the substance available for adsorption.

Adsorption Adsorption is the adhesion of atoms, ions, or molecules from a gas, liquid, or dissolved solid to a surface. This process creates a film of the adsorbate on the surface of the adsorbent.

Advanced oxidation processes **(AOPs)** Advanced oxidation processes (AOPs) are highly efficient novel methods that accelerate the oxidation and the degradation of a wide range of organic and inorganic substances that are resistant to conventional treatment methods.

Adverse effect (adverse outcome) Change in biochemistry, physiology, growth, development morphology, or behavior of an organism, including the effects of aging, that results in impairment of functional capacity or impairment of capacity to compensate for additional stress or increase in susceptibility to other environmental influences.

Aerosol Mixture of small particles (solid, liquid, or a mixed variety) and a carrier gas (usually air).

Alcohol ethoxylates (AEs) A surfactant found in products such as laundry detergents, surface cleaners, cosmetics, agricultural products, textiles, and paint.

Alcohol ethoxysulfates (AES) A surfactant found in products such as laundry detergents, surface cleaners, cosmetics, agricultural products, textiles, and paint.

Algicide Substance intended to kill algae.

Aliphatic hydroxyl Organic acids with carbon atoms arranged in branched or unbranched open chains rather than in rings.

Alkyldimethylamines (ADMAs) With chain lengths C10–C16 is the most commercially used amine oxide.

Alkyldimethylbenzylammonium chloride (ADBAC) A type of cationic surfactant. It is an organic salt classified as a quaternary ammonium compound.

Amphiphile A chemical compound possessing both hydrophilic (water-loving, polar) and lipophilic (fat-loving) properties. Such a compound is called amphiphilic or amphipathic.

Anthropogenic chemicals Anthropogenic chemicals are widely used in agriculture, industry, medicine, and military operations. Examples include pesticides such as atrazine, pentachorophenol (PCP), 1,3-dichloropropene, and DDT, explosives such as trinitrotoluene (TNT), solvents such as trichloroethylene, and dielectric fluids such as PCBs.

Antiandrogens Antiandrogens, also known as androgen antagonists or testosterone blockers, are a class of drugs that prevent androgens like testosterone and dihydrotestosterone (DHT) from mediating their biological effects in the body.

Antifogs *Anti-fog* agents, also known as *anti-fogging* agents and treatments, are chemicals that prevent the condensation of water in the form of small droplets on a surface which resemble fog.

Antimycotic (n, adj): fungicide (Substance) intended to kill a fungus or to inhibit its growth.

Antiresistant Substance used as an additive to a pesticide formulation in order to reduce the resistance of insects to the pesticide (e.g., an antimetabolite that inhibits metabolic inactivation of the pesticide).

Aphicide Substance intended to kill aphids.

Aphid Common name for a harmful plant parasite in the family Aphididae, some species of which are vectors of plant virus diseases.

Aquaculture Aquaculture, also known as aquafarming, is the farming of fish, crustaceans, molluscs, aquatic plants, algae, and other organisms. Aquaculture involves cultivating freshwater and saltwater populations under controlled conditions, and can be contrasted with commercial fishing, which is the harvesting of wild fish. Mariculture refers to aquaculture practiced in marine environments and in underwater habitats.

Arboricide Substance intended to kill trees and shrubs.

Ascaricide Substance intended to kill roundworms (Ascaridae).

Azelastine hydrochloride (AZH) Azelastine hydrochloride is chemically known as *4-(4-Chlorobenzyl)-2- [(4RS)-1-methylhexahydro-1H-azepin-4-yl] phthalazin-1(2H)-one hydrochloride*. AZH occurs as a white or almost white, crystalline powder. AZH is a potent, second generation, selective, histamine antagonist (histamine-H1-receptor antagonist) used as first line therapy of mild intermittent, moderate/severe intermittent and mild persistent rhinitis (new classification system for rhinitis). AZH has been formulated both as a nasal spray and as eye drops. Azelastine eye drops are indicated for the local treatment of seasonal and perennial allergic conjunctivitis.

B

Bactericide Substance intended to kill bacteria such as benzalkonium chloride, 1-dodecyl-3-methylimidazolium bromide, didecyldimethylammonium bromide, trihexyl(tetradecyl)phosphonium bromide, and trihexyl(tetradecyl)phosphonium chloride.

Benzyldimethyl hexadecylammonium chloride (BDHAC) Benzyldimethyl hexadecylammonium chloride is used in leather processing, textile dyeing. A mildew preventive in silicone-based water repellents. Compatible with many nonionic detergents. Active in moderately alkaline solutions.

Beta blockers Beta blockers are among the most important drugs used by cardiologists.

Bioaccumulation Bioaccumulation is the gradual accumulation of substances, such as pesticides or other chemicals in an organism. *Bioaccumulation* occurs when an organism absorbs a substance at a rate faster than that at which the substance is lost by catabolism and excretion.

Biocide (n)/biocidal (adj) Substance intended to kill living organisms.

Biocide A **biocide** is defined in the European legislation as a chemical substance or microorganism intended to destroy, deter, render harmless, or exert a controlling effect on any harmful organism.

Bioconcentration factor (BCF) The ratio of the chemical concentration in an organism or biota to the concentration in water.

Bioconcentration Bioconcentration is a term that was created for use in the field of aquatic toxicology. *Bioconcentration* can also be defined as the process by which a chemical concentration in an aquatic organism exceeds that in water as a result of exposure to a waterborne chemical.

Biodegradation Biodegradation is the naturally-occurring breakdown of materials by microorganisms such as bacteria and fungi or other biological activity.

Biopesticide Biological agent with pesticidal activity, e.g., the bacterium Bacillus thuringiensis when used to kill insects.

Biostatic Adjective applied to a substance that arrests the growth or multiplication of living organisms.

Biota All biological organisms as a totality.

Biotransformation Biotransformation means chemical alteration of chemicals such as nutrients, amino acids, toxins, and drugs in the body.

Bisphenol A (BPA) An organic synthetic compound with the chemical formula $(CH_3)_2C(C_6H_4OH)_2$ belonging to the group of diphenylmethane derivatives and bisphenols, with two hydroxyphenyl groups. It is a colorless solid that is soluble in organic solvents, but poorly soluble in water (0.344 wt.% at 83°C).

Boron-doped diamond (BDD) Boron-doped diamond is an excellent electrode material with superior material characteristics. BDD has the largest electrochemical potential window in aqueous solutions compared to traditional electrode materials such as gold, platinum and glassy carbon. Chemical reactions that occur within the electric potential range from − 1.2 V to + 2.5V can be investigated.

Boron-doped diamond (OT-Au-BDD) Boron-doped diamond an octanethiolate gold particles modified boron-doped diamond.

C

Carbamate pesticides Carbamate pesticides are derived from carbamic acid and kill insects in a similar fashion as organophosphate *insecticides*. They are widely used in homes, gardens, and agriculture. Like the organophosphates, their mode of action is inhibition of cholinesterase enzymes, affecting nerve impulse transmission.

Carcinogen (n)/carcinogenic (adj) Agent (chemical, physical, or biological) that is capable of increasing the incidence of malignant neoplasms, thus causing cancer.

Catchment quality control (CQC) Catchment quality control is a lowland river management concept for predicting the presence of organic micro-pollutants. Initially, CQC dealt with industrial or point discharges and full details of the procedure are given. More recently, attention has been given to non-point or diffuses discharges such as those of household and allied chemicals.

Cationic surfactants (CS^+) Cationic surfactants are amphiphilic compounds, which can dissociate in water with the formation of surface-active cations. These molecules have the capacity of self-assembling, being widely used in biotechnology.

Cetylpyridinium bromide (CPB) One of the antibacterial agents widely used as a disinfectant and also in the treatment of mouth, throat, skin, and eye infections (US Pharmacopeia) and can be immobilized on several carrier systems such as natural zeolite.

Cetylpyridinium chloride (CPC) Cetylpyridinium chloride is a cationic quaternary ammonium compound used in some types of mouthwashes, toothpastes, lozenges, throat sprays, breath sprays, and nasal sprays. It is an antiseptic that kills bacteria and other microorganisms.

Cetyltrimethylammonium bromide (CTAB) ([(C16H33)N(CH3)3]Br; cetyltrimethylammonium bromide; hexadecyltrimethylammonium bromide; CTAB) is a quaternary ammonium surfactant.

Chemical oxygen demond (COD) indicative measure of the amount of *oxygen* that can be consumed by reactions in a measured solution.

Cholinesterase inhibitors *Cholinesterase inhibitors*, also known as anti-*cholinesterase*, are chemicals that prevent the breakdown of the neurotransmitter acetylcholine or butyryl choline. This increases the amount of the acetylcholine or butyryl choline in the synaptic cleft that can bind to muscarinic receptors, nicotinic receptors and others.

Chronic toxicity Chronic toxicity is the development of adverse effects as the result of long-term exposure to a toxicant or other stressor. It can manifest as direct lethality but more commonly refers to sublethal endpoints such as decreased growth, reduced reproduction, or behavioral changes such as impacted swimming performance.

Concentration–response relationship (exposure–response relationship) Association between exposure concentration and the incidence of a defined effect in an exposed population.

Critical micille concentration (CMC) Defined as the concentration of surfactants above which micelles form and all additional surfactants added to the system go to micelles. The CMC is an important characteristic of a surfactant.

D

DEHP Di(2-ethylhexyl) phthalate.

Deinking The industrial process of removing printing ink from paperfibers of recycled paper to make *deinked* pulp.

Denitrification A microbially facilitated process where nitrate (NO_3^-) is reduced and ultimately produces molecular nitrogen (N_2) through a series of intermediate gaseous nitrogen oxide products.

Destabilization/coagulation A colloidal suspension results in joining of minute particles by physical and chemical processes.

Diclofenac **A** nonsteroidal anti-inflammatory drug (NSAID). This medicine works by reducing substances in the body that cause pain and inflammation.

Dissipation Reduction in the amount of apesticide or other compound that has been applied to plants, soil, or another environmental compartment.

Dissolved organic matters (DOM) Defined as the organic matter fraction in solution that passes through a 0.45-μm filter.

Dodecyltrimethylammonium bromide (DTAB) A quarternary *ammonium* cation having one dodecyl and three methyl substituents around the central nitrogen. It has a role as a surfactant.

Dodecyltrimethylammonium chloride (DTAC) It is a cationic alkyltrimethylammonium chloride surfactant that can undergo micellization in aqueous solution.

E

Ecofriendliness literally means earth-*friendly* or not harmful to the environment. This term most commonly refers to products that contribute to green living or practices that help conserve resources like water and energy.

Ecogenetics (n)/ecogenetic (adj) Study of the influence of hereditary factors on the response of individuals or populations to environmental factors.

Ectoparasiticide Substance intended to kill parasites living on the exterior of the host.

Electrochemical membrane bioreactor (EMBR) A reliable and promising technology for wastewater treatment and reclamation.

Electrochemical oxidation Known as anodic oxidation, is a technique used for wastewater treatment, mainly for industrial effluents, and is a type of advanced oxidation process (AOP).

Electron impact (EI) The most common ionization technique used for mass.

Emerging micropollutants (EMPs) The EMPs comprise of a wide range of natural and synthetic organic compounds, which include pharmaceuticals and personal care products (PPCPs), detergents, steroid hormones, industrial chemicals, pesticides, and many other contaminants of emerging concern.

Endocrine-disrupting chemicals (EDCs) Endocrine-disrupting chemicals (EDCs) are substances in the environment (air, soil, or water supply), food sources, personal care products, and manufactured products that interfere with the normal function of your body's *endocrine* system.

Environmental hazard An environmental hazard is a substance, a state or an event which has the potential to threaten the surrounding natural environment/or adversely affect people's health, including pollution and natural disasters such as storms and earthquakes.

Environmental Protection Agency (EPA) An independent agency of the United States federal government for environmental protection.

Environmental Risk Assessment (ERA) A process of predicting whether there may be a risk of adverse effects on the environment caused by a chemical substance.

Erythromycin Used to treat a wide variety of bacterial infections. It may also be used to prevent certain bacterial infections.

Estrone (E1): estradiol (E2) Also spelled oestradiol, is an estrogen steroid hormone and the major female.

European Medicines Agency (EMA) An agency of the European Union (EU) in charge of the evaluation and supervision of medicinal products.

European Medicines Agency (EMEA) It is an agency of the European Union (EU) in charge of the evaluation and supervision of medicinal products. Prior to 2004, it was known as the European Agency for the Evaluation of Medicinal Products or European Medicines Evaluation Agency (EMEA).

European Union (EU) It is a political and economic union of 27 member states that are located primarily in Europe. Its members have a combined area of 4,233,255.3 km^2 (1,634,469.0 sq mi) and an estimated total population of about 447 million.

Extracellular polymeric substances (EPS) Natural *polymers* of high molecular weight secreted by microorganisms into their environment. ... EPSs are mostly composed of polysaccharides (exopolysaccharides) and proteins, but include other macromolecules such as DNA, lipids, and humic *substances* extractable fraction.

Extraneous residue limit (ERL) Term referring to a pesticide residue or contaminant arising from environmental sources (including former agricultural uses) other than the use of a pesticide or contaminant substance directly or indirectly on the commodity.

F

Fatty acid alcohol/fatty acid esters (FAEs) A type of ester that result from the combination of a fatty acid with an alcohol. When the alcohol component is glycerol, the fatty acid esters produced can be monoglycerides, diglycerides, or triglycerides.

Fatty amines Any amine attached to a hydrocarbon chain of eight or more carbon atoms in length.

Fenton reagent A solution of hydrogen peroxide (H_2O_2) with ferrous iron (typically iron(II) sulfate, $FeSO_4$) as a catalyst that is used to oxidize contaminants or waste waters. Fenton's reagent can be used to destroy organic compounds.

Flocculation Refers to the process by which fine particulates are caused to clump together into a floc. The floc may then float to the top of the liquid (creaming), settle to the bottom of the liquid (sedimentation), or be readily filtered from the liquid.

Food and Drug Administration (FDA) Responsible for protecting the public health by ensuring the safety, efficacy, and security of human and veterinary drugs, biological products.

Fungicide Substance intended to kill fungi.

G

Gametocide Substance intended to kill gametes.

Global climate change Occurs when changes in Earth's climate system result in new weather patterns that remain in place for an extended period of time.

Gold particle-modified BDD (Au-BDD) Boron-doped diamond (BDD) has been modified with bare gold(Au) particles.

Graminicide Pesticide (herbicide) intended to kill weedy grasses (Gramineae).

H

Half-lives The half-life of a reaction, $t_{1/2}$, is the amount of time needed for a reactant concentration to decrease by *half* compared to its initial concentration.

Henry's Law A gas law that states that the amount of dissolved gas in a liquid is proportional to its partial pressure above the liquid.

Herbicide Substance intended to kill plants.

Hexadecyltrimethylammonium bromide (CTAB) An antiseptic agent also used in DNA extraction.

HOCl (hypochlorous acid) A weak *acid* that forms when chlorine dissolves in water, and itself partially dissociates, forming hypochlorite, ClO^-. HClO and ClO^- are oxidizers, and the primary disinfection agents of chlorine solutions.

HRT Hydraulic retention time.

Hydrolysis Hydrolysis is a chemical process in which a molecule of water is added to a substance. Sometimes this addition causes both substance and water molecule to split into two parts.

I

Insect repellents Insect repellents are used to repel mosquitoes, ticks, flies, and other biting insects.

Insecticide Substance intended to kill insects.

International Program of Chemical Safety (IPCS) The International Programme on Chemical Safety (IPCS) was formed in 1980 and is a collaboration between three United Nations bodies, the World Health Organization, the International Labour Organization and the United Nations Environment Programme, to establish a scientific basis for safe use of chemicals and to strengthen national capabilities and capacities for chemical safety.

L

Larva Recently hatched insect, fish, or other organism that have physical characteristics different from those seen in the adult, requiring metamorphosis to reach the adult body structure.

Larvicide Substance intended to kill larvae.

Lethal dose (LD) In toxicology, the lethal dose (LD) is an indication of the lethal toxicity of a given substance or type of radiation. Because resistance varies from one individual to another, the "lethal dose" represents a dose (usually recorded as dose per kilogram of subject body weight) at which a given percentage of subjects will die.

Linear alkyl benzene sulfonate (LAS) Prepared industrially by the sulfonation of linear alkylbenzenes (LABs), which can themselves be prepared in several ways.

M

Manual GCSOLAR program Contains a set of routines that compute direct photolysis rates and half-lives of pollutants in the aquatic environment.

MBR-GAC Integration of bioaugmented membrane bioreactor (MBR) method with granular activated carbon method in wastewater treatment.

MBR-NF Integration of bioaugmented membrane bioreactor (MBR) method with nanofiltration method in wastewater treatment.

MBR-PAC Integration of bioaugmented membrane bioreactor (MBR) method with powder activated carbon.

MBR-RO Integration of bioaugmented membrane bioreactor (MBR) *with reverse osmosis in wastewater treatment.*

Median lethal dose In toxicology, the median lethal dose, LD_{50} (abbreviation for "lethal dose, 50%"), LC_{50} (lethal concentration, 50%) or LCt_{50} is a measure of the lethal dose of a toxin, radiation, or pathogen. The value of LD_{50} for a substance is the dose required to kill half the members of a tested population after a specified test duration.

Membrane bioreactor (MBR) Membrane bioreactor (MBR) is the combination of a membrane process like microfiltration or ultrafiltration with a biological wastewater treatment process, the activated sludge process.

Membrane filtration process Membrane filtration process is a physical separation method characterized by the ability to separate molecules of different sizes and characteristics. Its driving force is the difference in pressure between the two sides of a special membrane.

Molluscicide (limacide = molluskicide) Substance intended to kill molluscs.

N

Naproxen Used to relieve pain from various conditions such as headache, muscle aches, tendonitis, dental pain, and menstrual cramps.

Natural organic matter (NOM) A broad term for the complex mixture of thousands of organic compounds found in water. These compounds are derived from decaying plant and animal matter.

Nematicide (nematocide) Substance intended to kill nematodes.

Nitrification The biological oxidation of ammonia to nitrite followed by the oxidation of the nitrite to nitrate. The transformation of ammonia to nitrite is usually the rate limiting step of nitrification. *Nitrification* is an important step in the nitrogen cycle in soil.

Nonsteroidal anti-inflammatory drugs (NSAIDs) members of a drug class that reduces pain, decreases fever, prevents blood clots, and in higher doses, decreases inflammation. Side effects depend on the specific drug but largely include an increased risk of gastrointestinal ulcers and bleeds, heart attack, and kidney disease.

NORMAN Network group The NORMAN network enhances the exchange of information on emerging environmental substances, and encourages the validation and harmonization of common measurement methods and monitoring tools so that the requirements of risk assessors and risk managers can be better met. It specifically seeks both to promote and to benefit from the synergies between research teams from different countries in the field of emerging substances.

O

Octanol–water partition coefficient (K_{OW}) The octanol/water partition coefficient (*K*ow) is defined as the ratio of a chemical's concentration in the octanol phase to its concentration in the aqueous phase of a two-phase octanol/water system. *K*ow = concentration in octanol phase/concentration in aqueous phase.

OMPs (organic micropollutants) They are ubiquitous in natural waters even in places where the human activity is limited. The presence of OMPs in natural water sources for human consumption encourages the evaluation of different water purification technologies to ensure water quality.

Organochlorine pesticides Organochlorine pesticides are chlorinated hydrocarbons used extensively from the 1940s through the 1960s in agriculture and mosquito control. Representative compounds in this group include DDT, methoxychlor, dieldrin, chlordane, toxaphene, mirex, kepone, lindane, and benzene hexachloride.

Organophosphate pesticides Organophosphates (also known as phosphate esters, or OPEs) are a class of organophosphorus compounds with the general structure $O = P(OR)_3$. They can be considered as esters of phosphoric acid. Like most functional groups organophosphates occur in a diverse range of forms, with important examples including key biomolecules such as DNA, RNA, and ATP, as well as many insecticides, herbicides, nerve agents, and flame retardants.

Ovicide Substance intended to kill eggs.

Ozonation A chemical water treatment technique based on the infusion of ozone into water. Ozone is a gas composed of three oxygen atoms (O3), which is one of the most powerful oxidants.

P

Parabens Parabens are a class of widely used preservatives in cosmetic and pharmaceutical products.

Peroxides A group of compounds with the structure $R - O - O - R$. The $O - O$ group in a *peroxide* is called the *peroxide* group or peroxo group.

Peroxydisulfate An oxyanion. It is commonly referred to as the persulfate ion or peroxodisulfate anions, but this term also refers to the peroxomonosulfate ion, So_5^{-2}.

Persistence, bioaccumulation potential, and toxicity (PBT) Persistent, bioaccumulative and toxic substances (PBTs) are a class of compounds that have high resistance to degradation from abiotic and biotic factors, high mobility in the environment and high toxicity.

Persistent organic pollutants (POPs) *Persistent organic pollutants (POPs)*, sometimes known as "forever chemicals" are **organic** compounds that are resistant to environmental degradation through chemical, biological, and photolytic processes.

Personal care products (PCPs) Personal care includes products as diverse as cleansing pads, colognes, cotton swabs, cotton pads, deodorant, eye liner, facial tissue, hair clippers, lip gloss, lipstick, lip balm, lotion, makeup, hand soap, facial cleanser, body wash, nail files, pomade, perfumes, razors, shaving cream, moisturizer, baby powder, toilet paper, toothpaste, facial treatments, wet wipes, and shampoo.

Pest Organism that may harm public health, attacks food and other materials essential to mankind, or otherwise affects human beings adversely.

Pesticide Substance intended to kill pests.

Pharmaceutically active compounds (PhACs) A group of compounds that include hormones, antibiotics and painkillers. Results of several recent studies suggest that these compounds are present in wastewater effluents and in polluted surface waters.

Photocatalysis Photocatalysis is a type of catalysis that results in the modification of the rate of a photoreaction—a chemical reaction that involves the absorption of light by one or more reacting species—by adding substances (catalysts) that participate in the chemical reaction without being consumed.

Photocatalytic degradation Promising technology due to its advantage of degradation on pollutants instead of their transformation under ambient conditions.

Photolysis Chemical decomposition induced by light or other radiant energy.

Phylogenetics Phylogenetics is the study of evolutionary relationships among biological entities—often species, individuals or genes (which may be referred to as taxa).

Polarity In chemistry, polarity is a separation of electric charge leading to a molecule or its chemical groups having an electric dipole moment, with a negatively charged end and a positively charged end. Polar molecules must contain polar bonds due to a difference in electronegativity between the bonded atoms.

Polyaromatic hydrocarbon (PAHs) hydrocarbons—organic compounds containing only carbon and hydrogen—that are composed of multiple aromatic rings (organic rings in which the electrons are delocalized).

Polystyrene sulfonate (PSS) A group of medications used to treat high blood potassium.

Positive-ion chemical ionization (PICI) new ionized species are formed when gaseous molecules interact with ions.

Powder activated carbon (PAC) small activated carbon particles, with a size that is predominantly less than 0.075 mm, produced by milling or pulverizing activated carbon.

ppb part per billion.

ppt part per trillion.

Predicted environmental concentration (PEC) The Predicted Environmental Concentration is an indication of the expected concentration of a material in the environment, taking into account the amount initially present (or added to) the environment, its distribution, and the probable methods and rates of environmental degradation and removal, either forced or natural.

Predicted no effect concentration (PNEC) Predicted no effect concentration is the concentration of a chemical which marks the limit at which below no adverse effects of exposure in an ecosystem are measured. PNEC values are intended to be conservative and predict the concentration at which a chemical will likely have no toxic effect. They are not intended to predict the upper limit of concentration of a chemical that has a toxic effect.

Preservative A substance or a chemical that is added to products such as food products, beverages, pharmaceutical drugs, paints, biological samples, cosmetics, wood, and many other products to prevent decomposition by microbial growth or by undesirable chemical changes.

Pressure retarded osmosis (PRO) A technique to separate a solvent (e.g., fresh water) from a solution that is more concentrated (e.g., seawater) and also pressurized.

Pseudopersistence Chemicals continually infused to the aquatic environment, essentially become "persistent" pollutants even if their half-lives, are short. Their supply is continually replenished.

Pyrethrins Pyrethrins are pesticides found naturally in some chrysanthemum flowers. They are a mixture of six chemicals that are toxic to insects. *Pyrethrins* are commonly used to control mosquitoes, fleas, flies, moths, ants, and many other pests.

Pyrethroid pesticides A pyrethroid is an organic compound similar to the natural pyrethrins, which are produced by the flowers of pyrethrums (*Chrysanthemum cinerariaefolium* and *C. coccineum*). *Pyrethroids* constitute the majority of commercial household insecticides.

Q

Quaternary ammonium compound (QAC) Also known as quats, are positively charged polyatomic ions of the structure $NR^{+}{}_{4}$, R being an alkyl group or an aryl group.

R

Reactive oxygen species (ROS) Chemically reactive chemical species containing oxygen. Examples include peroxides, superoxide, hydroxyl radical, singlet oxygen, and alpha-oxygen.

Reverse electrodialysis (RED) The salinity gradient energy retrieved from the difference in the salt concentration between seawater and river water.

Rodenticide Substance intended to kill rodents.

S

Silicone surfactants A group of small-molecule and polymeric surfactants that find a wide variety of applications due to their unusual properties.

Ski waxes A material applied to the bottom of snow runners, including skis, snowboards, and toboggans, to improve their coefficient of friction performance under varying snow conditions.

Sludge retention time (SRT) Sludge retention time is the average time the activated-sludge solids are in the system. The SRT is an important design and operating parameter for the activated-sludge process and is usually expressed in days.

Solubilizer An ingredient that helps solubilize (make soluble) an ingredient which is otherwise insoluble in a medium.

Sonolysis Or sonochemical technique is consistent with the production of iron oxide nanoparticles by using the physical impacts of ultrasound that are being produced by acoustic cavitation.

Sorption Sorption is a physical and chemical process by which one substance becomes attached to another.

Spermicides (nonoxynol-9) Spermicides (nonoxynol-9) sometimes abbreviated as N-9, is an organic compound that is used as a surfactant. It is a member of the nonoxynol family of nonionic surfactants. N-9 and related compounds are ingredients in various cleaning and cosmetic products. It is widely used in contraceptives for its spermicidal properties.

Stabilization ponds Or waste stabilization lagoons are ponds designed and built for wastewater treatment to reduce the organic content and remove pathogens from wastewater.

Sulfamethoxazole Sulfamethoxazole is an antibiotic. It is used for bacterial infections such as urinary tract infections, bronchitis, and prostatitis and is effective against both gram negative and positive bacteria such as Listeria monocytogenes and E.

Sulfamethoxazole (SMX) An antibiotic. It is used for bacterial infections such as urinary tract infections, bronchitis, and prostatitis and is effective against both gram negative and positive bacteria such as Listeria monocytogenes and *E. coli*.

SURFace ACTive AgeNT SURFace ACTive AgeNT or surfactant is a substance which lowers the surface tension of the medium in which it is dissolved, the interfacial tension with other phases, and is positively adsorbed at the liquid-vapour interface and other interfaces.

Surface tension The tendency of liquid surfaces to shrink into the minimum surface area possible.

Surfactants Compounds that lower the surface tension (or interfacial tension) between two liquids, between a gas and a liquid, or between a liquid and a solid. Surfactants may act as detergents, wetting agents, emulsifiers, foaming agents, and dispersants.

Synergism (in pesticide use) Action of a substance that is itself non-toxic, but, when co-administered with a pesticide, increases the pesticide's efficacy or overcomes pesticide resistance in the target organism.

T

Total organic carbon (TOC) The amount of carbon found in an organic compound and is often used as a non-specific indicator of water quality or cleanliness of pharmaceutical manufacturing equipment.

Trace organic contaminants (TrOCs) The term trace organic contaminant (TrOC) refers to a diverse and expanding array of natural as well as anthropogenic substances including industrial chemicals, chemicals used in households, compounds and their metabolites excreted by people, and by-products formed during wastewater and drinking-water treatment processes.

Triclocarban Triclocarban (sometimes abbreviated as TCC) is an antibacterial chemical once common in, but now phased out of, personal care products like soaps and lotions.

Triclosan Triclosan is an antibacterial and antifungal agent. It is a polychloro phenoxy phenol. It is widely used as a preservative and antimicrobial agent in personal care products such as soaps, skin creams, toothpaste, and deodorants as well as in household items such as plastic chopping boards, sports equipment, and shoes.

Trimethoprim (TMP) An antibiotic used mainly in the treatment of bladder infections.

Triton X-100 (TX-100) a nonionic surfactant that has a hydrophilic polyethylene oxide chain (on average it has 9.5 ethylene oxide units) and an aromatic hydrocarbon lipophilic or hydrophobic group.

U

Ultrafiltration (UF) A variety of membrane filtration in which forces like pressure or concentration gradients lead to a separation through a semipermeable membrane.

Ultrasonic solvent extraction (USE) Use of ultrasound in solid–liquid extraction includes the faster . . . Microwave-assisted extraction reduces time and solvent consumption.

V

Volatile organic carbones (VOCs) Organic chemicals that have a high vapor pressure at ordinary room temperature. Their high vapor pressure results from a low boiling point, which causes large numbers of molecules to evaporate or sublimate from the liquid or solid form of the compound and enter the surrounding air, a trait known as volatility.

Volatilization Volatilization is the process whereby a dissolved sample is vaporized.

W

Wastewater treatment plants (WWTPs) A wastewater treatment plant is a facility in which a combination of various processes (e.g., physical, chemical, and biological) are used to treat industrial wastewater and remove pollutants.

Z

Zwitterionic (amphoteric) Surfactants that carry both a positive and a negative charge.

Index

Note: Page numbers followed by "*f*" and "*t*" refer to figures and tables, respectively.

D

E

F

G

Q

R

S

X

Z